The Scientific Revolution
An Annotated Bibliography

for Philip
with many thanks
for
copy-editing
and
proofreading
Saj
7.12.96

The Scientific Revolution
An Annotated Bibliography

by

S.A. Jayawardene

LOCUST HILL PRESS
West Cornwall, CT
1996

Library of Congress Cataloging-in-Publication Data

Jayawardene, S. A.
    The scientific revolution : an annotated bibliography / by S.A.
Jayawardene.
        383p.   cm.
    Includes bibliographical references and index.
    ISBN 0-933951-71-X (lib. bdg. : alk. paper)
        1. Science--History--Bibliography. 2. Technology--History-
-Bibliography.   I. Title.
Q125.J39 1996
[Z7405.H6]
016.509--DC20   95-43553
                                                                        CIP

Printed on acid-free, 250-year-life paper
Manufactured in the United States of America

To the memory of my parents

5 | 2015
Fenny Smith [Rankin] told me that
SAJ returned to Sri Lanka in December 2014
and died shortly after (January 2015)
[1923 - 1915]

Contents

Introduction • *xi*

Abbreviations • *xv*

List of Illustrations • *xvii*

Research Aids • *3*
    Guides to Reference Books • *3*
    Bibliographies • *4*
    Bibliographies—History • *5*
    Bibliographies—History of Science • *6*
    Theses and Dissertations • *9*
    Libraries • *11*
    Dictionaries and Encyclopedias • *16*
    Historians of Science • *17*
    Chronologies • *18*

From Manuscript to Print. Humanism • *23*
    Bibliographies. Catalogues of Early Printed Books • *23*
    Printing and the Book Trade • *31*
    Public and Private Libraries (pre–1700) • *37*

Education • *52*
    Bibliographies • *52*
    General Works • *54*
    Monographs and Articles • *55*
    England • *59*
    Scotland • *65*
    France • *66*
    Germany • *68*
    Italy • *68*
    Netherlands • *69*

Science, Society, and Religion • *70*
    Scientific Societies • *80*

Royal Society of London • *84*
    Bibliographies • *84*
    Monographs and Articles • *84*

Philosophy • *87*
    Bibliographies • *87*
    General Works • *90*
    Monographs and Articles • *91*

Science • *101*
    Bibliographies • *101*
    Monographs and Articles • *105*

Occult Science • *131*
    Bibliographies • *131*
    Monographs and Articles • *135*

Astrology • *147*
    Bibliographies • *147*
    Monographs and Articles • *148*

Mathematics • *153*
    Bibliographies • *153*
    Monographs and Articles • *158*

Astronomy • *180*
    Bibliographies • *180*
    Monographs and Articles • *183*

Physics • *198*
    Bibliographies • *198*
    Monographs and Articles • *198*

Music • *207*

Chemistry • *210*
    Bibliographies • *210*
    Monographs and Articles • *212*

Earth Sciences • *216*
    Bibliographies • *216*
    Monographs and Articles • *218*

Marine Science. Geography. Surveying. Cartography • *223*
    Bibliographies • *223*
    Monographs and Articles • *228*

Technology • *241*
    Bibliographies • *241*
    General Works • *244*
    Monographs, Collections of Essays, and Articles • *246*
Instruments • *254*
    Bibliographies • *254*
    Monographs and Articles • *255*
Natural History • *264*
    Bibliographies • *264*
    Monographs and Articles • *269*
Medicine • *287*
  General • *287*
    Bibliographical Guides • *287*
    Bibliographies • *288*
    Catalogues • *290*
    Monographs and Articles • *294*
  Anatomy. Physiology • *300*
    Bibliographies • *300*
    Monographs and Articles • *302*
  Surgery • *308*
    Bibliographies • *308*
    Monographs and Articles • *309*
  Obstetrics. Gynaecology. Embryology • *310*
    Bibliographies • *310*
    Monographs and Articles • *311*
  Materia Medica • *313*
  Psychology. Psychiatry. Neurology • *316*
    Bibliographies • *316*
    Monographs and Articles • *317*
  Epidemic Diseases • *318*
    Bibliographies • *318*
    Monographs and Articles • *319*
  Paracelsian Medicine • *321*
    Bibliographies • *321*
    Monographs and Articles • *322*
  Medical Education • *323*
  Medical Biography • *327*
    Bibliographies and Biographical Dictionaries • *327*
    Biographies • *328*

The Scientific Revolution • *331*
        Bibliographies and Source Books • *331*
        Monographs and Articles • *332*
Author Index • *349*
Subject Index • *361*

Introduction

This bibliography is a key to the printed literature of the history of science over the 250 years that span the so-called Scientific Revolution—from the Renaissance to the early years of the Enlightenment. The scientific and cultural events that are covered took place during a period of varied and far-reaching political, economic, and spiritual activity throughout the world, such as the fall of Constantinople, the European voyages of discovery and the Atlantic slave trade, the inception of the colonial era (from the Treaty of Tordesillas to the Treaty of Utrecht), the Spanish and Roman Inquisitions, the Christian missions to unbelievers in the New World and the Old, the Reformation, the Council of Trent, and the Revocation of the Edict of Nantes.

The bibliography is arranged by subject area. The first six cover: research aids, chronologies, humanism, printing, education and social relations of science. These are followed by sections covering philosophy, the sciences, technology and medicine. The final section is on the Scientific Revolution in general. Considering the wide range of topics covered by the bibliography, no attempt has been made to provide a bibliographical essay on the historiography of science of the period. However, most sections of the work contain bibliographical items and the concluding section on the Scientific Revolution has a list of survey articles and recent studies—including Westfall's teaching guide (item 1568) and Floris Cohen's most recent *Historiographical Inquiry* (item 1644).

The arrangement of the bibliography is chronological within each section. Entries having the same publication date are arranged alphabetically. With few exceptions, no title has been entered under more than one subject category. Current bibliographies published after 1990 have not been examined systematically, and the inclusion of works published after that date is the result of my seeing a book review or of browsing among new accessions in libraries. It has not been possible to keep pace with the flood of recent publications on topics related to the Scientific Revolution. It is hoped that the *ISIS Current Bibliography* and the other current bibliographies listed will help to find them.

The material in the bibliography consists mainly of bibliographies, encyclopedias, library catalogues, monographs, source books, articles in periodicals, exhibition catalogues, proceedings of symposia, and essay reviews. As the bibliography is restricted to secondary sources, collected works of scientists have been included only in exceptional cases. Unpublished theses are excluded. In selecting the material I have made use of the standard reference works on the history of science and the resources of the Science Museum Library, the British Library, the Wellcome Institute for the History of Medicine, the Warburg Institute, and the libraries of the Institute of Education, the Senate House of the University of London and the Botany Department of the Natural History Museum, London. Some three thousand volumes were examined in the course of this work, and I am grateful to the staff of these libraries for the facilities offered to me.

The bulk of the material listed here has been examined by me personally. When I have not seen an item, I have said so. Each item has a full bibliographical description, and, with few exceptions, an annotation. Details of book reviews have been added where I thought they would be useful. Some annotations include citations taken from the book or article itself, or from the review cited.

There are two indexes—author and subject. In order to accommodate the wide range of topics covered by the bibliography, it has been necessary to be selective in the choice of entries for the subject index.

During the years that this work was in preparation I have accumulated an immense debt of gratitude to many persons—from Robert Multhauf who initiated the bibliography, to colleagues, friends, and family. Some helped me with comments, bibliographical information, illustrations, and encouragement. Others fetched books from the stacks of the libraries I was working in, sent me their publications, obtained books on loan, checked the indexes, copyedited the manuscript at various stages, and helped with the preparation of camera-ready copy. They will no doubt recognize their contributions in these pages and will, I hope, forgive me for not thanking them individually here. However, I would like to place on record my thanks to the late Ellen Wells for her patience, help, and encouragement from across the Atlantic and on her visits to London.

S. A. Jayawardene
Kew
Surrey (England)
May 1995

Abbreviations

AIHS	*Archives internationales d'histoire des sciences*
DNB	*Dictionary of National Biography* (London, 1885–1900)
DSB	*Dictionary of Scientific Biography* (New York, 1970–80)
MS(S)	manuscript(s)
NYTBR	*New York Times Book Review*
STC	Short-title catalogue
TLS	*Times Literary Supplement*

Illustrations

		page
1.	Bull of Pope Alexander VI, *Inter caetera* (4 May 1493)	*14–15*
2.	Portrait of Cardinal Bessarion	*34*
3.	University Library, Leiden, in the seventeenth century	*38*
4.	Title page of the English translation of Elie Benoît's *Histoire de l'Edit de Nantes*	*78*
5.	Galileo's *Dialogo* on the Index of Prohibited Books	*122*
6.	Horoscope of Wallenstein cast by Kepler	*151*
7.	Bombelli and the reconstruction of the *Ponte Santa Maria*	*167*
8.	Leibniz on the accusations of plagiarism	*172–73*
9.	Luca Pacioli's *Summa de arithmetica*	*178*
10.	*The Lusiad* by Luis de Camoens	*237*
11.	Bunch of coconuts from the *Hortus (Indicus) Malabaricus*	*285*
12.	Title page of Vesalius's *De humani corporis fabrica*	*304*
13.	Urine examination and treatment of syphilitic couple	*320*

The illustrations are reproduced by kind permission of:

   (1) University of London Library (nos. 1, 3, 4, 8)
   (2) Bodleian Library, Oxford (nos. 2, 5, 9, 10, 11, 12, 13)
   (3) Archives of the Kepler Commission, Munich (no. 6)
   (4) Vatican Archives (no. 7)

I am grateful to Constance Blackwell and the Foundation for Intellectual History for providing a grant to meet the cost of the illustrations.

The Scientific Revolution
An Annotated Bibliography

Research Aids

GUIDES TO REFERENCE BOOKS

1. American Library Association. *Guide to Reference Books.* 11th ed. Robert Balay, ed. Chicago, 1996. 2,020 pp.

Considered the bible of reference librarians in the U.S. Some 16,000 entries, classified in five main groups (49 sections and some 400 subsections): general reference works; the humanities; social and behavioural sciences; history and area studies; science, technology, and medicine. Annotated, with introductory notes to each section. Name, title, and subject index in one sequence. Includes some works published in 1993. Emphasis is on reference books for scholarly research but includes "representative works intended for the general reader."

Available on CD-ROM.

2. *Walford's Guide to Reference Material.* Sixth ed. London: Library Association, 1993–95. Vol. 1: *Science and Technology.* Edited by Marilyn Mullay and Priscilla Schlicke. 1993. 943 pp. [7,007 entries] Vol. 2: *Social and Historical Sciences, Philosophy and Religion.* Edited by Alan Day and Joan M. Harvey. 1994. 1,156 pp. [8,183 entries] Vol. 3: *Generalia, Language and Literature, The Arts.* Edited by Anthony Chalcraft, Ray Prytherch, and Stephen Willis. 1995. 1,148 pp. 8,774 entries.

British counterpart of Balay. Lacks introductory notes, but annotations are more detailed and include quotations from reviews. Many subsumed entries. International in scope but with more British material. Entries arranged according to the Universal Decimal Classification. Vols. 1 and 2 have author/title and subject indexes. Vol. 3 has an author/ title index and a cumulated subject index to all three volumes.

Similar in scope to Walford and Balay, but including a large number of titles other than reference books (histories, textbooks, etc.), and serving as a textbook of bibliography for students (by virtue of its method of presentation) is L. N. Malclès, *Les sources du travail bibliographique* (Geneva: Droz, 1950–58), 3 vols. Still useful though outdated.

3. Totok, Wilhelm, and Rolf Weitzel. *Handbuch der bibliographischen Nachschlagewerke. Bd. 1: Allgemeinbibliographien und allgemeine Nachschlagewerke. Bd. 2: Fachbibliographien und Fachbezogene Nachschlagewerke.* Sechste, erweiterte, vollig neu bearbeitete Auflage hrsg. von Hans-Jurgen Kernchen und Dagmar Kernchen. Frankfurt-am-Main: Klostermann, 1984–85.

Combines the best features of the guides of Balay, Walford, and Malclès. International in scope but with German slant. Selective, classified and annotated, with connecting narrative. Name and subject indexes. 2nd ed. reviewed in *Bulletin des bibliothèques de France* 1959, pp. 393–405.

4. Manfré, Guglielmo. *Guida bibliografica per gli studenti di lettere e filosofia e di magistero.* Naples: Edi-Guida, 1978. 870 pp.

Bibliographical guide intended for students in the Faculties of Arts and Education in Italian universities. Some 6,500 entries (mostly Italian) classified under 90 headings, annotated and with connecting narrative. Indexes: names and titles of works; subjects. Useful for research on Italian topics (e.g. universities, pp. 663–76).

BIBLIOGRAPHIES

5. Besterman, Theodore. *A World Bibliography of Bibliographies and of Bibliographical Catalogues, Calendars, Abstracts, Digests, Indexes, and the Like.* 4th ed. Lausanne: Societas Bibliographica, 1965–66. 5 vols.

Classified bibliography of some 117,000 bibliographies (books only), published from 1470 to 1963, under 16,000 headings. Includes catalogues of books and manuscripts. Coverage includes: academic writings; alchemy; almanacs; astrology; astronomy; cartography; chemistry; ecclesiastical history; education; geography; mathematics; medicine. No annotations, but the number of items in each bibliography is stated. Col. 17–48: Introduction. 3rd ed. reviewed in *Bulletin des bibliothèques de France* 1956, pp. 835–36.

For an annotated bibliography of bibliographies published before 1866, see J. Petzholdt, *Bibliotheca bibliographica* (Leipzig, 1866; reprinted Nieuwkoop, 1972).

6. Toomey, Alice F. *A World Bibliography of Bibliographies, 1964–1974. A List of Works Represented by Library of Congress Printed Catalog Cards.* Totowa, N.J.: Rowman and Littlefield, 1977. 2 vols.

Ten-year supplement to Besterman.

7. *Bibliographic Index: a Cumulative Bibliography of Bibliographies 1937– .* New York: Wilson: 1938– .

A subject list of bibliographies, including those appearing as parts of books, pamphlets, and periodicals. Some 2,800 periodicals were regularly examined in 1990. Useful as a supplement to Besterman for publications after 1963. Presently published every four months and cumulated annually.

SUBJECT INDEXES

8. Peddie, R. A. *Subject Guide to Books Published before 1880.* London: Grafton, 1933–48. 4 vols.

Classified bibliography in four series (with supplements), each containing over 50,000 titles. Arrangement alphabetical by subjects (some 8,000 in each volume) and chronological within each class. Subjects with extensive coverage (more than 100 titles) of pre-1700 literature: agriculture; alchemy; anatomy; architecture; arithmetic; astrology; astronomy; comets; fortification; geography; herbals; the horse; materia medica; military science; music; navigation; pharmacopeias; pharmacy; plague; surgery; syphilis; witchcraft.

BIBLIOGRAPHIES—HISTORY

9. American Historical Association. *Guide to Historical Literature.* New York: Macmillan, 1961. 962 pp.

Classified bibliography in forty sections with subdivisions. Some 20,000 entries with annotations. Still useful though outdated. Reviewed in *American Historical Review* 67 (1961), 76.

10. Read, Conyers, ed. *Bibliography of British History. Tudor Period, 1485–1603.* 2nd ed. Oxford: Clarendon Press, 1959. 624 pp.

Classified, selective bibliography. Topics include: discovery; exploration and colonization; military and naval history; cultural and social history. Partly annotated. 6,543 entries.

11. Levine, Mortimer. *Tudor England 1485–1603.* Conference on British studies, Bibliographical handbooks. Cambridge: University Press, 1968. 115 pp.

More selective than Read (item 10). 2,360 entries. Classified under fourteen headings (including: bibliographies; science and technology; military and naval affairs; religious history; intellectual history). Some annotations.

12. Davies, Godfrey, and Mary Frear Keeler, eds. *Bibliography of British History. Stuart Period, 1603–1714.* 2nd ed. Oxford: Clarendon Press, 1970. 734 pp.

Arrangement similar to Read (item 10). Pp. 249–51: Seamanship and naval warfare; construction and ships. Pp. 368–442: Cultural history.

13. Sachse, William L. *Restoration England, 1660–1689.* Conference on British studies, Bibliographical handbooks. Cambridge: University Press, 1971. 115 pp.

Select bibliography of 2,350 entries, partly annotated. Includes chapters on: agricultural history; science and technology; military and naval history; intellectual history.

14. Boyce, Gray Cowan, comp. *Literature of Medieval History 1930–1975. A Supplement to Louis John Paetow's* A Guide to the Study of Medieval History. New York: Kraus, 1981. Vols. 3, 4 and 5.

Pp. 1654–2330: Medieval culture 1100–1500. Includes sections on: New Aristotle; natural sciences; universities. Vol. 5: Index of personal names (authors and subjects).

15. *International Bibliography of Historical Sciences.* 1 (1926)– . Washington: International Committee of Historical Sciences, 1930– .

Classified. National histories not included. Sections K to M: Modern times (general; history of religion; intellectual history; economic and social history)—some 3,500 entries in Vol. 55 (1986).

BIBLIOGRAPHIES—HISTORY OF SCIENCE

16. Sarton, George. *Horus. A Guide to the History of Science. A First Guide for the Study of the History of Science with Introductory Essays on Science and Tradition.* Waltham, Mass.: Chronica Botanica, 1952. 316 pp.

Outdated but still useful, especially the chapters on: historical methods; historical tables; atlases and gazeteers; encyclopedias; biographical dictionaries; treatises and handbooks on the history of science; scientific instruments; history of special sciences; history of science in special countries. Introductory notes to each chapter and informative comments on authors. The list of journals in the history of science (Ch. 20) has now been supplemented by Corsi and Weindling (see item 21). Reviews: *Isis* 44 (1953), 91–93; *Scripta Mathematica* 19 (1953), 256–57; *Journal of Documentation* 9 (1953), 70–72.

17. Lopez Piñero, J. M., M. Peset Reig, and L. Garcia Ballester. *Bibliografia histórica sobre la ciencia y la técnica en España*. Cuadernos hispanicos de historia de la medicina y de la ciencia, 7, 13. Valencia: Catedra y Instituto de Historia de la medicina, 1968–73. 2 vols.

Bibliography of some 3,000 items of which half are biographical. The rest are classified under twenty headings. No chronological classification.

18. Ferguson, E. S. *Bibliography of the History of Technology*. Cambridge, Mass.: Society for the History of Technology, 1968. 347 pp.

Intended as an introduction to primary and secondary sources in the history of technology. However, chapters I to XI are of use to historians of science. See item 1135.

19. Russo, François. *Elements de bibliographie de l'histoire des sciences et des techniques*. 2nd ed. Paris: Hermann, 1969. 214 pp.

Essential reference book for the historian of science. In four parts: I and II. Generalities (History of science as a discipline; its organization; bibliographies, reference books and periodicals). III. History of science—by period. IV. History of science—by discipline. Part III is arranged for each period under: general studies; bibliographies; learned societies; dictionaries and encyclopedias; source books; works of individual authors and commentaries. Many entries annotated. Part III, C (pp. 50–102) relates to the period 1500–1800 and contains particulars of some one hundred and fifty scientists of the sixteenth and seventeenth centuries. Reviews: *Isis* 61 (1970), 266–67; *Centaurus* 15 (1970), 190–91.

20. Jayawardene, S. A., comp. *Reference Books for the Historian of Science: a Handlist*. Science Museum Library Occasional Publications, 2. London: Science Museum, 1982. 229 pp.

Vade-mecum of some one thousand titles, covering a wide range of research tools—from the *Dictionary of Scientific Biography* to *The Chicago Manual of Style*. In general, does not include books on the history of science apart from bibliographies. Reviewed in *Annals of Science* 40 (1983), 208–09.

21. Corsi, Pietro, and Paul Weindling, eds. *Information Sources in the History of Science and Medicine*. London: Butterworth Scientific, 1983. 531 pp.

Critical surveys (23 chapters, with bibliographies) of the literature, methods, and concepts of the history of science. Areas covered include: Renaissance science; scientific instruments; mathematics; physical science; chemistry; medicine. There are also chapters devoted to bibliographical sources; periodical literature; research methods and sources. Appendix contains a list of journals in the history of science. Reviewed in *Annals of Science* 40 (1983), 669–70 and in *Isis* 75 (1984), 567–69.

Italian translation (Rome: Theoria, 1990, NOT SEEN) reviewed in *Rivista di storia della medicina* (NS), 2 (1992), 86.

22. *Isis Cumulative Bibliography. Bibliography of the History of Science Formed from ISIS Critical Bibliographies 1–90, 1913–65.* Edited by Magda Whitrow. London: Mansell, 1971–84. 6 vols.

The result of weeding, tidying, and sorting the *Critical Bibliographies* 1–90. Vols. 1 and 2: Personalities and Institutions. Vol. 3: Subjects. Vol. 4: Civilizations and Periods. Prehistory to Middle Ages. Vol. 5: Civilizations and Periods. Fifteenth to Nineteenth Centuries. Vol. 6: Author Index. Reviews: *Annals of Science* 30 (1973), 123; 34 (1977), 432; 40 (1983), 310; *Technology and Culture* 19 (1978), 266–67; *Isis* 70 (1979), 160–63.

Book reviews and annotations in the original *Critical Bibliographies* are cited.

23. *Isis Cumulative Bibliography 1966–1975, 1976–1985. A Bibliography of the History of Science Formed from ISIS Critical Bibliographies 91–100, 101–110.* John Neu, ed. London: Mansell, 1979–85; Boston: G.K. Hall, 1989. 4 vols.

Continuation of above. Each cumulation has an index of book reviews.

24. Weiss, Burghard. *Wie finde ich Informationen zur Geschichte der Naturwissenschaften und der Technik?* Zweite . . . Auflage, Berlin: Arno Spitz, 1990. 454 pp. [First published 1985]

Research manual for students of the history of science and technology, with a classified bibliography of some 2,000 titles, mostly histories of science and technology. Partly annotated. With introductory chapters on (1) the state of the subject as an academic discipline in Germany, and (2) library resources and data banks. Reviews: *Isis* 76 (1985), 597–98; *Archives internationales d'histoire des sciences* 40 (1990), 407–8; *Sudhoffs Archiv* 74 (1990), 248–49.

25. "Critical Bibliography of the History of Science and Its Cultural Influences" (*formerly* "Bibliographie analytique des publications relatives à l'histoire de la science"), nos. 1–113. Published in *Isis*, vols. 1 (1913) to 79 (1988). Continued and issued separately as item 26.

26. *Isis Current Bibliography of the History of Science and Its Cultural Influences, 1989– .* John Neu, ed. Philadelphia: History of Science Society, 1989– .

Standard bibliography of the history of science begun by George Sarton in 1912 and edited by him until 1952. Not strictly critical but analytical with occasional notes and citations of book reviews. Mostly confined to literature in West European languages. Classification is chronological, ethnographical, and systematic, no item being entered under more than one heading. (Classification system has undergone some changes since 1955.) Four main divisions: A: History of science: general references and tools. B: Science and its history from special points of views (includes: scientific institutions; scientific instruments; scientific education). C: Histories of the special sciences (subdivided by subject). D: Chronological classification (includes sections for Islam, India, The Far East, Renaissance and Reformation, Seventeenth Century—further subdivided by subject). From 1968 book reviews are separately listed. For 1989, the sections for Renaissance and Reformation, and the Seventeenth Century contain some 500 entries, and the index of book reviews ca. 1,000 titles.

Three series of cumulations published, 1913–1985. See items 22 and 23.

27. *Bulletin signalétique 522: Histoire des sciences et des techniques.* Paris: CNRS, 1961– (Quarterly).

Current bibliography of articles and book reviews, arranged by subject and period. Brief notes. Annual cumulated subject and author indexes. Wide coverage.

27A. *Bibliografia italiana di storia della scienza* 1 (1982)– . Florence: Istituto e Museo di Storia della Scienza, 1985– .

Annual bibliography of writings in the history of science published in Italy. Classified with indexes (author, title and subject). Includes book reviews.

THESES AND DISSERTATIONS

General

28. *Comprehensive Dissertation Index (CDI), 1861–1972.* Ann Arbor, Mich.: Xerox University Microfilms, 1973. 37 vols.

Keyword index to dissertations accepted for academic doctoral degrees granted by US educational institutions (some 417,000 dissertations). On average there are about six keywords to a title of a dissertation. Vols. 1–32: Subjects; Vols. 33–37: Authors.

CDI Ten-Year Cumulation 1973–1982 (38 vols.) listing some 351,000 dissertations was published in 1984. There are annual supplements from 1983.

29. *Dissertation Abstracts International,* Vol. 30– . Ann Arbor, Mich.: University Microfilms, 1969– .

Vols. 1–11, published as *Microfilm Abstracts* (1938–51), and vols. 12–29 as *Dissertation Abstracts* (1952–69). There is a *Retrospective Index* (1970), 9 vols. in 11, to vols. 1–29. These vols. are restricted to US educational institutions.

History

30. Kuehl, Warren F. *Dissertations in History. An Index to Dissertations Completed in History Departments of United States and Canadian Universities 1873–1960, 1961–June 1970, 1970–June 1980.* Lexington: University of Kentucky Press; Santa Barbara, Calif.: ABC-Clio, 1965–85. 3 vols.

Vols. 1 and 2 are each arranged alphabetically by author, with subject indexes (13,579 dissertations). Vol. 3 contains 9,905 entries, arranged systematically by country, with author and subject indexes.

31. Bell, S. P. *Dissertations on British History, 1815–1914: an Index to British and American Theses.* Metuchen, N.J.: Scarecrow Press, 1974. 232 pp.

Contains some 2,300 titles dealing with the history of Great Britain and Ireland. Arranged alphabetically by author in four sequences (political, economic, ecclesiastical, and education). Author and subject indexes.

32. Jacobs, P. M., comp. *History Theses, 1901–70: Historical Research for Higher Degrees in the Universities of the United Kingdom.* London: University of London Institute of Historical Research, 1976. 456 pp.

List of 7,633 theses in broad subject groups. Author and subject indexes.

33. Horn, Joyce M., comp. *History Theses 1971–80. Historical Research for Higher Degrees in the Universities of the United Kingdom.* London: University of London Institute of Historical Research, 1984. 294 pp.

List of 4,416 theses arranged in broad subject groups. Subject and author indexes.

History of Science

34. Fisher, N. W., and W. H. Brock. "A List of Ph.D. Theses in the History of Science and Related Areas in British Universities, 1945–74." *British Journal for the History of Science* 8 (1975), 267–78.

Some 250 theses arranged by author. No index.

35. British Society for the History of Science. *A List of Theses in the History of Science in British Universities in Progress or Recently Completed.* Session 72/73– . London, 1973– (Annual).

Theses listed under institution. Subject index.

LIBRARIES

36. *World Guide to Special Libraries.* 2nd ed. *Handbook of International Documentation and Information*, 17. Munich: K. G. Saur, 1990. 2 vols. 1,196 pp.

Pp. 3–1030: Special libraries, arranged by subject, alphabetically by country within each subject group. Approximately 32,500 libraries under some one thousand subject groups, with cross references (list of subjects, pp. xxvii–xxxiv). Gives for each library: address, telephone number, telex number, fax number, details of holdings, accessible data banks. Pp. 1038–1196: Alphabetical index of libraries.

37. Williams, Moelwyn I., ed. *A Directory of Rare Book and Special Collections in the United Kingdom and the Republic of Ireland.* London: Library Association, 1985. 664 pp.

Arranged by city. Includes for each collection: a brief history; chronological summary of the contents; details of published literature. Name and subject index. Reviewed in *The Library* (6), 8 (1986), 390–92.

38. Taylor, Archer. *Book Catalogues: Their Varieties and Uses.* 2nd ed., revised by Wm. P. Barlow, Jr. Winchester, Hants: St. Paul's Bibliographies, 1986. 284 pp.

Study of catalogues which list printed books owned by private persons, institutions, booksellers, and publishers. Intended for students of W. European culture of the sixteenth, seventeenth, and eighteenth centuries. Pp. 1–91: The varieties of book catalogues. Pp. 92–171: The uses of catalogues. Pp. 172–222: Bibliographies of catalogues. Pp. 223–66: A list of catalogues (of private libraries) that have been recommended between the years 1670 and 1824 by bibliographers for reference use. Pp. 271–84: Indexes: (1) Dealers, institutions, owners, and publishers whose books have been listed in MS or printed catalogues, (2) The kinds of books and subjects listed in catalogues [includes: architecture, astrology, botany, the cabbala, folklore, humanities, machines, magic, mathematics, medicine, military science, natural history, occultism, philosophy, science], (3) Catalogues and compilations based on catalogues, (4) Compilers, editors, collectors, etc. Reviewed in *The Library* (5), 12 (1957), 211–13 ("Brings under scholarly control for the first time the jungle-like subject of book catalogues"— Sears Jayne). 2nd ed. reviewed in *The Library* (6), 11(1989), 169–70.

39. British Museum (Library). *Subject Index of the Modern Works (Books) Added to the Library, 1880/85—1971/75.* London, 1886–1986. 61 vols.

Title varies. (The library departments of the British Museum were incorporated into the British Library from 1 July 1973.) Subjects do not include: personal names (as they are entered in the General Catalogue); belles-lettres; drama; essays; fiction; poetry.

40. British Library. *Subject Catalogue of Printed Books: 1975–1985.* London, 1985. Microfiche.

Supersedes the *Subject Index of Modern Books* (item 39).

40A. Cole, George Watson, comp. *A Catalogue of Books Relating to the Discovery and Early History of North and South America, Forming a Part of the Library of E. D. Church.* New York: Dodd, Mead and Co., 1907. 5 vols.

1,385 entries, annotated and arranged chronologically from the beginning to 1884. Full entries with historical and bibliographical information, facsimiles of title pages, and locations for some fifty libraries. The collection is now in the Huntington Library, Huntington, California. Vol. 5, pp. 2579–635: Index (author, title, subject).

41. London Library. *Subject Index of the London Library.* By C. T. Hagberg Wright. London, 1909–55. 4 vols.

Subject index of the world's largest subscription library (temporary membership available to visiting scholars). All subjects in the humanities represented. For details of holdings, see item 37, pp. 251–53. Useful bibliographical tool. "Valued for its specific headings, adequate cross references and analytical entries" (Walford, *Guide to Reference Material*, 4th ed., 1987, Vol. 3, p. 52). The index is continued on cards.

42. Alden, John, and Dennis C. Landis, eds. *European Americana: A Chronological Guide to Works Printed in Europe Relating to the*

Americas, 1493–1776. (The John Carter Brown Library). New York: Readex Books, 1980– . Vol. 1: *1493–1600*. 1980. 467 pp. Vol. 2: *1601–1650*. 1982. 954 pp.

Union catalogue, arranged chronologically (and alphabetically for each year). Full entries with references to standard bibliographical works. Locations given for some three hundred North American and European libraries (and also Peru). Some annotations. Indexes include a general index of authors, titles, and subjects (Vol. 1, pp. 349–467; Vol. 2, pp. 675–954). Some 4,300 pre-1700 titles. Reviewed in *Papers, Bibliographical Society of America* 78 (1984), 91–95.

43. John Crerar Library, Chicago. *Author-Title Catalog*. Boston, Mass.: G. K. Hall, 1967. 35 vols.

———. *Classified Subject Catalog*. Boston, Mass.: G. K. Hall, 1967. 42 vols.

Library of scientific, technical and medical literature (founded 1895). Some 1,100,000 vols. (563,000 cards). Subject catalog arranged by Dewey decimal classification (subject index in Vol. 42).

44. Roller, Duane H. D., and Marcia M. Goodman. *The Catalogue of the History of Science Collections of the University of Oklahoma Libraries*. London: Mansell, 1976. 2 vols.

Card catalogue of some thirty-five thousand titles (books and microforms)—primary and secondary sources—in one alphabetical sequence. No subject index. Reviews: *Annals of Science* 34 (1977), 619; *Isis* 70 (1979), 278; *Journal for the History of Astronomy* 8 (1977), 215.

45. Bibliothèque Sainte-Geneviève, Paris. *Catalogue des ouvrages imprimés au XVIe siècle. Sciences—Techniques—Médecine*. Redigés par Jacqueline Linet et Denise Hillard. Paris: K. G. Saur, 1980. 493 pp.

The medieval abbey library's collections were dispersed in the sixteenth century. The present collections date from the seventeenth century. Some 2,000 titles.

46. Clark, Alan J., comp. *Book Catalogue of the Library of the Royal Society*. Frederick, Md.: University Publications of America, 1982. 5 vols.

Alphabetical catalogue. Some 62,500 entries (books and pamphlets in stock in November 1981).

201

167 EXEMPLAR BVLLAE SEV
DONATIONIS, AVTORITATE
CVIVS, EPISCOPVS ROMANVS
Alexander eius nominis fextus, con-
cefsit et donauit Caftellæ regibus
et fuis fuccefforibus, regiones
et Infulas noui orbis in
Oceano occidentali Hif-
panorum nauigationi-
bus repertas.·.

¶ THE COPPIE OF THE BULL   171
OR DONATION, BY TH[E]AU-
TORITIE WHEROF, POPE
Alexander the fyxte of that name,
gaue and graunted to the kynges of
Caftyle and theyr fucceffours the
Regions and Ilandes founde in
the Wefte Ocean fea by
the nauigations of the
Spanyardes.

LEXANDER EPISCOPVS, feruus feruorum Dei, Charifsimo in Chrifto filio Ferdinando Regi, et Charifsimæ in Chrifto filiæ Elizabeth Reginæ Caftellæ, Legionis, Aragonum, Siciliæ, et Granatæ, illuftribus, falutem et Apoftolicam benedictionem.

Inter cætera Diuinæ maieftati beneplacita opera et cordis noftri defiderabilia, illud profecto potifimum exiftit vt fides catholica et Chriftiana religio noftris præfertim temporibus exaltetur ac vbilibet amplietur ac dilatetur, animarumque falus procuretur, ac barbaræ nationes deprimantur et ad fidem ipfam reducantur. Vnde cum ad hanc facram Petri fedem Diuina fauente clementia (meritis licet imparibus) euocati fueremus, cognofcentes vos tanquam veros catholicos reges et principes: quales femper fuiffe nouimus, et a vobis præclare gefta, toti pene orbi notifsima demonftrant, nedum id exoptare, fed omni conatu, ftudio, et diligentia, nullis laboribus, nullis impenfis, nullifque parcendo periculis, etiam proprium fanguinem effundendo efficere, ac omnem animum veftrum, omnefque conatus ad hoc iam dudum dedicafse, quemadmodum recuperato regni Granatæ a tyrannis de Sarracenorum hodiernis temporibus per vos, cum tanta Diuini nominis gloria facta, teftatur. Digne ducimur non immerito, et debemus illa vobis etiam fponte, ac fauorabiliter concedere, per quæ huiufmodi fanctum ac laudabile ab immortali deo acceptum propofitum, indies feruentiori animo ad ipfius dei honorem et Imperij Chriftiani propagationem, profequi valeatis. Sane accepimus quod vos qui dudum animum propofueratis aliquas

Lexander byfhoppe, the feruaunte of the feruantes of God: To owre mofte deare beloued fonne in Chrift Kynge Ferdinande, And to owre deare beloued doughter in Chryfte Elyzabeth Queene of Caftyle, Legion, Aragon, Sicilie, and Granata, moft noble Princes, Gretynge and Apoftolical benediction.

Amonge other woorkes acceptable to the diuine maieftie and accordynge to owre hartes defyre, this certeinely is the chiefe, that the Catholyke fayth and Chriftian religion, fpecially in this owre tyme may in all places bee exalted, amplified, and enlarged, wherby the health of foules may be procured, and the Barbarous nations fubdued and brought to the fayth. And therefore wheras by the fauoure of gods clemencie (although not with equall defertes) we are cauled to this holy feate of Peter, and vnderftandynge you to bee trewe Catholyke Princes as we haue euer knowen you, and as youre noble and woorthy factes haue declared in maner to the hole worlde in that with all your ftudie, diligence, and induftrye, you haue fpared no trauayles, charges, or perels, aduenturynge euen the fhedynge of your owne bludde, with applyinge yowre hole myndes and endeuours here vnto, as your noble expeditions achyued in recoueryng the kyngdome of Granata from the tyrannie of the Sarracens in thefe our dayes, doo playnely declare your factes with fo great glorye of the diuine name. For the whiche as we thinke you woorthy, fo owght we of owre owne free wyl fauorably to graunt all thynges whereby you maye dayely with more feruent myndes to the honoure of god and enlargynge the Chriftian empire, profecute your deuoute and laud-

201

Extract from the bull of Pope Alexander VI, *Inter caetera* (4 May 1493), granting to the kings of Castile the regions and lands discovered and to be discovered in the Western Ocean by the navigations of the Spaniards—west of a line of demarcation drawn from north to south, one hundred leagues west of the Azores and the Cape Verde Islands. This line was shifted further west by

203

facri latuacri fufceptionem, qua mandatis Apoftolicis obligati eftis, et per vifcera mifericordiæ Domini noftri Iefu Chrifti attente requirimus, vt cum expeditionem huiufmodi omnino profequi et affumere prona mente orthodoxæ fidei zelo intendatis, populos in huiufmodi Infulis et terris degentes, ad Chriftianam religionem fufcipiendum inducere velitis et debeatis, nec pericula nec labores vllo vnquam tempore vos deterreant, firma fpe fiduciaque conceptis quod Deus omnipotens conatus veftros fœliciter profequetur. Et vt tanti negotij prouintiam Apoftolicæ gratiæ largitate donati, liberius et audacius affumatis, motu proprio non ad veftram vel

169 alterius pro vobis fuper hoc nobis oblatæ petitionis inftantiam, fed de noftra mera liberalitate, et ex certa fcientia, ac de Apoftolicæ poteftatis plenitudine, omnes Infulas et terras firmas inuentas et inueniendas, detectas et detegendas verfus Occidentem et Meridiem, fabricando et conftruendo vnam lineam a polo Arctico, fcilicet Septemtrione, ad polum Antarcticum, fcilicet Meridiem, fiue terræ firmæ et infulæ inuentæ et inueniendæ fint verfus Indiam aut verfus aliam quamcunque partem quæ linea diftet a qualibet Infularum quæ vulgariter nuncupantur de los Azores et Cabo verde centum leucis verfus Occidentem et Meridiem.

Itaque omnes Infulæ et terræ firmæ repertæ et reperiendæ, detectæ et detegendæ a præfata linea verfis Occidentem et Meridiem, quæ per alium Regem aut Principem Chriftianum non fuerint actualiter poffeffæ vfque ad diem natiuitatis Domini noftri Iefu Chrifti proxime præteritum, a quo incipit annus præfens Milleffimus Quadringenteffimus Nonogeffimus tercius, quando fuerint per nuncios et capitaneos veftros inuentæ aliquæ prædictarum Infularum, Autoritate omnipotentis Dei nobis in beato Petro conceffa, ac vicariatus Iefu Chrifti qua fungimur in terris, cum omnibus illatum dominijs, ciuitatibus, caftris, locis, et villis, iuribufque et iurifdictionibus ac partinentijs vniuerfis, vobis heredibufque et fuccefforibus veftris (Caftellæ et Legionis regibus) in perpetuum tenore præfentium donamus, concedimus, et affignamus: Vofque et hæredes ac fucceffores præfatos illarum Dominos, cum plena, libera, et omnimoda poteftate, autoritate, et iurifdictione, facimus, conftituimus, et deputamus. Decernentes nihilo minus per huiufmodi donationem, conceffionem, et affignationem noftram, nullo Chriftiano Principi qui actualiter præfatas Infulas et terras firmas poffederit vfque ad prædictum diem natiuitatis Domini noftri Iefu Chrifti ius quæntum, fublatum intelligi pofse aut auferri debere.

Et infuper mandamus vobis in virtute fanctæ obedi-

Apoftolicall obedience, and erneftely require yowe by the bowels of mercy of owre Lorde Iefu Chrift, that when yowe intende for the zeale of the Catholyke faythe to profecute the fayde expedition to reduce the people of the forefayde landes and Ilandes to the Chriftian religion, yowe fhall fpare no labours at any tyme, or bee deterred with any perels, conceauynge firme hope and confidence that the omnipotent godde wyll gyue good fucceffe to yowre godly attemptes. And that beinge autoryfed by the priuilege of the Apoftolycall grace, yowe may the more freely and bouldly take vpon yowe th[e]enterpryfe of fo greate a matter, we of owre owne motion, and not eyther at yowre requeft or at the inftant peticion of any other perfon, but of owre owne mere liberalitie and certeyne fcience, and by the fulneffe of Apoftolycall power, doo gyue, graunt, and affigne to yowe, yowre heyres and fucceffours, al the firme landes and Ilandes found or to be found, difcouered or to be difcouered toward the Weft and South, drawyng a line from the pole Artike to the pole Antartike (that is) from the north to the Southe: Conteynynge in this donation, what fo euer firme landes or Ilandes are founde or to bee founde towarde *India*, or towarde any other parte what fo euer it bee, beinge diftant from, or without the forefayd lyne drawen a hundreth leaques towarde the Wefte and South from any of the Ilandes which are commonly cauled *De los Azores* and *Cabo Verde*.

All the Ilandes therfore and firme landes, founde and to be founde, difcouered and to be difcouered from the fayde lyne towarde the Weft and South, fuch as haue not actually bin heretofore poffeffed by any other Chriftian kynge or prynce vntyll the daye of the natiuitie of owre Lorde Iefu Chryfte lafte pafte, from the which begynneth this prefent yeare beinge the yeare of owre Lorde. M. CCCC. lxxxxiii. when fo euer any fuch fhalbe founde by your meffingers and capytaines, Wee by the autoritie of almyghtie God graunted vnto vs in faynt Peter, and by the office which we beare on the earth in the fteede of Iefu Chrifte, doo for euer by the tenoure of thefe prefentes, gyue, graunte, affigne, vnto yowe, yowre heyres, and fucceffoures (the kynges of Caftyle and Legion) all thofe landes and Ilandes, with theyr dominions, territories, cities, caftels, towres, places, and vyllages, with all the ryght, and iurifdictions therunto perteynynge: conftitutynge, affignynge, and deputynge, yowe, yowre heyres, and fucceffours the lordes thereof, with full and free poure, autoritie, and iurifdiction. Decreeinge neuertheleffe by this owre donation, graunt, and affignation, that from no Chriftian Prince whiche actually hath poffeffed the forefayde Ilandes and firme landes vnto the day o. the natiuitie of owre lorde beforefayde theyr ryght obteyned to bee vnderftoode hereby to be taken away, or that it owght to be taken away.

Furthermore wee commaunde yowe in the vertue

203

the Treaty of Tordesillas (1494) between Spain and Portugal. From *The First Three English Books on America*, being chiefly translations, ... by Richard Eden, ed. by Edward Arber, Birmingham, 1885, pp. 201, 203 (University of London Library). **Entry 40A.**

DICTIONARIES AND ENCYCLOPEDIAS

47. Zischka, Gert A. *Index lexicorum. Bibliographie der lexikalischen Nachschlagewerke.* Vienna: Verlag Bruder Hollinek, 1959. 290 pp.

Classified bibliography of some seven thousand encyclopedias and dictionaries, arranged in twenty-one main sections. International in scope; no language dictionaries. Introduction contains a history of encyclopedias. Index of names and subject headings.

48. Collison, Robert. *Encyclopedias: Their History Throughout the Ages. A Bibliographical Guide with Extensive Historical Notes to the General Encyclopedias Issued Throughout the World from 350 B.C. to the Present Day.* 2nd ed. New York and London: Hafner Publishing Co., 1966. 334 pp.

Historical survey of encyclopedias (some 600 titles) in eight chapters. Ch. 1: The Beginnings. II: The Middle Ages. III: Bacon to the Encyclopédistes. IV: The Encyclopedia Britannica. V: Brockhaus. VII: The Nineteenth Century. VIII: The Twentieth Century. Pp. xiii–xvi: Chronology. Appendix 4: List of encyclopedias not mentioned in the text.

49. Tonelli, Giorgio. *A Short-Title List of Subject Dictionaries of the Sixteenth, Seventeenth and Eighteenth Centuries as Aids to the History of Ideas.* Warburg Institute Surveys, 4. London: The Warburg Institute, 1971. 64 pp.

Includes about one hundred major dictionaries of the sixteenth and seventeenth centuries—indexed under: encyclopedias; linguistic; theological; philosophical (including natural philosophy); mathematical; alchemical and chemical; medical; legal. Locations given for libraries in Rome, London, Göttingen, Munich, and Paris.

50. Phillips, Lawrence B. *A Dictionary of Biographical Reference Containing One Hundred Thousand Names Together with a Classified Index of the Biographical Literature of Europe and America.* London: Sampson Low & Son, and Marston, 1871. 1,020 pp.

A "students' dictionary" giving brief biographical details, with references to thirty-nine works in which fuller information can be found. Of these works only one is found in Hyamson (item 51).

51. Hyamson, Albert M. *A Dictionary of Universal Biography of All Ages and of All Peoples.* 2nd ed., entirely rewritten. London: Routledge and Kegan Paul, 1951. 680 pp. [First published 1916]

Finding list of some 110,000 biographies in eighteen biographical dictionaries, four encyclopedias, and the *Annual Register* (1850–1949). Profession, years of birth and death given. Reviewed in *Book Review Digest* 1916 and 1952.

52. Williams, E. N. *The Penguin Dictionary of English and European History, 1485–1789*. Penguin Reference Books. London: Penguin Books, 1980. 509 pp.

Provides substantial articles on the important personalities and topics in the mainstream of political history. With numerous cross references and a comprehensive index.

53. Hale. J. R., ed. *A Concise Encyclopedia of the Italian Renaissance*. New York: Oxford Univ. Press, 1981. 360 pp.

"*Aide-mémoire*, vade-mecum, interpreter" for scholar, student, traveller, and enquiring general reader. Consists of some eight hundred articles (themes, topics, and biographies). Subject index (pp. 6–13) arranged under: painting, sculpture, architecture; historic centres; history and politics; social history; economic history; religion; literature; learning and philosophy; travel and exploration; music; science. Pp. 352–57: Table of succession of emperors, popes, and princes, 1300–1620.

54. *Dictionary of Scientific Biography*. Charles C. Gillispie, editor-in-chief. New York: Charles Scribner's Sons, 1970–90. 18 vols.

Biographical articles of varying length on some five thousand scientists, from Thales to Bernal. Vols. 15, 17, and 18 are supplements. Vols. 16 and 18 (pp. 1017–76) contain indexes (names and subjects in one sequence) to vols. 1–15 and vols. 17–18 respectively. The indexes include persons who are not the subjects of biographies. Where relevant, topics indexed are subdivided chronologically (medieval, Renaissance, early/late seventeenth century, etc.). Vol. 15 (pp. 531–818) contains topical articles—six on the "scientific outlook and accomplishments of certain ancient civilizations" and one on early Japanese science. Some twelve hundred contributors. Vol. 16, pp. 479–503: List of scientists by field. Review symposium by Donald Fleming, Joseph Needham, Edward Grant and Jacques Roger in *Isis* 71 (1980), 633–52. See also item 470.

HISTORIANS OF SCIENCE

55. Jayawardene, S. A., and Jennifer Lawes. "Biographical Notices of Historians of Science: a Checklist." *Annals of Science* 36 (1979), 315–94.

Lists some three thousand notices relating to about eight hundred historians of science. Bibliographies indicated. For notices after 1979 see *Isis Critical Bibliography*, Section 39k.

56. History of Science Society. *1986 Guide to the History of Science.* Compiled by P. Thomas Carroll. Philadelphia, 1987. 304 pp. [Triennial publication of the Society]

Pp. 15–96: Guide to the profession (graduate study and research; scholarly journals; societies and organisations). Pp. 97–293: Directory of members (pp. 255–72: subject interests of members).

CHRONOLOGIES

57. Darmstaedter, Ludwig. *Handbuch zur Geschichte der Naturwissenschaften und der Technik in chronologischer Darstellung.* 2nd ed. Berlin: Springer, 1908. 1,262 pp.

Events of scientific and technological importance arranged in one sequence. Pp. 72–159: sixteenth and seventeenth centuries.

58. Aiton, Arthur S., and Louis C. Karpinski. "Chronology of Events of Scientific Importance in North and South America in the Sixteenth Century." *Archeion* 22 (1940), 382–97.

59. Pelseneer, Jean. "Tables chronologiques des principaux faits concernant les sciences qui ont eu lieu en Belgique au XVIe siècle." *Archeion* 23 (1941), 395–409; 24 (1942), 216–17, 218, 223.

Pp. 218–223: Name index.

60. Bologa,Valeriu L. "Die Wissenschaft bei den rumänishen Völkern." *Archeion* 23 (1941), 64–73.

Chronology of sixteenth-century scientific events in Rumania.

61. Datta, Bibhutibhusan. "Chronology of the History of Science in India during the XVIth Century." *Archeion* 23 (1941), 78–83.

62. Hansen, Axel. "Die Wissenschaft in Dänemark im XVI Jahrhundert." *Archeion* 23 (1941), 73–78.

63. Hoeven, J. van der. "Tables chronologiques des principaux faits concernant les sciences qui ont eu lieu aux Pays-Bas au XVIe siècle." *Archeion* 23 (1941), 410–21.

64. Mikami, Yoshio. "A Chronology of the XVIth Century: China and Japan." *Archeion* 23 (1941), 211–26.

65. Gliozzi, Mario. "Tavole di cronologia scientifica italiana dal 1501 al 1600." *Archeion* 24 (1942), 23–81, 217–23; 25 (1943), 54–55.

Chronology of scientific activity and publications. With index of names. Vol. 25, pp. 54–55 contain comments by Lynn Thorndike.

66. Brunet, Pierre. "Table chronologique du XVIe siècle concernant la France." *Archeion* 24 (1942), 198–216, 218–23.

Chronology of scientific activity and publications. Pp. 218–23: Name index.

67. Vera, Francisco. "Tablas cronológicas de España para el siglo XVI." *Archeion* 24 (1942), 403–37.

68. Vetter, Quido. "Tablas cronológicas para ceskoslovensko." *Archeion* 25 (1943), 40–54.

Chronology with name index.

69. Mayer, Alfred. *Annals of European Civilization, 1501–1900.* London: Cassell, 1949. 458 pp.

Pp. 1–30: Index of names. Pp. 31–42: Index of places. Pp. 45–161: Annals 1501–1700. Pp. 309–457: Summaries (41 subjects including: academies, architecture, astronomy, biology, botany, chemistry, discoveries, geography, mathematics, medicine, zoology).

70. Stephanides, M. "Tables chronologiques d'histoire des sciences du XVIe siècle pour ce qui regarde le monde grec." *Archives internationales d'histoire des sciences* 2 (1949), 1144–50.

Includes scientific works of Greek antiquity published during the sixteenth century.

71. Williams, Neville. *Chronology of the Expanding World: 1492–1762.* London: Barrie and Rockliffe, 1969. 700 pp.

Events of the year (month by month) and achievements (arranged under: politics, science, philosophy, etc.) on opposite pages. Pp. 1–423 cover the years to 1700.

72. Langer, William L., comp. *An Encyclopedia of World History, Ancient, Medieval, and Modern, Chronologically Arranged.* 6th ed. London: George G. Harrap, 1972. 1569 pp.

Pp. 383–91: The great discoveries. Pp. 395–587: The early modern period. Pp. 1379–569: Index.

73. Freeman-Grenville, G.S.P. *Chronology of World History. A Calendar of Principal Events from 3100 B.C. to A.D. 1973.* 2nd ed. London: Collings, 1975. 753 pp.

"Skeletal outline of the principal events and dates. . . ." Pp. 336–433 cover the period 1448–1709 under: W. and N. Europe; Central and S. Europe; Africa; The East and Australasia; The Americas; Religion and culture. Pp. 645–753: Index.

74. Grun, B. *The Time-Tables of History: A Chronology of World Events Based on Werner Stein's Kulturfahrplan.* London: Thames and Hudson, 1975. 661 pp.

Covers world events from 5000 B.C. to A.D. 1974 in seven columns: history, politics; literature, theater; religion, philosophy, learning; visual arts; music; science, technology, growth; daily life. Pp. 209–321 cover the years 1447–1702.

75. Powell, Ken, and Chris Cook. *English Historical Facts, 1485–1603.* London: Macmillan, 1977. 228 pp.

Important historical facts assembled under ten headings, with introductory essays and notes. Pp. 83–134: The Church. Pp. 135–54: Education. Pp. 206–20: Selected Tudor biographies. Bibliographical note.

76. Pascoe, L. C., ed. *Encyclopaedia of Dates and Events.* (Teach Yourself Books). 3rd ed. London: English Universities Press, 1979. 830 pp.

Events arranged in four tables: political, constitutional, and religious history; language, literature, and drama; art, architecture, music; economics, exploration, science, and technology. Pp. 683–830: name and subject index. Pp. 244–386 cover the period 1450–1700.

77. Cook, Chris, and John Wroughton. *English Historical Facts, 1603–1688.* London: Macmillan, 1980. 231 pp.

Important historical facts assembled under ten headings, with introductory essays and notes. Pp. 57–82: Selected biographies. Pp. 100–40: The Church. Pp. 182–225: Education and learning. Bibliography.

78. Delorme, Jean. *Les grands dates des temps modernes.* (Que sais-Je? 1147) Paris: Presses Universitaires de France, 1982. 127 pp.

Historic events of the period 1494–1789. Annotated.

79. Babuscio, Jack, and Richard Minta Dunn. *European Political Facts, 1648–1789.* New York: Facts on File Publications, 1984. 387 pp.

Arranged in the following chapters: Heads of state and key ministers; Political chronology; The Enlightenment (Glossary of people and terms of the Enlightenment; Chronology of the Enlightenment); Defence and warfare (principal European armed conflicts, land battles and sea battles); Treaties and diplomacy; The Church; Population; Colonies and dependencies. Index.

80. Parkinson, Claire L. *Breakthroughs. A Chronology of Great Achievements in Science and Mathematics, 1200–1900.* London: Mansell Publishing, 1985. 576 pp.

Provides "concise, clear statements on events or accomplishments contributing to the development of Western science." Pp. 26–134 cover the period 1451–1700. Pp. 522–76: Name and subject indexes. With a bibliography of sources.

81. Steinberg, S. H. *Historical Tables, 58 B.C.–A.D. 1985.* 11th ed., updated by John Paxton. London: Macmillan Reference Books, 1986. 277 pp.

Pp. 96–157 cover the period 1440–1700. Events classified under: Western Europe; Central, Northern and Eastern Europe; countries overseas; ecclesiastical history [until 1656]; constitutional and economic history [after 1656 Constitutional history; economic history and natural science]; cultural life.

82. Gascoigne, Robert Mortimer. *A Chronology of the History of Science, 1450–1900.* New York: Garland Publishing, 1987. 585 pp.

Chronicles the work of some one thousand persons in about five thousand entries. Part 1 (The Cognitive Dimension) covers scientific events and publications, classified under nineteen subjects, arranged in decades. Part 2 (The Social Dimension) is subdivided into twelve geographical regions, arranged in decades. Pp. 537–44: Teaching institutions (indexed). Pp. 544–85: Index (names, subjects, societies, etc.).

83. Cook, Chris, and John Stevenson. *British Historical Facts, 1688–1760.* London: Macmillan, 1988. 252 pp.

Important historical facts in the following chapters: The monarchy; chronology; administrations and political biographies; holders of major public office; Parliament; elections; religion; treaties and diplomacy; the armed forces; the colonies; law and order; social developments; economy and finance; local government.

84. Hellemans, Alexander, and Bryan Bunch. *The Timetables of Science. A Chronology of the Most Important People and Events in the History of Science.* New York: Simon and Schuster, 1988. 656 pp.

Pp. 90–145: The Renaissance and the Scientific Revolution (1453–1659). Pp. 146–87: The Newtonian epoch (1660–1734). For each of these periods there is an introduction (overview), and the tables are interspersed with short essays on notable topics. Pp. 609–56: Name and subject indexes.

85. Manley, Deborah, ed. *The Guinness Book of Records 1492; The World Five Hundred Years Ago.* Enfield, Middlesex: Guinness Publishing, 1992. 192 pp.

"The world of five hundred years ago seen "through a periscope" by some fifty specialists. "Thousands of records [events] are recorded, in everything from astrology to zoos, from Greenland to the Cape of Good Hope, from China to Peru." In eleven chapters. Richly illustrated. With index and reading list.

From Manuscript to Print. Humanism

BIBLIOGRAPHIES. CATALOGUES OF EARLY PRINTED BOOKS

86. Bigmore, E.C., and C.W.H. Wyman. *A Bibliography of Printing.* With notes and illustrations. London: Bernard Quaritch, 1880–86. 3 vols.

Bibliography of the history of the printed book. With biographical notes and some detailed annotations. 7,500 entries, alphabetically arranged. No index.

87. Arber, Edward. *The Term Catalogues, 1668–1709 A.D.; With a Number for Easter Term, 1711 A.D. A Contemporary Bibliography of English Literature in the Reigns of Charles II, James II, William and Mary, and Anne.* Edited from the very rare quarterly lists of new books and reprints . . . issued by the Booksellers of London. London, 1903–1906. 3 vols.

Each volume has two indexes (title; name and subject). Subjects indexed include: algebra, almanacks, arithmetic, astrology, astronomical maps, astronomy, bibliography, botany, chemistry, cookery, cosmography, ephemerides, fortification, geography, geometry, globes, Gregorian calendar, gunnery, horology, Julian calendar, logarithms, longitude, maps, mathematics, mechanics, medical works, mensuration, meteorology, meteors, mines, mining, Napier's rods, navigation, nautical almanacs, optics, perspective, philosopher's stone, physics, physiology, Rudolphine Tables, surveying, trigonometry, veterinary science, weights and measures, zoology.

88. Growoll, A., and Wilberforce Eames. *Three Centuries of English Booktrade Bibliography. An Essay on the Beginnings of Booktrade Bibliography since the Introduction of Printing and in England since*

1595. *Also a List of Catalogues, et cetera Published for the English Booktrade from 1595–1902.* London: The Holland Press, 1964. 195 pp. [First published 1903]

Pp. 107–32: Booktrade catalogues published from 1595 to 1714.

89. Steinschneider, Moritz. *Die europäischen Übersetzungen aus dem arabischen bis Mitte des 17. Jahrhunderts.* Graz: Akademische Drück- und Verlagsanstalt, 1956. 84 + 108 pp.

In three parts: (1) works of known translators; (2) works whose translators are not known or uncertain; (3) anonymous works. Works whose authors and translators are uncertain. First published in *Sitzungsberichten der K. Akademie der Wissenschaften in Wien, philos.-hist. Kl.,* 149 (1904) and 151 (1905).

90. Richardson, E. C. *A List of Printed Catalogues of Manuscript Books.* (*A Union World Catalogue of Manuscript Books,* Vol. 3). Edited by the American Library Association. New York: Wilson, 1935. 386 pp.

Rough checklist (arranged by city) based chiefly on printed sources, compiled for the projected *Union World Catalog of Manuscript Books* which was not realised. Reviewed in *Revue d'histoire ecclésiastique* 31 (1936), 621–30.

91. Spargo, J. W. "Some Reference Books of the Sixteenth and Seventeenth Centuries: A Finding List." *Papers, Bibliographical Society of America,* 31 (1937), 133–75.

Annotated bibliography of some 120 works, not limited to English. With index of subjects. Complements item 125. "It is through books of this kind that we can most easily and accurately estimate the power of some of the most influential elements in the thought of past ages. Through them we are able to grasp how the learned and perhaps the not-so-learned worked, what books they used, what gave them their opinions on this matter or that; and thus we can follow some of the ramifications in the development of their ideas" (author). See also items 94 and 121.

92. Klebs, Arnold C. "Incunabula scientifica et medica: Short Title List." *Osiris* 4 (1938), 1–359.

Alphabetical list of some 1,000 titles, with references to standard bibliographical works (e.g. *Gesamtkatalog der Wiegendrücke;* Hain's *Repertorium bibliographicum;* Osler's *Incunabula medica*).

93. Sarton, George. "The Scientific Literature Transmitted Through the Incunabula." *Osiris* 5 (1938), 41–247.

Pp. 43–93: Introduction, in Sarton's inimitable style, to scientific incunabula. Pp. 94–181: Sixty facsimiles with explanatory notes. Pp. 182–245: Appendices (the most popular authors; the bestsellers, anonyma; chronological list).

94. Taylor, Archer. *Renaissance Guides to Books. An Inventory and Some Conclusions.* Berkeley: University of California Press, 1945. 130 pp.

Reviews the "kinds of bibliographies made during the Renaissance" and shows "what amazing equipment the Renaissance scholar possessed." Pp. 85–117: Brief guide to bibliographies published before 1700. See also items 91, 121, 125.

95. Starnes, De Witt T. *Renaissance Dictionaries, English-Latin and Latin-English.* Austin: University of Texas Press, 1954. 427 pp.

Survey of E–L and L–E lexicography of the period 1440–1740. Analysis of 22 works from the *Promptorium parvulorum* (ca. 1440) to Robert Ainsworth's *Thesaurus linguae latinae compendiarius* (1736). Pp. 393–94: Bibliography of references to the history of L–E lexicography. Pp. 394–99: Short title list of L–E and E–L dictionaries (1500–ca. 1800) in American libraries. Pp. 403–28: Name, title, and subject index.

96. Sears, Jayne. *Library Catalogues of the English Renaissance.* Berkeley: University of California Press, 1956. 226 pp. [Reissued in 1983 with new preface and notes (pp. viii–xiv) by St. Paul's Bibliographies, Foxbury Meadows, Godalming, Surrey].

A survey of extant catalogues (848) of English libraries (institutional and private). These catalogues "provide the best available subject index to the reading materials available in Renaissance England." Pp. 206–24: Name and subject index. (Subjects include: arithmetic, astrology, astronomy, cosmography, geography, geometry, mathematics, medicine, philosophy.) Reviews: *The Library* (5), 13 (1958), 64–66; *TLS* 8 July 1983, p. 738; *Journal of Library History* 20 (1985), 320; *Indexer* 14 (1984), 75.

97. Kristeller, Paul Oskar, F. Edward Cranz, and Virginia Brown, eds. *Catalogus translationum et commentariorum. Medieval and Renaissance Latin Translations and Commentaries. Annotated Lists and Guides.* Washington, D.C.: Catholic University of America Press, 1960–92. 7 vols.

Lists and describes the Latin translations of ancient Greek authors and the Latin commentaries on ancient Latin (and Greek) authors up to the year 1600. It is intended to illustrate the impact of Greek and Roman literature on the Middle Ages and the Renaissance. Includes sixteenth-century eds. of:

Alexander Aphrodisiensis, Hermetica philosophica, Aristarchus, Autolycus, Hypsicles, Arator, Pappus, Strabo, Theophrastus, Thessalus astrologus, Vitruvius, Dioscorides, Paulus Aegineta, Cato Censor, C. Plinius Secundus, Marcus Terentius Varro, Pomponius Mea, Flavius Renatus Vegetius.

98. Cabeen, David C., and Jules Brody, gen. eds. *A Critical Bibliography of French Literature*. Vol. III. *The Seventeenth Century*. Nathan Edelman, ed. Vol. IIIA. *The Seventeenth Century. Supplement*. Gaston Hall, ed. Syracuse, N.Y.: Syracuse Univ. Press, 1961–83. 638 pp.; 460 pp.

Select annotated bibliography. Nos. 528–78, 5079–88A: Education (background material); 4155–4455, 7705–7902: Philosophical and scientific literature. Also contains separate sections for Pascal and Descartes.

99. Cosenza, Mario Emilio. *Biographical and Bibliographical Dictionary of the Italian Humanists and of the World of Classical Scholarship in Italy, 1300–1800*. 2nd ed. Boston, Mass.: G.K. Hall, 1962–67. 6 vols. [1st ed. of 1954 is on microfilm (Graphic Microfilm Corporation)].

A reproduction of some 70,000 handwritten cards, with biobibliographical notes from Cosenza's file of humanists. "Covers all scholars of Italian origin who devoted their studies to any phase of Greek and Roman civilization. Includes scholars in the fields of literature, art, science, social customs, archeology, epigraphy, mathematics, law, and medicine." Also includes foreign scholars who spent a considerable time in Italy, and all patrons of scholarship from monarchs to booksellers and collectors. Vol. 5: Synopsis and bibliography; Vol. 6: Supplement.

100. Wightman, W. P. D. *Science and the Renaissance. Vol. 1: An Introduction to the Study of the Emergence of the Sciences in the Sixteenth Century. Vol. 2: An Annotated Bibliography of the Sixteenth-Century Books Relating to the Sciences in the Library of the University of Aberdeen*. Edinburgh: Oliver and Boyd, 1962. 2 vols. 327 pp.; 293 pp.

Study based on a collection of some 750 sixteenth-century books in the library of King's College, Aberdeen. No subject index. Reviews: *Renaissance News* 16 (1963), 215–17; *Bibliothèque d'humanisme et Renaissance* 25 (1963), 433–36; *History of Science* 3 (1964), 131–34.

101. Schüling, Hermann. *Bibliographischer Wegweiser zu dem in Deutschland erschienenen Schrifttum des 17. Jahrhunderts*. Berichte und Arbeiten aus der Universitätsbibliothek Giessen, 4. Giessen, 1964. 176 pp.

Sources for a German national bibliography of the seventeenth century. (Georgi's *Allgemeines europäisches Bücherlexikon* [1742–58]) contains only about a third of the German book production.) 1,123 items. (Book-fair catalogues, retrospective bibliographies, selective bibliographies, library catalogues, regional bibliographies, bibliographies of publishers and printers, indexes of prohibited books, catalogues of academic theses, catalogues of newspapers, etc.)

102. Kristeller, Paul Oskar. *Iter Italicum. A Finding List of Uncatalogued or Incompletely Catalogued Humanistic Manuscripts of the Renaissance in Italian and Other Libraries.* 6 vols. London: Warburg Institute, 1965–1992. 6 vols.

Vols. 1 and 2: Italy, Vatican City; Vol. 3: Australia to Germany; Vol. 4: Great Britain to Spain; Vols. 5 and 6: Sweden to Yugoslavia, Utopia [unidentified private collectors]; supplements: Italy, Vatican, Austria to Spain. Primarily confined to philosophical, scholarly, and literary MSS in Latin that were copied during the years 1300 to 1600, containing the writings of scholars who were active during that period. Some scientific writings are included where there seemed to be a connection with the main stream of the humanistic and philosophic culture of the Renaissance period.

103. Gerlo, Aloïs. *Bibliographie de l'humanisme belge, précédée d'une bibliographie générale concernant l'humanisme européen.* Brussels: Presses Universitaires de Bruxelles, 1965. 248 pp.

Systematic bibliography. Pp. 1–82: European humanism; pp. 127–224: Belgian humanists—biobibliography. See also items 112 and 122.

104. Kristeller, Paul Oskar. *Latin Manuscript Books Before 1600. A List of the Printed Catalogues and Unpublished Inventories of Extant Collections.* 3rd ed. New York: Fordham Univ. Press, 1965. 284 pp.

Annotated bibliography of catalogues of pre-1600 MSS. In three sections: Bibliography and statistics of libraries and their collections of MSS; Works describing MSS of more than one city; Printed catalogues and handwritten inventories of individual libraries, by cities. Pp. 233–84: supplementary material.

105. *Bibliographie internationale de l'humanisme et de la Renaissance.* Travaux parus en 1965– . Geneva: Droz, 1966– .

Annual bibliography of writings on humanism (fifteenth and sixteenth centuries) and the Renaissance. Arranged in two sequences: personalities and anonymous works of the Renaissance; subjects. Author index.

106. Adams. H. M. *Catalogue of Books Printed on the Continent of Europe, 1501–1600, in Cambridge Libraries.* Cambridge: University Press, 1967. 2 vols.

More than 30,000 entries covering the University and College libraries. "The early books in the Cambridge College collections are to a large extent the working tools of scholars. Taken together, they reflect the dominant interests of those engaged in teaching and research in the university" (J.W. Jolliffe). Reviewed in *The Library* 24 (1969), 63–65 and *TLS* 28 Mar. 1968, p. 234.

107. Carter, John, and Percy H. Muir. *Printing and the Mind of Man. A Descriptive Catalogue Illustrating the Impact of Print on the Evolution of Western Civilization During Five Centuries.* With an introductory essay by Denys Hay. London: Cassell and Co., 1967. 280 pp.

Extensively annotated catalogue of some four hundred books—of which 170 are pre-1700 (70 scientific)—which are of prime importance because of the ideas they brought to the world for the first time. The books formed part of an exhibition illustrating the history of the printing industry. Reviewed in *TLS* 22 June 1967, p. 561 (see also p. 579).

108. Stillwell. M. B. *The Awakening Interest in Science During the First Century of Printing, 1450–1550. An Annotated Checklist of First Editions Viewed from the Angle of Their Subject Content.* New York: Bibliographical Society of America, 1970. 399 pp.

List of nine hundred important and typical texts arranged alphabetically by author under: astronomy, mathematics, medicine, natural science, physics, and technology. Each section is preceded by a subject analysis. Includes indexes and chronological tables. Annotations include references to bibliographies and histories of science. Reviewed in *TLS* 4 Dec. 1970, p. 1428.

109. Pollard, A. W., and G. R. Redgrave. *A Short-Title Catalogue of Books Printed in England, Scotland, and Ireland and of English Books Printed Abroad, 1475–1640.* 2nd ed. by W. A. Jackson, F. S. Ferguson, and K.F. Pantzer. London: Bibliographical Society, 1972–91. 3 Vols.

Comprehensive record of English books. Arranged alphabetically by author. Locations (in representative British and American libraries) given. Vol. 3 contains an index of printers and publishers by Katherine Pantzer and a chronological index by Philip R. Rider. Reviewed in *The Library* (6), 9 (1987), 289–92; 14 (1992), 146–48.

110. Wing, D.G. *Short-Title Catalogue of Books Printed in England, Scotland, Ireland, Wales, and British America and of English Books Printed in Other Countries 1641–1700.* Second ed., rev. and enl. New York: The Modern Language Association of Americas, 1972–88. 3 vols. [First published 1945–51]

Continuation of Pollard and Redgrave's *STC* (item 109). Convenient locations given. Reviews: *TLS* 26 Jan. 1973, p. 100; 17 Dec. 1982, p. 1403; *The Library* (6), 11 (1989), 383–88.

111. Watson, George, ed. *The New Cambridge Bibliography of English Literature.* Cambridge: Cambridge Univ. Press, 1972–77. 5 vols.

Vol. 1: 600–1660; Vol. 2: 1660–1800; Vol. 5: Index. Vols. 1 and 2 contain chapters on travel, philosophy, science, and education. Vol. 5 is a general index to vols. 1–4, listing primary authors and major anonymous works as well as certain subject headings from the bibliography. First published as *The Cambridge Bibliography of English Literature* (1940–57, 4 vols. + supp.).

112. Gerlo, Aloïs, and Hendrik D. L. Vervliet. *Bibliographie de l'humanisme des anciens Pays-Bas. Avec un répertoire bibliographique des humanistes et poètes neo-latins.* Brussels: Presses Universitaires de Bruxelles, 1972. 546 pp.

Revised ed. of 103 but without the chapter on European humanism. Pp. 233–491: biobibliography of humanists.

113. *ABHB: Annual Bibliography of the History of the Printed Book and Libraries.* Volume 1– : Publications of 1970. The Hague: Martinus Nijhoff, 1973– .

". . . aims at recording all books and articles of scholarly value which relate to the history of the printed book, to the history of the arts, crafts, techniques and equipment, and of the economic, social and cultural environment, involved in its production, distribution, conservation, and description." Enumerative bibliographies are not generally included as they are recorded in other bibliographies (*Bibliographic Index* and *Bibliographische Berichte*). Classified under 12 main headings (including: booktrade; bibliophily; and libraries). Topical subheadings include: philosophy; mathematics and pure sciences; applied sciences (including medicine); geography. Vol. 17A is a cumulated subject index for 1970–1986.

114. Allison, A. F., and V. F. Goldsmith. *Titles of English Books (and of Foreign Books Printed in England). An Alphabetical Finding-List by Title of Books Published under Author's Name, Pseudonym or Initials.* Volume 1: *1475–1640.* Volume 2: *1641–1700.* Folkestone (Kent): Dawson, 1976–77. 2 vols.

Complements items 109 and 110.

115. Munby, A. N. L., and L. Coral, eds. *British Book Sale Catalogues, 1676–1800. A Union List*. London: Mansell, 1977. 146 pp.

Some four hundred catalogues of sales prior to 1700. Press marks of British Library holdings given. Reviewed in *The Library* (5), 33 (1978), 336–38.

116. Jayawardene, S. A. "Western Scientific Manuscripts Before 1600: a Checklist of Published Catalogues." *Annals of Science* 35 (1978), 143–72.

386 items. Includes some 125 separately published catalogues.

117. Halkett, Samuel, and John Laing. *A Dictionary of Anonymous and Pseudonymous Publications in the English Language*. [Vol. 1] *1475–1640*. 3rd (rev. and enl.) ed. John Horden, ed. London: Longman, 1980. 271 pp.

New ed. being published in three chronological segments, volumes for 1641–1700 and 1701–1800 to follow. This volume contains some four thousand anonyma and pseudonyma. Entries include: title; place and date of printing; references to *STC* and other standard bibliographies. Annotation includes name and dates of writer where known. Reviewed in *Library Quarterly* 53 (1983), 71–72.

118. Wells, Ellen B. "Scientists' Libraries: A Handlist of Printed Sources." *Annals of Science* 40 (1983), 317–89.

Lists catalogues of and publications relating to books and MSS owned by scientists (880) of whom about 85 flourished during the sixteenth and seventeenth centuries.

119. Brooks, Richard A., gen. ed. *A Critical Bibliography of French Literature*. Vol. II (Revised). *The Sixteenth Century*. Edited by Raymond C. La Charité. Syracuse, N.Y.: Syracuse Univ. Press, 1985. 847 pp.

Select annotated bibliography. Nos. 180–204: Education (background material); 745–85: Science and travel (background material); 2186–2320: Scientific poetry; 4177–4255: Agriculture; astronomy and pseudo-science; biological sciences; geography, exploration and travel; mathematics; medicine; technology.

120. Kohl, Benjamin G. *Renaissance Humanism, 1300–1550. A Bibliography of Materials in English*. New York: Garland, 1985. 354 pp.

Writings of humanists in English translation and studies in English on Renaissance humanism. 3,088 entries. No annotations. Author and subject index. Intended for the student whose first (and often only) language is English.

121. Cochetti, Maria. *Repertori bibliografici del Cinquecento.* Con una premessa di Alfredo Serrai. Il bibliotecario, 3. Rome: Bulzoni, 1987. 152 pp.

Annotated bibliography of 106 published bibliographies of the sixteenth century. Includes: eight works on medicine, two on the mechanical arts, and one each on natural philosophy, zoology and botany. Eleven standard reference works are cited and the library where each book was consulted is named. Indexes. See also items 91, 94 and 125.

122. De Schepper, Marcus, and Chris L. Heesakkers, eds. *Bibliographie de l'humanisme des anciens Pays-Bas. Avec un répertoire bibliographique des humanistes et poètes néo-latins: supplément 1970– 1985* avec compléments à l'édition de A. Gerlo et H. D. L. Vervliet (Bruxelles, 1972). Brussels: Koninklijke Academie voor Wetenschappen, Letteren en Schone Kunsten van Belgie, 1988. 439 pp. [NOT SEEN]

See items 103 and 112. Reviewed in *The Library* (6), 12 (1990), 246–48.

PRINTING AND THE BOOK TRADE

123. Thompson, James Westfall. *The Frankfort Book Fair. The Franco-fordiense Emporium of Henri Estienne.* Edited with historical introduction. Original Latin text with English translation on opposite pages and notes. Chicago: The Caxton Club, 1911. 204 pp. [Reprinted in 1968 by Burt Franklin]

Pp. 3–123: Introduction: the beginnings of the German book trade. The origin and character of the Frankfurt Book Fair. Famous Frankfurt printers. The origin of the Mess-Katalog. The decline of the Frankfurt Book Fair and the growth of the Leipzig book trade. Pp. 125–80: "Encomium of the Frankfort Fair" by Henri Estienne. Pp. 195–204: Index.

124. Bay, J. Christian. "Conrad Gesner (1516–1565), the Father of Bibliography. An Appreciation." *Papers, Bibliographical Society of America* 10 (1916), 53–88.

Biographical sketch based on Swiss and German sources with emphasis on his *Bibliotheca Universalis.*

125. Besterman, Theodore. *The Beginnings of Systematic Bibliography.* 2nd ed., revised. London: Oxford Univ. Press, 1936. 81 pp.

The manuscript age. Johann Tritheim (1462–1516), father of bibliography. The earliest medical and legal bibliographies. Conrad Gesner, universal bibliographer. The beginnings of national bibliography. Cornelius à Beughem (1638–1710), poly-bibliographer. National and subject bibliography in the seventeenth century. List of bibliographies printed to the end of the sixteenth century (45 items). Classified table of bibliographies to the end of the sixteenth century (pp. 30–31).

126. Malclès, Louise-Noëlle. *La bibliographie.* 2nd ed. Paris: Presses Universitaires de France, 1962. 136 pp.

Pp. 15–60: Excellent survey of the history of bibliography in the sixteenth and seventeenth centuries. Pp. 95–103: List of specialized retrospective bibliographies published in the nineteenth century (some 40 scientific).

127. Bennett, H. S. *English Books and Readers, 1558 to 1603. Being a Study of the Book Trade in the Reign of Elizabeth I.* Cambridge: University Press, 1965. 320 pp.

Surveys the printed matter (other than ephemera) "in the hope of showing the many ways in which the printer put at the disposal of readers a wealth of matter touching almost every side of their daily life, both intellectual and practical." Pp. 112–258: The variety of books (subjects include education; medicine; husbandry, agriculture; arithmetic, astronomy and popular science; geography; witchcraft). See 132.

128. Pollard, Graham, and Albert Ehrman. *The Distribution of Books by Catalogue from the Invention of Printing to A.D. 1800 Based on Material in the Broxbourne Library.* Cambridge: printed for presentation to members of the Roxburghe Club, 1965. 427 pp.

Series of historical essays about book catalogues: fifteenth-century broadsides and advertisements concerning the book trade; catalogues of sixteenth-century printers; German book-fair catalogues; the "Latin trade" in England (i.e., books imported in the sixteenth and seventeenth centuries— the classics, medicine, science, theology, eds of the Fathers, polyglot bibles . . .); seventeenth-century catalogues; booksellers' catalogues; antiquarian catalogues; inventories; auction sales; institutional library catalogues; private libraries open to scholars. Pp. 283–352: Bibliography. Pp. 353–427: Name and subject index.

129. Berry, W. Turner, and H. Edmund Poole. *Annals of Printing. A Chronological Encyclopaedia from the Earliest Times to 1900.* London: Blandford Press, 1966. 315 pp.

Pp. 10–141 cover the years 1450–1700. Pp. 297–315: Name and subject index. With bibliographical references. Reviewed in *The Library* (5), 23 (1968), 151.

130. Fischer, Hans. "Conrad Gesner (1516–1565) as Bibliographer and Encyclopedist." *The Library* (5), 21 (1966), 269–81.

Discusses Gesner's *Bibliotheca universalis,* his *Historia animalium,* the *Historia plantarum,* and his eds. of Galen.

131. Labowsky, L. "Bessarione." *Dizionario biografico degli italiani.* Vol. 9, pp. 686–96. Rome: Enciclopedia Trecanni, 1967.

Emigré Greek humanist, metropolitan of Nicaea, Bessarion joined the Roman Catholic Church and was made a Cardinal (1439) and patriarch of Constantinople (1463). His collection of MSS was bequeathed to the Republic of Venice, and formed the nucleus of the Marciana (Library of St. Mark). Among his protégés were Peuerbach and Regiomontanus. See items 154 and 196.

132. Bennett, H. S. *English Books and Readers, 1475 to 1557. Being a Study in the History of the Book Trade from Caxton to the Incorporation of the Stationers' Company.* 2nd ed. Cambridge: University Press, 1969. 537 pp. [First published 1952]

Topics covered include: literacy and educational possibilities of the period; patronage; the demand for books; translations and translators. Pp. 65–151 contain a discussion of the variety of books (subjects include medicine, manuals for farmers, arithmetic, astronomy and popular science, prognostication, geography).

133. Clair, Colin. *A Chronology of Printing.* London: Cassell, 1969. 229 pp.

Pp. 8–94 cover the period 1450–1700. Pp. 193–228: Name and subject index. Reviewed in *TLS,* 24 July 1969, p. 802.

134. Bennett, H. S. *English Books and Readers, 1603 to 1640. Being a Study in the History of the Book Trade in the Reigns of James I and Charles I.* Cambridge: University Press, 1970. 253 pp.

Examines "the number and variety of books which were published between the death of Elizabeth I and 1640, the eve of the Civil War" and investigates "the different kinds of audiences the books hoped to interest." Pp. 87–198: The variety of books (includes: education; medicine; horses and horsemanship; gardening; husbandry; arithmetic; astronomy and popular science; geography, travel and adventure). Review: *The Library* (5), 26 (1971), 274–75. Sequel to 127.

gia Xyſto detuliſſe,quo repente renunciato,adoratoque,Beſſarion dixiſſe fertur: Hæc tua,
Nicolae, intempeſtiua ſedulitas, & tiaram mihi, & tibi galerum eripuit . Nec multo pòſt
cum honore legationis in Galliam eſt ablegatus, quòd Xyſtus noua licentia Pontificatum
nomine principatus gerendum ratus, liberè, & grauiter, religioſèque ſententias dicentis
vultum non perferret. Sed Beſſarion,dum è Gallia rediens côcepto morbo Rauennæ ſub.
ſtitiſſet,ſeptuageſimo ſeptimo ætatis anno moritur. Funus autem Romæ celebratum in
templo Apoſtolorum,vbi marmoreum tumulum, viuens ſibi cum hac Græca inſcriptio-
ne extruxerat.hanc ſic vertit Maioranus Salentinus.

Beſſarion feci hunc tumulum,qui conderet oſſa | *Venerat vnde olim ſpiritus aſtra petet.*

MAIORANI.

Viuens adhuc feci hoc ſepulchrum corpori | *Me docta genuit Græcia,at honores dedit*
Beſſarion,ad Deum remittens ſpiritum. | *Vrbs Roma,qua mihi perpetuò eſt patria.*

IANI VITALIS,

Non tibi ſit laudi ſanctum celebraſſe Platonem, | *Verùm quòd per te migrauit Græcia Romam:*
Caſtaq̃, Socraticæ frena pudicitiæ: | *Et didicit Latios Attica Muſa ſonos.*
Non quòd virtutũ exemplũ,quod lumen honoris, | *Per te hinc Romanas miratur Tybris Athenas*
Quòd Sol extincta religionũ eras: | *Argolicam & Romam Græcia Beſſarion.*

LATOMI.

Woodcut portrait of Cardinal Bessarion by Tobias Stimmer. From P. Giovio,
Elegia virorum literis illustrium, Basle, 1577, p. 30 (Bodleian Library, Oxford,
K.5.16.Art.). **Entries 131, 154, 196, 206.**

135. Thornton, John L., and R. I. J. Tully. *Scientific Books, Libraries and Collectors. A Study of Bibliography and the Book Trade in Relation to Science.* 3rd rev. ed. London: Library Association, 1971. 508 pp. *Supplement 1969–75.* London, 1978. 172 pp. [First published 1954]

Pp. 1–141: Scientific literature from before the invention of printing to 1700. Pp. 339–53: Private scientific libraries. Pp. 381–465: Bibliography. Updated by supplement. Reviewed in *TLS,* 3 Dec. 1971, p. 1533 and *Isis* 63 (1972), 567–68.

136. McLean, Antonia. *Humanism and the Rise of Science in Tudor England.* London: Heinemann, 1972. 258 pp.

On the invention of printing and its influence on education, science and medicine. Reviewed in *Historia Mathematica* 1 (1974), 348–50.

137. Febvre, Lucien, and Henri-Jean Martin. *The Coming of the Book: the Impact of Printing 1450–1800.* London: NLB, 1976. 378 pp. [Translation of the 1958 French ed.]

Aims to "establish how and why the printed book was something more than a triumph of ingenuity, but was also one of the most potent agents at the disposal of Western civilization in bringing together the scattered ideas of representative thinkers" and to show that it was "one of the most effective means of mastery over the whole world." Reviewed in *TLS* 14 October 1977, p. 1191.

138. Grendler, Paul F. *The Roman Inquisition and the Venetian Press, 1540–1605.* Princeton, N.J.: Princeton Univ. Press, 1977. 374 pp.

Documented internal history of the Index of Prohibited Books as it affected the book trade of Venice—"one of the great European publishing centers," which accounted for about half the book trade of sixteenth-century Italy. (Chapters: The Venetian bookmen; the growth of censorship; the Counter-Reformation; the clandestine book trade; Venice and Rome part company; the Republic protects the press; the waning of the Index; the impact of Index and Inquisition on Italian intellectual life.) Pp. 304–74: Inventories of prohibited books, bibliography, and index.

139. Eisenstein, Elizabeth. *The Printing Press as an Agent of Change. Communications and Cultural Transformations in Early Modern Europe.* Cambridge: Cambridge Univ. Press, 1979. 2 vols. 794 pp.

Studies the shift from script to print in W. Europe, and its consequences. "The communication shift altered the way Western Christians viewed their sacred book and the natural world. It made the word of God appear more multiform and his handiwork more uniform. The printing press laid the basis for both literal fundamentalism and for modern science." Pp. 520–682: Technical literature goes to press; Resetting the stage for the Copernican revolution; Sponsorship and censorship of scientific publication. Pp. 709–67: Bibliographical index. Reviews: *Isis* 71 (1980), 474–77; *Papers, Bibliographical Society of America* 75 (1981), 228–30; *The Library* (6), 3 (1981), 261–63. For abridgement, see 142.

140. Lowry, Martin. *The World of Aldus Manutius: Business and Scholarship in Renaissance Venice.* Oxford: Blackwells, 1979. 350 pp.

On the "more mundane but also more pressing details of how a fifteenth-century publishing house was run and what made it successful." Pp. 309–31: Bibliography. A revised Italian ed. (NOT SEEN) was published in 1984 (Il Veltro, Rome).

141. Chrisman, Miriam Usher. *Lay Culture, Learned Culture: Books and Social Change in Strasbourg, 1480–1599.* New Haven: Yale Univ. Press, 1982. 401 pp.

Study of book publication in Strasbourg and its impact on the community. Pp. 1–75: The book trade (the printers; the intellectual community; the reading and book-buying public). Pp. 123–41: Scientific publication, 1480–1520. Pp. 170–91: Scientific publication, 1530–1548. Pp. 207–30: Vernacular literature and popular science.

142. Eisenstein, Elizabeth. *The Printing Revolution in Early Modern Europe.* Cambridge: Cambridge Univ. Press, 1983. 296 pp.

Abridged version of 139. Pp. 277–89: Selected (annotated) reading list. Reviewed in *Bulletin of the History of Medicine* 59 (1985), 539–40.

143. McNally, Peter F., ed. *The Advent of Printing: Historians of Science Respond to Elizabeth Eisenstein's* The Printing Press as an Agent of Change. Graduate School of Library and Information Studies, McGill University, Montreal, Occasional Paper 10. Montreal, 1987. 37 pp.

Proceedings of a seminar held by the Roundtable on Bibliography, McGill University, 9 March 1984. Of the five papers presented, three are relevant: Peter B. McNally, "'. . . The Eye of the Beholder': Opinions Concerning Elizabeth Eisenstein and *The Printing Press* . . ." (pp. 1–7. With a list of 52 reviews); Philip M. Teigen, "A Prolegomenon to the Interpretation of *The Printing Press* . . ." (pp. 8–14); William R. Shea, "The Printing Press as an Agent of Change at the Time of the Scientific Revolution" (pp. 15–19). See item 139.

144. Martin, Henri-Jean. *Le livre français sous l'Ancien Régime. Histoire du livre.* Paris: Promodis, Editions du Cercle de la Librairie, 1987. 303 pp.

Survey of the book trade in France from 1500 to 1800. A collection of articles published between 1956 and 1986. Reviewed in *The Library* (6), 10 (1988), 168–70.

145. Besson, Alain, ed. *Thornton's Medical Books and Collectors. A Study of Bibliography and the Book Trade in Relation to the Medical Sciences.* 3rd, revised ed. Aldershot, Hants.: Gower Publishing, 1990. 417 pp. [First published 1949, with introduction by Sir Geoffrey Keynes. 2nd ed. 1966]

A survey of medical literature from Ancient Egypt to modern times. This ed. has been entirely rewritten by a team of ten specialists; two chapters in the previous ed. (history of publishers; medical societies) not retained. Pp. 1–29: Medical books before the invention of printing; pp. 30–42: Medical incunabula; pp. 43–82: Medical books of the sixteenth century; pp. 83–115: Seventeenth-century medical books; pp. 301–41: Medical libraries of today; pp. 342–57: Bibliography; pp. 358–417: Index.

PUBLIC AND PRIVATE LIBRARIES (PRE-1700)

146. Bodleian Library, Oxford. *The First Printed Catalogue of the Bodleian Library 1605. A Facsimile. Catalogus librorum Bibliothecae publicae quam vir ornatissimus Thomas Bodleius Eques auratus in Academia Oxoniensi nuper instituit.* Oxford: Clarendon Press, 1986. 655 pp.

Pp. 1–180: Theology; pp. 181–218: Medicine; pp. 219–74: Law; pp. 275–415: The Arts; pp. 417–25: Commentaries on Aristotle; pp. 427–655: Appendices. Pp. 652–55: Indexes to writings on Hippocrates and Galen. [Unpaginated]: Author index.

147. University Library, Leiden. *Catalogus Bibliothecae Publicae Lugduno-Batavae.* Leiden: Elsevier, 1640. 216 + 21 pp.

Seventeenth-century catalogue (classified) of books and MSS (university founded in 1575). Subject groups include: medicine, cosmography, geography, mathematics, philosophy. The catalogue has special sections for the Scaliger and Golius collections of Oriental books and MSS.

University Library, Leiden, in the seventeenth century. Engraving by Woudanus (Municipal Archives, Leiden). From John Willis Clark, *The Care of Books*, Cambridge, 1901 (University of London Library). *Entries 147, 192.*

148. Bateman, Christopher. *Bibliotheca Scarburghiana; or, a Catalogue of the Incomparable Library of Sir Charles Scarburgh, Knt., M.D. Containing (almost) a Collection of Greek Books in All Faculties; with a Large Collection of Mathematicks and Physicks . . . Which Will Be Sold on Friday the Eighth Day of February 1694/5.* London, 1695. 115 pp.

Some 2,500 books including about 600 on mathematics and 450 on medicine. A second sale catalogue was issued for the 18th February, *Bibliotheca mathematica et medica Scarburghiana* (1695), containing 837 titles of mathematical and 440 medical works.

149. Bernard, Francis, *M.D. A Catalogue of the Library of the Late Learned Dr. Francis Bernard, Fellow of the College of Physicians, and Physician to S. Bartholomew's Hospital. Being a Large Collection of the Best Theological, Historical, Philological, Medicinal and Mathematical Authors, in the Greek, Latin, Italian, Spanish, French, German, Dutch and English Tongues, in All Volumes; Which Will Be Sold by Auction at the Doctor's Late Dwelling House in Little Britain: the Sale to Begin on Tuesday, October 4, 1698. London, 1698.* 450 pp.

Catalogue of some fifteen thousand titles, including: 938 on mathematics, and 4,484 on medicine.

150. Ballard, Thomas. *Catalogue of the Libraries of the Learned Sir Thomas Brown [1605–82], and Dr Edward Brown, His Son, Late President of the Royal College of Physicians. Consisting of Many Very Valuable and Uncommon Books, in Most Faculties and Languages, Chiefly in Physick, Chirurgery, Chymistry, Divinity, Philology, History, and Other Polite Parts of Learning. . . . Which Will Begin to Be Sold by Auction, . . . on Monday the 8th Day of January 1710/11.* London, 1711. 58 pp.

Catalogue of some 2,400 books: theology (134); history, philology, geography, etc. (519); medicine and philosophy (442); mathematics (98); French language (329); Italian language (94); Spanish language (19); German and Flemish (42); English language (583).

151. Cooke, Jacob. *Bibliotheca Bernardiana; or, a Catalogue of the Library of the Late Charles Bernard, Esq., [1650–1711] Serjeant Surgeon to Her Majesty. Containing a Curious Collection of the Best Authors in Physick, History, Philology, Antiquities, etc. . . . Which Will Begin to Be Sold by Auction on Thursday the 22nd of March 1710–11.* London, 1711. 221 pp.

3,511 titles, under twelve headings. Subjects include medicine (991), geography (300) and mathematics (113).

152. Boerhaave, Hermann (1668–1738). *Bibliotheca Boerhaaviana, sive catalogus librorum instructissimae bibliothecae viri summi D. Hermanni Boerhaave, . . . quorum publica fiet auctio . . . 8 junii et seqq. diebus 1739.* Leiden, 1739. 68 pp.

Catalogue of some 3,300 books, of which 829 are medical and 978 scientific.

153. Favaro, Antonio. "La libreria di Galileo Galilei, descritta ed illustrata." *Bullettino di bibliografia e di storia delle scienze matematiche e fisiche,* 19 (1886), 219–93; 20 (1887), 372–76.

An attempt at "reconstructing" Galileo's library. 521 items. Name index.

154. Omont, H. "Inventaire des manuscrits grecs et latins données a Saint-Marc de Venise par le Cardinal Bessarion (1468)." *Revue des bibliothèques* 4 (1894), 129–87.

List of 482 Greek and 264 Latin MSS. Includes 218 MSS (144 Greek, 74 Latin) of medicine, mathematics, astronomy, music, philosophy. The Greek MSS include works of: Apollonius, Archimedes, Diophantus, Dioscorides, Euclid, Galen, Hermes Trismegistus, Hero, Hippocrates, Ptolemy. Pp. 131–81: Inventory of 1468. Pp. 182–87: Subject and name indexes, index of Greek MSS, index of Latin MSS. See item 206.

155. Biblioteca Apostolica Vaticana. *Codices urbinates graeci.* Recensuit Cosimus Stornajolo. Rome, 1895. 354 pp.

165 MSS. Indexes of authors, titles and topics. Pp. ix–cxcix: History of the collection and the old inventory (see note below).

156. Biblioteca Apostolica Vaticana. *Codices urbinates latini.* Recensuit Cosimus Stornajolo. Rome, 1902–21. 3 vols. 1,779 MSS. Indexes in each vol.

Catalogues (155 and 156) of the *fondo urbinate* of the Vatican Library. These MSS previously formed part of one of the finest Renaissance libraries—that of the Dukes of Urbino (1463–1657)—rich in works of mathematics, astronomy, and medicine. Luca Pacioli and Federico Commandino are among those who are believed to have used the library. See item 786 (index under "Urbino, Ducal Library").

157. Lawler, John. *Book Auctions in England in the Seventeenth Century (1676–1700). With a Chronological List of the Book Auctions of the*

Period. The Book Lover's Library, ed. Henry B. Wheatley. London: Elliot Stock, 1898. 241 pp. [1968 reprint by Gale Research, Detroit]

Notes on 134 book auctions (several collections of medical and scientific books). Pp. 225–41: Index of names and titles. Pp. ix–xliv: Introduction.

158. Sayle, C. "The Library of Thomas Lorkyn [1528–1591]." *Annals of Medical History* 3 (1921), 310–23.

List of 272 medical works. Lorkyn was Regius professor of medicine at Cambridge. He gave his library to the University.

159. Tricot-Royer, J. "La bibliothèque de Vopiscus Fortunatus Plempius [1601–1671], professeur de médecine au XVIIe siècle." *Recueil des mémoires couronnées et autres mémoires de l'Académie royale de médecine de Belgique* 22 (1925), (5), 1–112. [Also published separately]

Catalogue of 766 medical books based on the 1673 sale catalogue. Pp. 3–12: Biobibliographical note and analysis of catalogue. Pp. 13–18: Author index. Plemp studied medicine at Leyden, Padua, and Bologna, practised medicine for ten years in Amsterdam, and afterwards was professor of medicine at Louvain.

160. De Vecchi, Bindo. "I libri di un medico umanista fiorentino del sec. XV. Dai 'Ricordi' di maestro Antonio Benivieni." *Bibliofilia* 34 (1933), 293–301.

Inventory of books (1487): philosophy (21), Greek (6), logic (7), rhetoric (32), theology (3), medicine (73), astrology (9), miscell. (24). Benivieni (1443–1502) is considered the founder of pathological anatomy.

161. Vorstius, Doris. *Grundzüge der Bibliotheksgeschichte*. Leipzig: Otto Harrassowitz, 1935. 96 pp.

Notes for students at library-school on private, public, monastic, and university libraries, from the Renaissance to the Enlightenment. Pp. 73–82: List of the most important libraries of the world (date of foundation and number of volumes). Pp. 82–90: Survey of the most important special libraries classified by subject.

162. Kibre, Pearl. *The Library of Pico della Mirandola*. New York: Columbia Univ. Press, 1936. 330 pp. [Reprinted 1966 by AMS Press, New York]

A study and inventory of the library of Giovanni Pico della Mirandola (1463–1494), containing some 1,190 vols. It was in the same class as the libraries of Bessarion, Giorgio Valla, and Federigo, Duke of Urbino (see items 155, 156). Pp. 11–22: History of the library. Pp. 23–36: Greek works (pp. 33–34: science). Pp. 86–111: Occult and natural science. Pp. 115–297: Inventory. Pp. 301–30: Name and subject index.

163. Goldschmidt, E. P. *Hieronimus Münzer und seine Bibliothek*. Studies of the Warburg Institute, 4. London, 1938. 154 pp.

Study of the life and work (pp. 13–104) and reconstructed catalogue of the library (pp. 115–45) of Münzer (1438–1508), physician, humanist, geographer, and bibliophile. After studying in Leipzig and Pavia (M.D. 1478), he practiced medicine in Nuremberg. See the author's article, "Hieronymus Muenzer and Other Fifteenth Century Bibliophiles," *Bulletin of the New York Academy of Medicine* 14 (1938), 491–508.

164. Liddell, J. R. "The Library of Corpus Christi College, Oxford in the Sixteenth Century." *The Library* (4) 18 (1938), 384–416.

Includes a catalogue of 371 items of which 1–66 are on science, medicine, and philosophy.

165. Almagià, Roberto. "La biblioteca d'un umanista del seicento (Luca Holstenio)." *Archeion* 22 (1940), 47–56.

Luc Holste (1595–1661), German humanist, geographer, and theological writer was librarian of the Vatican from 1641 to 1661. His library of some 3,500 books was rich in works of philosophy, medicine, mathematics, and geography (pp. 51–54).

166. Sanchez Canton, F. J. *La libreria di Juan des Herrera*. Consejo Superior de Investigaciónes Científicas Instituto Diego Velasquez. Madrid, 1941. 46 pp.

Catalogue of 450 books (subjects include: art and architecture; military science; cosmography; scientific instruments; medicine; astrology; magic; philosophy; travel) based on the inventory of 1597. Herrera (1530–1597) was Philip II's architect and director of the Academy of Mathematics, Madrid (founded 1582).

167. Barnard, E. A. B., and L. B. Newman. "John Deighton of Gloucester, Surgeon." *Transactions, Bristol and Gloucestershire Archaeological Society* 64 (1943), 71–88.

List of surgical instruments and analysis of library (121 books). Based on an inventory of 24 Feb. 1639/40.

168. Bishop, W. J. "Some Medical Bibliophiles and Their Libraries." *Journal of the History of Medicine* 3 (1948), 229–62.

Notes on some 50 medical men, from Ugolino de Montecantini (1415) to William Hunter (1783), of whom about 25 lived during the 16th–17th centuries. Some of them have bequeathed their books to the Royal College of Physicians.

169. Milkau, Fritz, ed. *Handbuch der Bibliothekswissenschaft.* Zweite, vermehrte und verbesserte Aufl., hrsg. von Georg Leyh. Vol. 3 (1): *Geschichte der Bibliotheken.* Wiesbaden: Otto Harrassowitz, 1953. 830 pp.

A survey of the history of libraries. Pp. 499–681: European libraries other than French, from the Renaissance to the beginning of the Enlightenment (by Alice Bömer). Pp. 682–708: French libraries from the Renaissance to the Revolution (by Ludwig Klaiber).

170. Hessel, Alfred. *A History of Libraries.* Translated with supplementary material by Reuben Peiss. New Brunswick: Scarecrow, 1955. 198 pp.

Pp. 39–61 cover the Renaissance, Reformation, and the seventeenth century. Pp. 142–80: Bibliography (some 750 references in one alphabetical sequence).

For a survey of Renaissance libraries, see Pearl Kibre, "The Intellectual Interests Reflected in Libraries of the Fourteenth and Fifteenth Centuries." *Journal of the History of Ideas* 7 (1946), 257–97.

171. Jayne, Sears, and Francis R. Johnson, eds. *The Lumley Library. The Catalogue of 1609.* London: British Museum, 1956. 372 pp.

2,609 titles arranged under: Theology (1–936), History (937–1527), Arts and philosophy (1528–2290), Medicine (2291–2426), Cosmography and geography (2485–2517), Law (2427–84, 2518–67), Music (2568–2609). Pp. 323–72: Index of proper names and short titles. The library was unequalled for its collection of books on the sciences (in catalogue under arts and philosophy). The nucleus of the library was the "theological arsenal" of Thomas Cranmer, Archbishop of Canterbury. Pp. 2–37 (Introduction) contain a history of the library. Review: *The Library* (5), 12 (1957), 63–65.

172. Wormald, Francis, and C.E. Wright, eds. *The English Library Before 1700. Studies in Its History.* London: Univ. of London, The Athlone Press, 1958. 273 pp.

Based on lectures given at the School of Library and Archives, University College, London. Pp. 148–75: The dispersal of the libraries in the sixteenth century (C. E. Wright). Pp. 213–35: The libraries of Cambridge, 1570–1700 (J. C. T. Oates). Pp. 236–55: Oxford libraries in the seventeenth and eighteenth centuries (J. N. L. Myers).

173. Dickins, Bruce. "Henry Gostling's Library. A Young Don's Books in 1674." *Transactions, Cambridge Bibliographical Society* 3 (1961), 216–24.

List of 114 books of which 23 are mathematical. Gostling (student at St. John's and Fellow of Corpus Christi) was Taxor of the University of Cambridge in 1670.

174. Wilkinson, R.S. "The Alchemical Library of John Winthrop, Jr. (1606–1676) and His Descendants in Colonial America." *Ambix* 11 (1963), 33–51; 13 (1966), 139–86.

Chemist and practitioner of chemical medicine, Winthrop emigrated in 1631 and became America's first astronomer and the first colonial member of the Royal Society. The catalogue of 275 books is restricted to works on alchemy and chemical medicine (and other relevant non-chemical works).

175. Devresse, Robert. *Le fonds grec de la Bibliothèque Vaticane des origines à Paul V*. Studi e testi, 244. Vatican City, 1965. 521 pp.

Edition of inventories of Greek MSS in the Vatican Library made from the time of its foundation during the pontificate of Nicholas V (1446–55) to the end of the sixteenth century. Some 1,200 MSS of which about 200 relate to scientific subjects (from Aristotle to mathematics and medicine). Author and subject index.

176. Harrison, J. R., and Peter Laslett. *The Library of John Locke [1632–1704]*. Oxford Bibliographical Society Publications, NS, Vol. 13. Oxford, 1965. 292 pp.

Alphabetical catalogue of 3,641 books: medicine (402); geography and exploration (275); philosophy (269); natural science (240). Pp. 1–61: Essay on John Locke and his books. Review: *The Library* (5), 21 (1966), 343–47. See also 181.

177. Rostenberg, Leona. *Literary, Political, Scientific, Religious and Legal Publishing, Printing and Bookselling in England, 1551–1700: Twelve Studies*. New York: Burt Franklin, 1965. 2 vols. 463 pp.

Study of twelve members of the Stationers' Company, publishers, printers, and booksellers—"their publications, their careers, the milieu in which they lived and worked, their influence upon the period and the period's influence upon them." Pp. 237–80: The New Science; John Martyn, "Printer to the Royal Society." Pp. 281–313: The Liberal Arts—Robert Scott, importer and university agent.

178. Anderson, Andrew H. "The Books and Interests of Henry, Lord Stafford (1501–1563)." *The Library* (5), 21 (1966), 87–114.

Catalogue (pp. 97–114) of 342 books on subjects ranging from theology (64) and law (48) to medicine (58), astronomy and mathematics (36), cosmography and military science (11).

179. Morgan, Paul. "George Hartgill: An Elizabethan Parson-Astronomer and His Library." *Annals of Science* 24 (1968), 295–311.

List of 101 books on mathematics, astronomy, and physick.

180. Oakeshott, Walter. "Sir Walter Ralegh's Library." *The Library* (5), 23 (1968), 285–327.

Study of shelf-list (pp. 296–327) of 515 volumes, majority of which have been identified. About 100 titles relate to science (geography, philosophy, psychology, medicine, chemical, cosmography, chronology). Five languages (English, French, Spanish, Italian, Latin) represented.

181. Ashcraft, Richard. "John Locke's Library: Portrait of an Intellectual." *Transactions, Cambridge Bibliographical Society* 5 (1969), 47–60.

Observations on the relationship between the catalogue of a seventeenth-century library and the intellectual life of the period. Complements the introductory essay by Harrison and Laslett (176). Study is based on the author's examination of more than 80 library catalogues of Locke's contemporaries.

182. Di Pietro, G., and M. L. Righini Bonelli. *Catalogo della Biblioteca Mediceo-Lorenese*. Florence: Olschki, 1970. 578 pp.

Catalogue of a collection of scientific books (chronologically arranged), the major part of which belonged to the Medici family. 1,768 titles of which 616 are pre-1700. Name index.

183. Hughes, Barnabas B. "The Private Library of Johann Scheubel, Sixteenth-Century Mathematician." *Viator* 3 (1972), 417–32.

Brief account with a reconstructed catalogue (109 titles).

184. Reti, Ladislao. *The Library of Leonardo da Vinci*. Los Angeles, Calif., 1972. 28 pp. [First published in the *Burlington Magazine* 110 (1968), no. 779]

Annotated catalogue of 116 items, based on a list in Codex Madrid II (MS 8936). "The entries for books on anatomy, medicine, natural history, arithmetic, geography, geometry, astronomy, philosophy, bear witness to Leonardo's wide and insatiable longing for scientific knowledge."

185. Ullman, Berthold L., and Philip A. Stadter. *The Public Library of Renaissance Florence: Niccolò Niccoli, Cosimo de' Medici and the Library of San Marco.* Padua: Antenore, 1972. 369 pp.

The San Marco Library, the first public library in Italy, was open to all scholars from its inception (1444). At the time of Bernard de Montfaucon's visit in 1700, it was second in Florence only to the Laurenziana. The catalogue of 1500 (pp. 107–267), lists 1,232 items, of which more than half have been identified. Items 572–810 include: philosophy, medicine, mathematics, astronomy. General index includes authors and subjects.

186. Forbes, Eric G. "The Library of the Rev. John Flamsteed, F.R.S. [1646–1719], First Astronomer Royal." *Notes and Records, Royal Society of London* 28 (1973), 119–43.

List of 260 books owned before 1685.

187. Rhodes, Dennis E. "An Unknown Library in Southern Italy in 1557." *Transactions, Cambridge Bibliographical Society* 6 (1973), 115–25.

Library of Luca Gaurico (1475–1558). Catalogue of some 150 books, mainly on astrology and astronomy, edited and partly annotated. See item 690.

188. Rose, Paul Lawrence. "Humanist Culture and Renaissance Mathematics. The Italian Libraries of the *Quattrocento*." *Studies in the Renaissance* 20 (1973), 46–105.

Examines the collection and translation of Greek mathematical MSS by fifteenth-century Italian humanists. A revised version was published as Chapter 2 of item 786.

189. Feisenberger, E. A., ed. *Sale Catalogues of Libraries of Eminent Persons.* Vol. 2. *Scientists.* London: Mansell, 1975. 296 pp.

Libraries of Elias Ashmole (1617–92), Robert Hooke (1635–1702), John Ray (1627–1705), Edmund Halley (ca. 1656–1743). 12–page introduction. Reviewed in *Isis* 67 (1976), 486–87.

190. Kolb, Robert. *Caspar Peucer's Library: Portrait of a Wittenberg Professor of the Mid-Sixteenth Century.* St. Louis: Center for Reformation Research, 1976. 76 pp.

Pp. 37–76: Inventory of 1,455 books arranged under: theology, law, medicine (259–680), history, philosophy (1043–1328), poetry, and lexica. Mathematics and astronomy included under philosophy. Peucer (1525–1602), son-in-law of Melanchthon, was professor of mathematics and medicine. Reviewed in *Annals of Science* 35 (1978), p. 209.

191. Callmer, Christian. *Königin Christina, ihre Bibliothekare und ihre Handschriften. Beiträge zur europäischen Bibliotheksgeschichte.* Acta Bibliothecae regiae Stockholmiensis, 30. Stockholm, 1977. 271 pp.

The story of the acquisition and dispersal of Queen Christina of Sweden's MSS. Pp. 43–93 contain biographical notes on her librarians (among them: Isaac Vossius, Nicolaus Heinsius, Gabriel Naudé). Pp. 256–59. Index of MSS. Reviewed in *Bibliothèque de l'Ecole des Chartes* 139 (1981), p. 85–87. The MSS in the Vatican Library have been catalogued by J. Bignami Odier and F. De Marco in *Les manuscrits de la reine de Suède au Vatican: Réédition du Catalogue de Montfaucon et côtes actuelles,* Studi e Testi 238 (Vatican City, 1964), 133 pp. (2,111 MSS, of which some 250 are scientific). For an account of the printed books, see article by Callmer, "Queen Christina's Library of Printed Books in Rome," *Queen Christina of Sweden: Documents and Studies,* edited by Magnus von Platen, *Analecta Reginensia* 1 (Stockholm: Norstedt, 1966), pp. 59–73.

192. Scaliger, Justus Joseph (1540–1609). *The Auction Catalogue [1609] of the Library of J. J. Scaliger.* A facsimile ed. with an introduction by H. J. de Jonge. Utrecht: HES Publishers, 1977. 9 + 78 pp.

French orientalist and distinguished classical scholar, son of Julius Caesar Scaliger (see *DSB*). He was professor at Geneva and, later, at Leiden. Among the many classical texts edited by him is the *Astronomicon* of Manilius. His *De emendatione temporum* (1583) is considered to be the first scientific work on chronology. His work on squaring the circle was refuted by Van Roomen and Viète. His collection of Oriental and Western MSS was bequeathed to the University of Leiden. The present catalogue lists some 1,700 books of which some 250 are on medicine, mathematics, and philosophy.

193. Czartoryski, Pawel. "The Library of Copernicus." *Studia Copernicana* 16 (1978), 355–96.

"An attempt at presenting Copernicus' library and his readings."

194. Grierson, Philip. "John Caius' Library." *Biographical History of Gonville and Caius College* 7 (1978), 509–25.

Inventory of 142 titles (many of them medical) with comments. Caius was Master of Gonville Hall (1559–73) and President of the College of Physicians (1555–60).

195. Harrison, John. *The Library of Isaac Newton.* Cambridge: Cambridge Univ. Press, 1978. 286 pp.

Pp. 1–27: Isaac Newton, user of books. Pp. 28–57: Dispersal of the library after his death. Pp. 58–78: The composition of the library. Pp. 79–265:

The catalogue, its compilation and content (1,752 titles). Reviewed in *Isis* 70 (1979), 619–20.

196. Labowsky, Lotte. *Bessarion's Library and the Biblioteca Marciana. Six Early Inventories.* Sussidi eruditi, 31. Rome: Edizioni di Storia e Letteratura, 1979. 547 pp.

Bessarion's collection of MSS, donated to the Marciana, played an important part in the transmission of Greek learning (philosophy, mathematics, astronomy, and medicine) to the West. Pp. 1–144: B.'s library and the first century of the Marciana. Pp. 145–427: Inventories of 1468, 1474, 1524, 1543, 1545/46, 1575. On B.'s role as humanist patron, see item 786. Review: *Papers, Bibliographical Society of America* 76 (1982), 362–63.

197. Gaskell, Philip. *Trinity College Library: The First 150 Years.* Cambridge: Cambridge Univ. Press, 1980. 275 pp.

The reason for the special interest in the history of this library is that "a number of Trinity men—more perhaps than of the members of any other college—contributed importantly to England's spiritual, intellectual, and scientific development from soon after the foundation of the college in 1546 until the end of the seventeenth century." The library has grown to be the greatest of all the Oxford and Cambridge college libraries. Pp. 147–212: A catalogue of the college library in 1600. Pp. 241–58: The science books in the class catalogue of c. 1645. Reviewed in *History of Universities* 2 (1982), 251–52.

198. Grendler, Marcella. "A Greek Collection in Padua. The Library of Gian Vincenzo Pinelli (1535–1601)." *Renaissance Quarterly* 33 (1980), 386–416.

Discusses Pinelli's collection of Greek books and MSS. He "made his home into an informal academy where locals and travelers on the road to Venice or Rome found intelligent conversation and lively exchange of ideas." He also had a collection of mathematical and astronomical instruments, globes, maps, fossils and metals, and even a botanical garden. Pp. 402–405 refer to the mathematical works.

199. Hess, Heinz Jürgen. "Bücher aus dem Besitz von Christian Huygens (1629–1695) in der Niedersächsischen Landesbibliothek Hannover." *Studia Leibnitiana* 12 (1980), 1–51.

List of 121 books from the library of Huygens acquired by Leibniz, with transcript of marginal notes on eleven of them.

200. Morgan, Paul, comp. *Oxford Libraries Outside the Bodleian. A Guide.* 2nd ed. Oxford: Bodleian Library, 1980. 264 pp.

Describes the collections of college libraries and others not dependent on the Bodleian. Twenty-one libraries belong to pre-1700 foundations.

201. Paredi, Angelo. *A History of the Ambrosiana*. Notre Dame, Ind.: Univ. of Notre Dame Press, 1983. 110 pp.

Brief survey of the history of the Milanese library (founded by Cardinal Federico Borromeo in 1607) containing the Codex Atlanticus of Leonardo da Vinci and a large number of scientific MSS. Pp. 99–110: Bibliography and catalogues.

202. Philip, Ian. *The Bodleian Library in the Seventeenth and Eighteenth Centuries*. The Lyell lectures, Oxford, 1980–81. Oxford: Clarendon Press, 1983. 139 pp.

Pp. 1–69: From the foundation to Thomas Hyde (Librarian, 1665–1700). Reviewed in *Library History* 6 (1982–84), 116–18.

203. Thompson, Lawrence S. "Renaissance Libraries." *Encyclopedia of Library and Information Science*, Executive ed., Allen Kent. New York: Marcel Dekker, 1983. Vol. 36, Suppl. 1, pp. 488–508.

Documented survey covering: Italy; The Vatican; France; Eastern Europe, England and Spain; The Germanies; Reformation libraries.

204. Elmer, Peter. *The Library of Dr. John Webster: The Making of a Seventeenth-Century Radical*. Medical History, Supplement no. 6. London: Wellcome Institute for the History of Medicine, 1986. 275 pp.

Biographical notice, a reconstruction of the library, and a survey of its contents. Pp. 52–234: Catalogue of library (1,501 titles: medicine 242; mathematics 79; natural science 326; theology 397; literature 148). Pp. 239–75: Index to the catalogue. Reviewed in *The Library* (6), 11 (1989), 71–72.

205. Leedham-Green, E. S. *Books in Cambridge Inventories: Book-Lists from Vice-Chancellor's Court Probate Inventories in the Tudor and Stuart Periods*. Cambridge: Cambridge Univ. Press, 1987. 2 vols.

Vol. 1: The inventories of students and graduates who died while at the university between 1535 and 1700. With biographical notices. 649 pp. Vol. 2: Catalogue. Pp. 1–821: Author/Title catalogue of books; Pp. 821–24: Maps, pictures, etc. Pp. 824–27: Instruments; Pp. 828–61: Subject index (includes: arithmetic, astronomy, divination, magic, geography, geometry, medicine, music, natural philosophy, useful arts). Reviewed in *The Library* (6), 11 (1989), 383–88.

206. Zorzi, Marino. *La Libreria di San Marco. Libri, lettori, società nella Venezia dei Dogi.* Milan: Mondadori, 1987. 597 pp.

Documented history of the Biblioteca Marciana, Venice. Pp. 23–242 deal with the history of the library from the time of Bessarion to the end of the seventeenth century. See also items 131, 154, 196.

207. Jolly, Claude, ed. *Histoire des bibliothèques françaises. Les bibliothèques sous l'Ancien Régime 1530–1789.* Paris: Promodis–Editions du Cercle de la Librairie, 1988. 547 pp.

Includes: Ecclesiastical libraries (pp. 11–73); the King's Library, 1490–1664; private libraries (De Thou, Richelieu, Mazarin, Seguier, Colbert). Pp. 188–91: Note on Protestant libraries. Reviewed in *The Library* (6), 12 (1990), 144–49.

208. Kiessling, Nicholas. *The Library of Robert Burton [1576–1640].* Oxford Bibliographical Society Publications, NS, Vol. 22. Oxford, 1988. 433 pp.

Catalogue of 1,738 books (and two MSS) mainly on theology, literature and history—the library of the author of *The Anatomy of Melancholy*, one of the largest undispersed collections in England dating from the pre-Civil War period. About 74 percent of the books were published during his lifetime. The majority of the books are now in the Bodleian and Christ Church libraries. Pp. 370–77: Subject index (about four hundred scientific titles). Reviewed in *The Library* (6), 12 (1990), 248–50.

On Burton, see Sir William Osler, "Robert Burton, the Man, His Book, His Library," *Proceedings, Oxford Bibliographical Society* 1 (1922–26), 159–246.

209. Nickson, M. A. E. "Hans Sloane, Book Collector and Cataloguer, 1682–1698." *British Library Journal* 14 (1988), (1), 52–89.

On Sloane's own catalogues of his library. His vast collection of books and MSS became the nucleus of the library of the British Museum. See also pp. 1–20 for note, "Hans Sloane, Scientist," by Maarten Ultee. Sloane (1660–1753) was made a F.R.S. in 1685 for his contributions to botany, and was elected President in 1727.

210. Rostenberg, Leona. *The Library of Robert Hooke: the Scientific Book Trade of Restoration England.* Santa Monica, Calif.: Modoc Press, 1989. 257 pp.

A study of the library of Robert Hooke in the context of the scientific book trade of Restoration England, based on an analysis of the 1703 sale catalogue, his *Diaries,* and the *Philosophical Transactions.*

210A. Mugnai Carrara, Daniela. *La biblioteca di Nicolò Leoniceno; Tra Aristotele e Galeno: cultura e libri di un medico umanista.* Florence: Leo S. Olschki Editore, 1991. 247 pp.

Reconstruction of the library of Nicolò Leoniceno of Ferrara (1428–1524), leading medical humanist of his day. Review: *The Library* (6), 14 (1992), 382–83.

Education

BIBLIOGRAPHIES

211. Buisson, Ferdinand Edouard. *Répertoire des ouvrages pédagogiques du XVIe siècle. (Bibliothèques de Paris et des départements.)* Paris: Le Musée Pédagogique, 1886. 733 pp. [Reprinted in 1968 by De Graaf, Nieuwkoop]

Bibliography of educational books in Latin and French by some one thousand authors (c. 5,000 titles) arranged alphabetically. With a subject index (twenty-five headings, subdivided by author). Subjects covered include: grammar; rhetoric; arithmetic, music; geometry; astronomy; cosmography; geography; natural history. Locations given for libraries in Paris.

212. Erman, Wilhelm, and Ewald Horn. *Bibliographie der deutschen Universitäten. Systematisch geordnetes Verzeichnis der bis Ende 1899 gedruckten Bücher und Aufsätze über das deutsche Universitätswesen.* Leipzig: Teubner 1904–1905. 3 vols.

Vol. 1: General (24 main subjects); Vol. 2: Individual universities (arranged by city, with subdivisions); Vol. 3: Index, errata, and supplement. Includes Swiss, Austrian, and Baltic universities. Some 40,000 entries. Reviewed in *Zentralblatt für Bibliothekswesen* 21 (1904), 287–88; 22 (1905), 588–99.

213. Bernard Quaritch Ltd. *A Catalogue of Rare and Valuable Early Schoolbooks (Fifteenth, Sixteenth and Seventeenth Centuries) and of Books Relating to Early Education.* Catalogue no. 464. London, 1932. 99 pp.

Catalogue of 342 items, annotated. Nos. 1–111: History of education; 112–314: Schoolbooks; 315–38: Books on manners and morals.

214. Commission Internationale pour l'Histoire des Universités. *Bibliographie internationale de l'histoire des universites*, 1–2. Etudes et travaux, 2, 5. Geneva: Librairie Droz, 1973–76. 2 vols.

Intended to complete the bibliography in H. Rashdall, *The Universities of Europe in the Middle Ages*, new ed. by F. M. Powicke and A. B. Emden (Oxford: Oxford Univ. Press, 1936), 3 vols., and cover the history from the origins to 1800. Nos. 1–2 cover: Spain, Louvain, Copenhagen, Prague, Portugal, Leiden, Pécs, Franeker, Basel.

215. Gabriel, A. L. *Summary Bibliography of the History of the Universities of Great Britain and Ireland up to 1800, Covering Publications between 1900 and 1968*. Notre Dame, Ind.: Medieval Institute, University of Notre Dame, 1974. 154 pp.

1,514 entries. No annotations. History of education in Europe (1–161); History of education in England (162–294); Cambridge, Oxford, and English colleges on the Continent (295–1352); Scottish universities and colleges (1353–1458); Wales (1459–64); Irish colleges (1465–1514).

216. Stark, Edwin, comp. *Bibliographie zur Universitätsgeschichte. Verzeichnis der im Gebiet der Bundesrepublik Deutschland 1945–1971 veröffentlichten Literatur*. Ed. Erich Hassinger. Freiburger Beiträge zur Wissenschafts- und Universitätsgeschichte, Bd. 1. Freiburg-Munich: Karl Aber, 1974. 316 pp.

Bibliography of studies of individual universities arranged alphabetically by city. Some three thousand items.

217. Stelling-Michaud, Sven. "La storia della università nel medioevo e nel Rinascimento: stato degli studi e prospettive di ricerca." *Le origini dell'università*. A cura di Girolamo Arnoldi. Bologna: Società editrice il Mulino, 1974, pp. 153–217.

Richly documented survey (290 footnotes) of recent scholarship. A revised version of: "L'histoire des universités au Moyen-Age et à la Renaissance au cours des vingt-cinq dernières anneés." *Comité International des Sciences Historiques, XIe Congrès, (Stockholm, 21–28 Août 1960), Rapports I* (Göteborg: Almqvist & Wiksell, 1960), pp. 97–143.

218. Fletcher, John M., ed. *The History of European Universities. Work in Progress and Publications*. 1 (1977)–5 (1981). Published for the International Commission for the History of Universities by the Department of Modern Languages, The University of Aston in Birmingham. Birmingham, 1978– 82. 5 pts.

Annual newsletter containing a list of work in progress and a current bibliography arranged by country. Each issue contains an index of countries and institutions. The bibliography is continued in *History of Universities* 7 (1988)– . See item 220.

219. Guenée, Simonne. *Bibliographie de l'histoire des universités françaises des origines à la Révolution.* Tome 1. *Généralités. Université de Paris.* Tome 2. *D'Aix-en-Provence à Valence et académies protestantes.* Paris: A. et J. Picard, 1978–81. 2 vols. 567 + 495 pp.

Selective bibliography (classified, with occasional annotations) covering scholarly works published up to 1975. Vol. 1 covers general works (2,210 entries) and the University of Paris (4,926 entries); Vol. 2 covers 54 provincial universities and colleges, and 13 Protestant academies. Reviews: *History of Universities* 2 (1982), 240; 5 (1985), 187–88.

220. *History of Universities,* Vol. 1 (1981)– .

Annual. Concerned mainly with the history of European and American universities before World War II. Includes book reviews. Vols. 7 (1988)– contain: "Publications on University History since 1977: a Continuing Bibliography," ed. John M. Fletcher.

221. Fletcher, John M., and Julian Deahl. "European Universities, 1300–1700: The Development of Research 1969–1979, with a Summary Bibliography." *Rebirth, Reform and Resilience: Universities in Transition, 1300–1700.* Editors: James M. Kittelson and Pamela J. Transue. Columbus: Ohio State Univ. Press, 1984, pp. 324–57.

Survey of historical research done during the years 1969–79 (pp. 324–37) with a bibliography (ca. 300 items).

GENERAL WORKS

222. Schmidt, K. A., ed. *Geschichte der Erziehung vom Anfang an bis auf unsere Zeit bearbeitet in Gemeinschaft mit einer Anzahl von Gelehrten und Schulmännern.* Stuttgart: F. G. Cotta'schen Buchhandlung, 1884–1901. 5 vols.

History of education from pre-Christian times to the present day. Vol. 2(2): The Renaissance and humanism; The Reformation and the sixteenth century. Vol. 3(1): Society of Jesus; Education in France and England in the sixteenth century. Vol. 3(2): Ratke; Comenius and his predecessors. Vol. 4(1): The Thirty Years' War; Pietism and education; John Locke; Education in seventeenth- and eighteenth-century France. With bibliographies and name index.

223. Watson, Foster, ed. *The Encyclopedia and Dictionary of Education. A Comprehensive, Practical and Authoritative Guide on All Matters Connected with Education, including Educational Principles and Practice, Various Types of Teaching Institutions, and Educational Systems Throughout the World.* London: Sir Isaac Pitman and Sons, 1921–22. 4 vols.

Contains articles (many of historical interest) by about nine hundred eminent authorities. Vol. 4 contains a 26–page classified index.

224. Monroe, Paul, ed. *A Cyclopedia of Education.* New York: The Macmillan Co., 1911–13. 5 vols.

c. 7,000 signed articles (some with bibliographies)—from ABACUS to ZWINGLI—on all aspects of education and its history. The 1968 reprint by Gale Research contains an analytic subject index (NOT SEEN).

MONOGRAPHS AND ARTICLES

225. Laurie, S. S. *Studies in the History of Educational Opinion from the Renaissance.* Reprint. London: Frank Cass, 1968. 261 pp. [First published 1903]

Thirteen lectures (out of 16) relate to the period from Trotzendorf (1490–1556) to John Locke (1632–1704).

226. Adamson, John William. *Pioneers of Modern Education, 1600–1700.* Cambridge: University Press, 1905. 285 pp.

John Brinsley, Wolfgang Ratke, Bacon, Comenius, Samuel Hartlib, Milton, Petty, John Dury, Charles Hoole, De la Salle, A. H. Francke. Author was professor of education at King's College, London.

227. Woodward, William Harrison. *Studies in Education During the Age of the Renaissance, 1400–1600.* With a foreword by Lawrence Stone. Classics in Education, 32. New York: Teachers College Press, 1967. 340 pp. [First published 1906]

Thirteen essays on the "origin and development of the idea of a liberal education—embracing character, manners and instruction—during the two important formative centuries of modern Europe." With a bibliographical note by Lawrence Stone.

Topics include: Erasmus (1466–1556); College de Guyenne; Cardinal Sadoleto (1477–1547); Juan Luis Vives (1492–1540); Melanchthon (1497–1560); Castiglione's *Il Cortigiano*; Thomas Elyot (c.1490–1546); Roger Ascham (1515–68).

228. Graves, Frank Pierrepont. *Peter Ramus and the Educational Reformation of the Sixteenth Century.* New York: Macmillan, 1912. 226 pp·

Ramus was "a practical reformer [of education], a writer of textbooks, the founder of a new and influential point of view in subject matter and method, a popular and successful teacher, and an active correspondent and personal acquaintance of the educational leaders of his day in all countries." Pp. 219–22: Bibliography.

229. Farrell, Allan P., S.J. *The Jesuit Code of a Liberal Education. Development and Scope of the Ratio Studiorum.* Milwaukee, Wis.: Bruce Publishing, 1938. 478 pp.

Part 1 reconstructs from primary sources the history of the formation of the *Ratio studiorum* (foundation of the Colleges of Messina and Palermo (1548–49); the early years of the Roman College (1551–65); the development of the Jesuit colleges in the rest of Italy, in Spain and Portugal; drafting of the *Ratio*). Part 2, pp. 219–362 contains an analysis of the *Ratio* (1586, 1591, 1599). Pp. 365–76 touch on the Jesuit interest in mathematical and scientific studies. Pp. 455–78: Bibliography and index.

230. Needham, Joseph, ed. *The Teacher of Nations. Addresses and Essays in Commemoration of the Visit to England of the Great Czech Educationalist Jan Amos Komensky, Comenius 1641–1941.* Cambridge: University Press, 1942. 99 pp.

Eleven papers on the life, work, and influence of Comenius (1592–1670), educational reformer. With a chronology and select bibliography.

231. Battersby, W. J. *De La Salle: a Pioneer of Modern Education.* London: Longmans, Green, and Co., 1949. 236 pp.

Jean Baptiste de la Salle (1651–1719) was the first to establish training colleges for teachers, as distinct from ecclesiastical seminaries. His greatest practical achievement was the establishment of a body of trained teachers for "poor schools" at a time when there were none. He was the first to establish secondary schools of a non-classical type intended for the new middle class. Pp. 227–30: Bibliography.

232. Weiss, Roberto. "Learning and Education in Western Europe from 1470 to 1520." *The New Cambridge Modern History*, Vol. 1, edited by G. R. Potter. Cambridge: Cambridge Univ. Press, 1957, pp. 95–126.

Describes how humanism gradually became an integral part of western European culture during this half-century.

233. Rioux, Georges. *L'oeuvre pédagogique de Wolfgangus Ratichius (1571–1635)*. Paris: J. Vrin, 1963. 315 pp.

Detailed study of the life and work of Wolfgang Ratke, German educationist. With bibliography and index of names.

234. Yates, Frances A. *The Art of Memory*. London: Routledge and Kegan Paul, 1966. 400 pp.

"The Greeks, who invented many arts, invented an art of memory which, like their other arts, was passed on to Rome whence it descended in the European tradition. This art seeks to memorise through a technique of impressing 'places' and 'images' on memory" (Preface). Surveys the history from Simonides of Ceos to Leibniz. Pp. 105–389 cover the 15th to 17th centuries.

235. Rusk, Robert R. *The Doctrines of the Great Educators*. Revised and enlarged fourth ed. London: Macmillan, 1969. 356 pp. [First published in 1918]

"An exposition of the doctrines of a limited number of representative educators" from Plato to Whitehead. Includes (pp. 52–156): Thomas Elyot (1490?–1546); Ignatius Loyola (1491–1551); Comenius (1592–1670); Milton (1608–74); Locke (1632–1704).

236. O'Malley, C. D., ed. *The History of Medical Education. An International Symposium Held February 5–9, 1968*. UCLA Forums in Medical Sciences, no. 12. Berkeley-Los Angeles: Univ. of California Press, 1970. 548 pp.

Nineteen papers of which eight (pp. 89–299) relate to medical education in western Europe during the Renaissance and seventeenth century. With bibliographies. Reviewed in *Isis* 62 (1971), 252.

237. Bowen, James. *A History of Western Education*. Vol. 2. *Civilization of Europe: Sixth to Sixteenth Century*. Vol. 3. *The Modern West: Europe and the New World*. London: Methuen, 1975–81. 2 vols.

Vol. 2, pp. 207–435: Humanism and higher learning in Italy of the Quattrocento (Ch. 7–8); Expansion of education and humanism in France, Germany, and England to 1500 (Ch. 9–10); Christian humanism—Erasmus and the ideal of piety (Ch. 11); Martin Luther and the Reformation in Germany

(Ch. 12); Extension of educational thought and practice in the sixteenth century (Ch. 13); The search for method—towards a *Ratio* (Ch. 14). Vol. 3, pp. 1–167: The educational heritage of the modern West (Ch. 1); The scientific revolution of the seventeenth century (Ch. 2); A "Reformation of schooles"—Utopia and reality in the seventeenth century (Ch. 3); Reformation to Enlightenment—the extension of schooling, the Catholic conservative tradition, the Protestant initiative (Ch. 4–5). Each volume has notes, bibliography and index. Reviewed in *TLS* 25 April 1975, p. 452 and *TLS* 9 April 1982, p. 30.

238. Boyd, Williams, and Edmund J. King. *The History of Western Education*. 11th ed. London: Adam and Charles Black, 1975. 517 pp.

Pp. 159–279: Humanistic education; The Reformation and education; The broadening of humanism; The seventeenth century. Bibliographies.

239. Schmitt, Charles B. "Philosophy and Science in Sixteenth-Century Universities: Some Preliminary Comments." *The Cultural Context of Medieval Learning. Proceedings of the First International Colloquium on Philosophy, Science, and Theology in the Middle Ages, September 1973.* John Emery Murdoch and Edith Dudley Sylla, eds. Boston Studies in the Philosophy of Science, 26. Dordrecht: Reidel, 1975, pp. 485–537.

Deals with several aspects of sixteenth-century university culture (including: teaching of philosophy; influence of humanism, the Reformation, and scientific change on the universities; Pisa and Oxford; developments in the teaching of medicine and mathematics).

240. Schmitt, Charles B., ed. "Continuity and Change in Early Modern Universities." *History of Universities* 1 (1981), 1–213.

Six papers focusing on the theme "Innovation and Tradition: the Universities of Early Modern Europe"—a colloquium held at the Warburg Institute in 1979. (Reaction of English and German universities to the Reformation and Counter-Reformation; The classroom in a Parisian college of the Renaissance; Spanish universities and the debate over the justice of the conquest of America; New Testament and the Dutch universities; Philosophy teaching in France; German universities after the Thirty Years War.) Includes two essay reviews: Recent studies on Italian universties; Oxford and Cambridge college histories.

241. Schmitt, Charles B. *The Aristotelian Tradition and Renaissance Universities*. London: Variorum Reprints, 1984. 362 pp.

Collection of 15 papers published between 1965 and 1983, organized around the two themes. Reviewed in *Isis* 76 (1985), 406–7.

242. Houston, R. A. *Literacy in Early Modern Europe. Culture and Education, 1500–1800*. London: Longman, 1988. 266 pp.

Examines the "central issues of education, literacy and culture between the Renaissance and the Industrial Revolution." Discusses: the nature of education; the extent of reading and writing among the people; the uses to which literacy was put. Pp. 235–62: Bibliography.

243. Hay, Denys. "Schools and Universities." *The New Cambridge Modern History*. Vol. 2; *The Reformation*. Second ed. Edited by G. R. Elton. Cambridge: Cambridge Univ. Press, 1990, pp. 452–77.

Discusses educational changes during the Reformation.

ENGLAND

244. Ball, W. W. Rouse. *Notes on the History of Trinity College, Cambridge*. London: Macmillan and Co., 1899. 183 pp.

Pp. 37–110 cover the history from the foundation up to 1700.

245. Watson, Foster. "The Curriculum and Text-Books of English Schools in the First Half of the Seventeenth Century." *Transactions, Bibliographical Society* (London) 6 (1900–1902), 159–267. [Reprinted separately in 1903]

Pp. 159–235: Elementary subjects (reading, writing, arithmetic, foreign languages, drawing, religious instruction, geography or cosmography, history, music); Grammar (Latin and Greek grammars); Letter writing; Rhetoric and themes; The oration. Pp. 237–67: Appendix (Illustrations of exercise work in the English grammar schools).

246. Watson, Foster, comp. *English Writers on Education, 1480–1603. A Source Book*. A facsimile reproduction with an introduction by Robert D. Pepper. Gainesville, Fla.: Scholars' Facsimiles and Reprints, 1967. 153 pp. [Originally published as *Notices of Some Early English Writers on Education, with Descriptions, Extracts, and Notes* (1902–1906) in Annual Reports of the US Commissioner of Education.]

A bibliographical study of English education. Incomplete. Describes the works of some fifty educationists (several non-English: Erasmus, Vives, Castiglione).

247. Watson, Foster. *The English Grammar Schools to 1660. Their Curriculum and Practice.* London: Frank Cass, 1968. 548 pp. [First published 1908]

History of teaching in the grammar schools since the invention of printing, based on documents and school textbooks. The expansion of the medieval quadrivium into the separate subjects of mathematics and the sciences had little influence on the grammar schools.

248. Watson, Foster. "Notes and Materials on Religious Refugees in Their Relation to Education in England before the Revocation of the Edict of Nantes, 1685." *Proceedings, Huguenot Society of London* (1909–11), 299–475. [Reprinted in 1911 as *English Refugees and English Education* by Spottiswoode and Co., London]

On the significance of the Huguenot strain in English education.

249. Watson, Foster. *The Beginnings of the Teaching of Modern Subjects in England.* London: Pitman and Sons, 1909. 555 pp.

"Traces the origins of the teaching of English and other modern languages; also of history, geography, drawing, mathematics, astronomy, and other scientific subjects, chiefly outside the grammar-school curriculum and chiefly in the 17th century."

250. Parker, Irene. *Dissenting Academies in England. Their Rise and Progress and Their Place among the Educational Systems of the Country.* Cambridge: University Press, 1914. 168 pp.

The education of the middle classes in "modern" subjects was "advocated chiefly by the Puritans in the first half of the 17th century; it was attempted by the Tutors of the Dissenting Academies."

251. Chaplin, Arnold. "The History of Medical Education in the Universities of Oxford and Cambridge, 1500–1850." *Proceedings, Royal Society of Medicine, Sec. Hist. Med.* 13 (1919–20), 83–107.

Oxford: Medical education under the "Statuta antiqua"; The "Caroline" or "Laudean" code of 1636; The history of the medical professorships and lectureships. Cambridge: "Statuta antiqua"; The Elizabethan statutes of 1570; Professorships and lectureships.

252. McLachlan, H. *English Education under the Test Acts, Being the History of the Non-Conformist Academies, 1662–1820.* Publications, University of Manchester, History ser., no. 59. Manchester: Manchester Univ. Press, 1931. 344 pp.

Pp. 45–117 describe Nonconformist academies founded in the seventeenth century. Pp. 300–3: Textbooks in the early academies. Reviewed in *TLS* 12 Nov. 1931, p. 878.

253. Johnson, Francis R., and Sanford V. Larkey. "Robert Recorde's Mathematical Teaching and the Anti-Aristotelian Movement." *Huntington Library Bulletin* 7 (1935), 59–87.

Describes Recorde's aims and methods in the teaching of the sciences and points out briefly their "relation to Renaissance pedagogical theories and to the general anti-Aristotelian movement."

254. Gunther, R. T. *Early Science in Oxford*. Vol. 11. *Oxford Colleges and Their Men of Science*. Oxford: Privately published, 1937. 429 pp.

Sets out the achievements of the "more scientifically minded Oxford men." Pp. 325–36: Old scientific books in college libraries.

255. Allen, Phyllis. "Medical Education in Seventeenth Century England." *Journal of the History of Medicine* 1 (1946), 115–43.

On changes in the classical curriculum in medicine. Topics include: conservatism of Oxford and Cambridge; the Dissenting Academies; Gresham College; College of Physicians; College of Barber-Surgeons.

256. Allen, Phyllis. "Scientific Studies in Seventeenth Century English Universities." *Journal of the History of Ideas* 10 (1949), 219–53.

On the gradual acceptance of scientific studies into the university curriculum during the Age of Newton.

257. Costello, William T. *The Scholastic Curriculum at Early Seventeenth-Century Cambridge*. Cambridge: Harvard Univ. Press, 1958. 221 pp.

On the training given at Cambridge to our literary ancestors of the Renaissance and the early seventeenth century. Reviewed in *TLS* 5 December 1958, p. 709, and *Isis* 51 (1960), 112–13.

258. Curtis, Mark H. *Oxford and Cambridge in Transition, 1558–1642. An Essay on Changing Relations between the English Universities and English Society*. Oxford: Clarendon Press, 1959. 314 pp.

Describes the changes which the universities of Oxford and Cambridge underwent during the Tudor and early Stuart periods. Revises some earlier interpretations on the basis of new evidence. Pp. 226–60: The universities and the advancement of learning. Pp. 290–314: Index of names and subjects. With notes but no bibliography. Review: *TLS* 20 Nov. 1959, p. 681.

259. Jarman, T. L. *Landmarks in the History of Education. English Education as Part of the European Tradition.* 2nd ed. London: John Murray, 1963. 325 pp. [First published 1951]

Traces the evolution of education in England as part of the historical development of European education. Pp. 308–25: Bibliography and index.

260. Stone, Lawrence. "The Educational Revolution in England, 1560–1640." *Past and Present* 28 (1964), 41–80.

Attempts to study the "scale of growth and the shifts in social distribution of education in England between 1560 and 1640."

261. Simon, Joan. *Education and Society in Tudor England.* Cambridge: University Press, 1966. 452 pp. [Paperback ed. published in 1979]

Scholarly survey dealing with all aspects of education, based on recent studies. Pp. 404–36: Sources and bibliography; pp. 437–52: Name and subject index. Reviewed in *English Historical Review* 82 (1967), 384.

262. Lawson, John. *Medieval Education and the Reformation.* London: Routledge and Kegan Paul, 1967. 115 pp.

An outline history of grammar schools and universities and their development (and expansion) under the impact of the Reformation.

263. Greaves, Richard L. *The Puritan Revolution and Educational Thought: Background for Reform.* New Brunswick, N.J.: Rutgers Univ. Press, 1969. 188 pp.

Study of the educational theories of the Puritan Revolution in England. Pp. 63–92: Educational reform in the sciences. Pp. 147–88: Notes, select bibliography, and index.

264. Axtell, James L. "Education and Status in Stuart England: The London Physician." *History of Education Quarterly* 10 (1970), 141–59.

On the education and status of the Stuart physician and the early years of the College of Physicians.

265. Debus, Allen G., ed. *Science and Education in the Seventeenth Century. The Webster-Ward Debate.* London: Macdonald, 1970. 307 pp.

Pp. 1–64: Introduction to the debate (1654) between John Webster, John Wilkins, Seth Ward and Thomas Hall on the reform of the English universities. Pp. 67–307 contain facsimile reproductions of the tracts: *Academiarum examen* (1653) by John Webster; *Vindiciae academiarum* (1654) by John Wilkins and Seth Ward; *Historio-mastix* (1654) by Thomas Hall. Reviews: *Isis* 64 (1973), 423–24; *Ambix* 19 (1972), 221–22.

266. Kearney, Hugh. *Scholars and Gentlemen: Universities and Society in Pre-industrial Britain, 1500–1700.* London: Faber and Faber, 1970. 214 pp.

Discusses the curriculum of the universities of England, Scotland and Ireland against the social background, using students' notebooks. Bibliography. Reviewed in *TLS* 16 April 1971, p. 440.

267. Webster, Charles, ed. *Samuel Hartlib and the Advancement of Learning.* Cambridge: University Press, 1970. 220 pp.

A selection of the writings of Hartlib and John Dury. Pp. 1–72: Introduction; pp. 202–20: Notes, bibliography, and index.

268. Shapiro, Barbara J. "The Universities and Science in Seventeenth-Century England." *Journal of British Studies* 10 (1971), (2), 47–82.

Examines the "state of the sciences in Oxford and Cambridge prior to, during, and after the Interregnum in order to suggest that universities had shown a continuous interest in science."

269. Frank, Robert G., Jr. "Science, Medicine, and the Universities of Early Modern England: Background and Sources." *History of Science* 11 (1973), 194–216, 239–69.

From the late sixteenth to the mid-eighteenth centuries, the universities of Oxford and Cambridge "occupied a solid and by no means modest place in the institutional landscape of English science and scientists."

270. Orme, Nicholas. *English Schools in the Middle Ages.* London: Methuen, 1973. 369 pp.

Survey of English schools between the years 1200 and 1530. Pp. 326–54: Bibliography; pp. 355–69: Index.

271. Cressy, David. *Education in Tudor and Stuart England.* London: Edward Arnold, 1975. 141 pp.

Presents a collection of documents chosen to "outline the school system of Tudor and Stuart England and to illustrate some of the social and political pressures bearing on education in the sixteenth and seventeenth centuries." The documents (156), with individual commentaries, are arranged in eight chapters: Perspectives on education; Control of education; The organization of schools; Schoolmasters; The curriculum; Educational opportunity; Education of women; The universities.

272. Hunter, Michael. *John Aubrey and the Realm of Learning*. London: Duchforth, 1975. 256 pp.

Study of Aubrey's writings and their background, especially his relationship to the seventeenth-century scientific movement. Reviewed in *Isis* 68 (1977), 483–84 and *TLS* 26 Mar. 1976, p. 351.

273. Webster, Charles. "The Curriculum of the Grammar Schools and Universities, 1500–1600: A Current Review of the Literature." *History of Education* 4 (1975) (1), 51–68.

On the development of grammar schools and the way in which the Quadrivium was used as a basis for the introduction of "modern" subjects into education. Bibliography (some 200 items) classified under: general; texts; the idea of education; the grammar schools; the universities and "modern" subjects).

274. Howson, Geoffrey. *A History of Mathematics Education in England*. Cambridge: Cambridge Univ. Press, 1982. 294 pp.

Claims to be the first book "which attempts to tell the story of the development of mathematics education in England." Covers the story from Bede to Elizabeth Williams (b. 1895). Documented. Pp. 6–28: Robert Recorde; pp. 29–44: Samuel Pepys; pp. 45–49: The Puritan revolution and its aftermath; the early dissenting academies. Name and subject indexes. Reviewed in *Isis* 75 (1984), 575.

275. O'Day, Rosemary. *Education and Society, 1500–1800. The Social Foundations of Education in Early Modern Britain*. London: Longman, 1982. 324 pp.

Pp. 282–316: Notes and references; Select bibliography for further reading.

276. Aston, T. H., gen. ed. *The History of the University of Oxford*. Vol. 3: *The Collegiate University*. Edited by James McConica. Oxford: Clarendon Press, 1986. 775 pp.

A unified study (10 chapters) by specialists, of the history of the university from the end of the Middle Ages to the end of the Tudor period. Pp. 213–56: The Faculty of Medicine (by Gillian Lewis); pp. 440–519: The provision of books (by N. R. Ker); pp. 733–75: Index. Documented but no bibliography.

277. Bill, E. G. W. *Education at Christ Church, Oxford, 1660–1800.* Oxford: Clarendon Press, 1988. 367 pp.

An attempt to "discover who went to Christ Church between the Restoration and the end of the eighteenth century, what they were taught, and who taught them."

278. Feingold, Mordechai. "The Universities and the Scientific Revolution: The Case of England." *New Trends in the History of Science. Proceedings of a Conference Held at the University of Utrecht.* Edited by R. P. W. Visser, H. J. M. Bos, L. C. Palm, and H. A. M. Snelders. Amsterdam: Rodopi B.V., 1989, pp. 29–52.

On the need for a major re-evaluation of the role of the universities in the Scientific Revolution. Commentary by H. Floris Cohen (pp. 49–52).

279. Orme, Nicolas. *Education and Society in Medieval and Renaissance England.* London: The Hambledon Press, 1989. 297 pp.

Complements the author's earlier studies: *English Schools in the Middle Ages* (1973) and *From Childhood to Chivalry* (1984). Aims to give an outline of the history of non-university education from the twelfth to the sixteenth centuries. See item 270.

280. Alexander, Michael Van Cleave. *The Growth of English Education 1348–1648. A Social and Cultural History.* University Park: Pennsylvania State Univ. Press, 1990. 286 pp.

Pp. 247–86: Notes, suggestions for additional reading, and index. Reviews: *Choice* 28 (1990), 357; *Journal of Higher Education* 62 (1991), 467.

SCOTLAND

281. Wilson, Duncan K. *The History of Mathematical Teaching in Scotland to the End of the Eighteenth Century.* London: University Press, 1935. 107 pp.

Pp. 6–30: Brief survey from the founding of St. Andrews University to 1700. Reviewed in *TLS* 24 Oct 1935, p. 674.

282. French, Roger. "Medical Teaching in Aberdeen: from the Foundation of the University to the Middle of the Seventeenth Century." *History of Universities* 3 (1983), 127–57.

Some information on books used in teaching in Aberdeen, where the first chair of medicine (in Britain) had been established in 1495.

FRANCE

283. Sédillot, L. Am. "Les professeurs de mathématique et de physique générale au Collège de France. (Première période: François I, 1530–1547. Deuxième période: Les derniers Valois, 1547–1589. Troisième période: 1589–1774." *Bullettino di bibliografia e di storia delle scienze matematiche e fisiche* 2 (1869), 343–68, 387–448, 461–510.

Administrative history of the college (continued up to 1869 in Vol. 3) with special reference to the chairs of mathematics and physics. Bio-bibliographical notes in the text.

284. Lefranc, Abel. *Histoire du Collège de France depuis ses origines jusqu'à la fin du premier Empire.* Paris: Librarie Hachette et Cie., 1893. 432 pp.

Pp. 39–252 cover the history of the College from Francis I (1529) to Louis XIV (1715).

285. Hamy, E. T. "Recherches sur les origines de l'enseignement de l'anatomie humaine et de l'anthropologie au Jardin des Plantes." *Nouvelles archives du Muséum d'histoire naturelle* (3), 7 (1895), 1–30.

Notes on teaching at the Jardin des Plantes (founded 1626) from 1635 to 1680.

286. Irsay, Stephen d'. *Histoire des universités françaises et étrangères des origines à nos jours.* Tome I. *Moyen âge et Renaissance.* Tome II. *Du XVIe siècle à 1860.* Paris: Editions Auguste Picard, 1933–35. 2 vols.

Vol. I, pp. 223–361 and Vol. II, pp. 1–104 relate to the period from the Renaissance to the end of the seventeenth century.

287. Dainville, François de. *La géographie des humanistes. (Les Jésuites et l'éducation de la société française.)* Paris: Beauchesne, 1940. 563 pp.

Historical study of the teaching of geography by the Jesuits (up to 1700).

288. Artz, Frederick Binkherd. *The Development of Technical Education in France, 1500–1850.* Cambridge, Mass.: The Society for the History of Technology and the MIT Press, 1966. 274 pp.

Pp. 1–59: The beginnings, 1500–1700. Review: *TLS* 26 Jan. 1967, p. 70.

289. Durkheim, Emile. *The Evolution of Educational Thought. Lectures on the Formation and Development of Secondary Education in France.* London: Routledge and Kegan Paul, 1977. 354 pp. [Translated from the 1938 French ed.]

Text of a course given to *agrégation* candidates at the University of Paris (1904–5). Pp. 63–173: Analysis of the educational system in France from the twelfth to the sixteenth century. Pp. 177–277: The Renaissance (Rabelais or the encyclopedic movement; Erasmus and the humanist movement); educational theory in the sixteenth century; The educational thought of the Renaissance; the Jesuits.

290. Dainville, François de (1909–1971). *L'éducation des jésuites (XVIe– XVIIIe siècles).* Textes réunies et présentés par Marie Madeleine Compère. Paris: Editions de Minuit, 1978. 570 pp.

Collection of twenty-nine articles of which six (pp. 311–423) relate to the teaching of science. With notes and a bibliography of the author's works.

291. Brockliss, L. W. B. "Aristotle, Descartes, and the New Science: Natural Philosophy at the University of Paris 1600–1740." *Annals of Science* 38 (1981), 33–69.

On the decline of Aristotelian physics at the University of Paris.

292. Parias, Louis-Henri, ed. *Histoire générale de l'enseignement et de l'éducation en France.* Tome 2. *De Gutenberg aux Lumières.* Par Francois Lebrun, Marc Vénard et Jean Quéniart. Paris: Nouvelle Librairie de France, 1981. 677 pp.

Study of pre-school and school education, 1480–1789. Pp. 631–40: Chronology and bibliography.

293. Compère, Marie Madeleine, and Dominique Julia. *Les collèges françaises: 16e–18e siècles. 1. Répertoire. France du Midi.* Paris: INRP-CNRS, 1984. 759 pp.

Information on some 370 colleges in 44 *départements*. Details include: brief history; bibliography; library catalogues.

294. Brockliss, L. W. B. *French Higher Education in the Seventeenth and Eighteenth Centuries: A Cultural History.* Oxford: Clarendon Press, 1987. 544 pp.

Study, based on student *cahiers* and professional textbooks, of the structure and content of the college and university curriculum. Pp. 337–440 cover philosophy (physics) and medicine. Pp. 486–523: Bibliography. Reviewed in *Isis* 79 (1988), 346–47 and *TLS* 30 Oct. 1987, p. 1183.

GERMANY

295. Günther, Siegmund. *Geschichte des mathematischen Unterrichts im deutschen Mittelalter bis zum Jahre 1525.* Monumenta Germaniae Pedagogica, hrsg. von Karl Kehrbach, Vol. 3. Berlin: Hofmann, 1887. 408 pp.

Pp. 207–370 cover the period 1450 to 1525.

296. Paulsen, Friedrich. *The German Universities and University Study.* London: Longmans, Green, and Co., 1906. 451 pp. [Translated from the German ed. of 1902]

Study of German universities, their structure, function, and their relation to the life of the community and to the central government. With bibliography.

297. Paulsen. Friedrich. *Geschichte des gelehrten Unterrichts auf den deutschen Schulen und Universitäten vom Ausgang des Mittelalters bis zur Gegenwart. Mit besonderer Rücksicht auf den klassischen Unterricht.* 3rd ed. Leipzig: Veit, 1919. 2 vols.

Vol. 1 (636 pp.) covers the period 1450 to 1740. Vol. 2 contains a list of works cited and three indexes (names, places, and subjects), covering both volumes.

298. Schmitz, Rudolf, ed. *Die Naturwissenschaften an der Philipps-Universität Marburg, 1527–1977.* Marburg: N. G. Elwert Verlag, 1978. 540 pp.

Pp. 3–32: Physics from Antonius Niger (1533–36) to Denis Papin (1688–1707); Pp. 75–106: Botany from Euricius Cordus (1527–33) to Daniel Nebel (1693–1707); Pp. 140–41: Botanic Garden (1527); Pp. 185–203: Chemistry from Dryander (1535–60) to Hartmann (1592–1627); Pp. 333–57: Pharmacy from Hartmann to Daniel Nebel. Pp. xv–xxxi: Chronology. Bibliographies.

ITALY

299. Crombie, A. C. "Mathematics and Platonism in Sixteenth-Century Italian Universities and in Jesuit Educational Policy." *Prismata: naturwissenschaftsgeschichtliche Studien. Festschrift für Willy Hartner.* Edited by Y. Maeyama and W. G. Saltzer. Wiesbaden: Steiner, 1977, pp. 63–94.

300. Schmitt, Charles B. "Filosofia e scienze nelle università italiane del XVI secolo." *Il Rinascimento: interpretazioni e problemi. A Eugenio Garin nel suo 70° compleanno.* Rome: Laterza, 1979, pp. 353–98. [English translation, London: Methuen, 1982]
Survey of recent studies.

301. Palmer, Richard. *The "Studio" of Venice and Its Graduates in the Sixteenth Century.* Contributi alla storia dell' Università di Padova, 12. Trieste: Edizioni Lint, 1983. 204 pp.

On the College of Physicians of Venice and its students. Includes a directory of six hundred students who graduated in the sixteenth century.

302. Grendler, Paul F. *Schooling in Renaissance Italy: Literacy and Learning, 1300–1600.* Johns Hopkins University Studies in Historical and Political Science, 107th Series, 1. Baltimore, Md.: Johns Hopkins Univ. Press, 1989. 477 pp.

Study of primary and secondary education in Renaissance Italy based on original textbooks, teachers' manuals, and archival records. Pp. 431–61: Bibliography; pp. 463–77: Index. Reviewed in *Isis* 82 (1991), 127–28. This work and others on the same topic are the subject of a critical survey by Robert Black in "Italian Renaissance Education: Changing Perspectives and Continuing Controversies," *Journal of the History of Ideas* 52 (1991), 315–34. Grendler's reply is on pp. 335–37.

302A. Findlen, Paula. *Possessing Nature. Museums, Collecting and Scientific Culture in Early Modern Italy.* Studies on the History of Society and Culture. Berkeley, Calif.: Univ. of California Press, 1994. 449 pp.

A well-documented study of the lost world of late Renaissance and Baroque Italian museums—from Ulisse Aldrovandi to Athanasius Kircher—and their place in the history of science (including their role in the development of natural history as a discipline). Pp. 241–87: Museums of medicine; pp. 409–49: Bibliography and index.

NETHERLANDS

303. Lunsingh Scheurleer, Th. H. and G. H. M. Posthumus Meyjes, eds. *Leiden University in the Seventeenth Century. An Exchange of Learning.* Leiden: Universitaire Pers, 1975. 496 pp.

Pp. 217–343: The teaching of medicine and science; Pp. 395–459: The library; Pp. 467–79: Select bibliography.

Science, Society, and Religion

304. Smiles, Samuel. *The Huguenots: Their Settlements, Churches, and Industries, in England and Ireland.* 6th ed. London: John Murray, 1889. 458 pp.

Pp. 242–60: Huguenot men of science and learning. Pp. 261–77: Refugee artisans.

305. Pastor, Ludwig, Freiherr von. *The History of the Popes, from the Close of the Middle Ages. Drawn from the Secret Archives of the Vatican and Other Original Sources.* London: Hodges; Kegan Paul; Routledge and Kegan Paul, 1891–1953. 40 vols. [Translated from the German]

Monumental study on the papacy, copiously annotated. Each volume has a detailed table of contents and an index of names. Vols. 2–32 cover the period from the pontificate of Nicholas V (1447–55) to that of Innocent XII (1691–1700).

306. White, Andrew Dickson. *A History of the Warfare of Science with Theology in Christendom.* New York: Dover, 1960. 2 vols. [First published in 1899]

On the changes in religious dogma necessitated by advances in man's knowledge of the universe. With bibliographical notes. Reviewed in *Hispanic American Historical Review* 41 (1961), 307.

307. Willey, B. *The Seventeenth Century Background. Studies in the Thought of the Age in Relation to Poetry and Religion.* Harmondsworth: Penguin Books, 1972. 283 pp. [First published 1934]

Provides a sketch of the intellectual background of the period.

308. Merton, Robert K. *Science, Technology and Society in Seventeenth Century England.* New York: Howard Fertig, 1970. 279 pp. [Originally published in *Osiris* 4 (1938), 360–632]

With a new preface and updated bibliography. Deals with the "special cultural and economic conditions which made modern science possible." Reviews: *Science and Society* 2 (1938), 566–71 (J. Needham); *Isis* 31 (1940), 438–41 (R. F. Jones). For assessments of the impact and influence of this work, see items 318, 350, 352.

309. Raistrick, Arthur. *Quakers in Science and Industry. Being an Account of the Quaker Contributions to Science and Industry During the Seventeenth and Eighteenth Centuries.* London: The Bannisdale Press, 1950. 361 pp.

Covers the years 1650 to 1800.

310. Kocher, Paul H. *Science and Religion in Elizabethan England.* San Marino, Calif.: The Huntington Library, 1953. 340 pp.

Seeks to analyze how Elizabethan Christianity and science "impinged upon each other, and in so doing affected other attitudes, in a number of key topics ranging widely from medicine to astrology, from Satan to the new world in the moon."

311. Raven, Charles E. *Natural Religion and Christian Theology.* The Gifford Lectures 1951, first series: Science and religion. Cambridge: University Press, 1953. 224 pp.

Ten lectures including: Gesner and the age of transition; Cudworth [Ralph, 1617–1688] and the age of genius; Newton and the age of the machine (pp. 80–144).

312. Russo, François, S.J. "Rôle respectif du catholicisme et du protestantisme dans le développement des sciences aux XVIe et XVIIe siècles." *Journal of World History* 3 (1957), 854–80.

Differs from Pelseneer's somewhat simplistic view (see item 314) that modern science was a result of the Reformation.

313. Westfall, Richard S. *Science and Religion in Seventeenth-Century England.* New Haven, Conn.: Yale Univ. Press, 1958. 235 pp.

Examines the way in which "natural science came to replace Christianity as the central point of intellectual life." Pp. 221–28: Bibliographical essay.

314. Pelseneer, Jean. "La réforme du XVIe siècle à l'origine de la science moderne." *La science au seizième siècle.* Alexandre Koyré, ed. Paris: Hermann, 1960, pp. 151–68.

The author, while acknowledging his Roman Catholic faith, argues in favor of the Protestant origins of modern science.

315. Hill, Christopher. *The Century of Revolution, 1603–1714.* London: Sphere Books, 1961. 301 pp.

"The object of this book is to try to understand the changes which set England on the path of Parliamentary government, economic advance and imperialist foreign policy, of religious toleration and scientific progress." Reviewed in *TLS* 1 Sep. 1961, p. 579.

316. Rowbottom, Margaret E. "Some Huguenot Friends and Acquaintances of Robert Boyle (1627–1691)." *Proceedings of the Huguenot Society, London* 20 (1961), (2), 177–94.

Peter du Moulin, Samuel Hartib, John Dury, Nicholas Le Fevre, Hugh Chamberlen, Denis Papin.

317. Rabb, T. K. "Puritanism and the Rise of Experimental Science in England." *Journal of World History* 7 (1962), 46–67.

Concludes that: "Puritanism cannot be regarded as a main factor or a tangible cause. Yet it would be hard to deny that its indirect help played a part of considerable importance in the developments of the time."

318. Hall, A. Rupert. "Merton Revisited: or, Science and Society in the Seventeenth Century." *History of Science* 2 (1963), 1–16.

Merton's *Science, Technology and Society in Seventeenth-Century England* (1938) viewed by an intellectual historian. See item 308.

319. Aston, Trevor, ed. *Crisis in Europe, 1560–1660. Essays from "Past and Present."* With an introduction by Christopher Hill. London: Routledge and Kegan Paul, 1965. 368 pp.

Collection of essays (thirteen contributors) covering the many aspects of the crisis and change that occurred in western and central Europe—in society, government, economics, religion, and education—during the seventeenth century.

320. Hill, Christopher. *Intellectual Origins of the English Revolution.* Oxford: Clarendon Press, 1965. 333 pp.

Ford lectures, Oxford. Covers: London science and medicine; Francis Bacon and the Parliamentarians; Ralegh—science, history, and politics; Sir Edward Coke—myth-maker. Appendix: A note on the universities. Reviews: *Isis* 57 (1966), 142–43; *TLS* 20 May 1965, p. 387. For criticism of some of the ideas expressed here see H. F. Kearney, "Puritanism, Capitalism and the Scientific Revolution," *Past and Present* 28 (1964), 81–101. Hill's reply is in Vol. 29 (1964), 88–97.

321. National Museum of Fine Arts, Stockholm. *Christina, Queen of Sweden—A Personality of European Civilisation. National Museum, Stockholm, 29 June–16 October, 1966.* Eleventh Exhibition of the Council of Europe. Nationalmusei Utställningskatalog 305. Stockholm, 1966. 622 pp.; 96 pls.

Catalogue of an exhibition intended to revive the memory of Christina of Sweden (1626–89), patron of learning, lover of the fine arts and "precursor of the great intellectual rulers of the Enlightenment." The catalogue is divided into 29 sections, each with an introductory note. Pp. 26–27: Calendar of events in Christina's life; pp. 30–39: Queen Christina; pp. 44–53: Christina and the scholars; pp. 364–75: Scholars in Rome; pp. 376–88: Academies and festivities; pp. 529–45: Manuscripts. For a recent study of Christina, with an extensive bibliography, see Susanna Ackerman, *Queen Christina and Her Circle: the Transformation of a Seventeenth-Century Philosophical Libertine*, Brill's Studies in Intellectual History, 21 (Leiden: E.J. Brill, 1991), 339 pp.

322. Shapiro, B. J. "Latitudinarianism and Science in Seventeenth-Century England." *Past and Present* 40 (1968), 16–41.

Argues that religious moderation was connected with the English scientific movement far more intimately than Puritanism.

323. Ben David, Joseph. *The Scientist's Role in Society. A Comparative Study.* Englewood Cliffs, N.J.: Prentice Hall, 1971. 207 pp.

Describes the "emergence and the development of the social role of the scientist, and of the organization of scientific work." Pp. 1–87: The sociology of science; Science in comparative perspective; The sociology of Greek science; The emergence of the scientific role; The institutionalization of science in seventeenth-century England.

324. Bangert, William V., S. J. *A History of the Society of Jesus.* St. Louis: Institute of Jesuit Sources, 1972. 558 pp.

A history of the Jesuits from their foundations in 1556 to the present. Documented, with bibliography and index. Pp. 1–362 cover the years 1491–1687.

325. Hooykaas, Reyer. *Religion and the Rise of Modern Science.* Edinburgh: Scottish Academic Press, 1972. 162 pp.

"The confrontation of Graeco-Roman culture with biblical religion engendered, after centuries of tension, a new science. This science preserved the indispensable parts of the ancient heritage (mathematics, logic, methods of observation and experimentation), but it was directed by different social and methodological conceptions, largely stemming from a biblical world view. Metaphorically speaking, whereas the bodily ingredients of science may have been Greek, its vitamins and hormones were biblical." In five chapters: God and Nature; Reason and experience; Nature and art; The rise of experimental science; Science and the Reformation.

326. Mathias, Peter, ed. *Science and Society, 1600–1900.* Cambridge: Cambridge Univ. Press, 1972. 166 pp.

Six essays, of which three relate to the seventeenth century (pp. 1–53). Reviewed in *Isis* 65 (1974), 400–02.

327. Smith, Alan G. R. *Science and Society in the Sixteenth and Seventeenth Centuries.* London: Thames and Hudson, 1972. 216 pp.

Discusses "the significance of those scientific discoveries which later historians have recognized as most important in establishing the new ideas about man and the universe which were emerging by 1700." Reviews: *American Historical Review* 80 (1975), 397–98; *History* 59 (1974), 268–69.

328. Wightman, W. P. D. *Science in a Renaissance Society.* London: Hutchinson, 1972. 191 pp.

Historical study of the interrelations of science and society during the period 1450–1620.

329. Hill, Christopher. *Change and Continuity in Seventeenth-Century England.* London: Weidenfeld and Nicolson, 1974. 370 pp.

A collection of articles written since 1960, whose main theme is the interrelationship between material and intellectual aspects of the English Revolution, between economics, politics and ideas. Reviews: *Isis* 68 (1977), 153–54; *TLS* 24 Oct. 1975, p. 1250.

330. Manuel, Frank E. *The Religion of Isaac Newton.* The Fremantle Lectures, 1973. Oxford: Clarendon Press, 1974. 141 pp.

Studies of the non-scientific thought of Newton, based mainly on the Newton MSS in Jerusalem. Four lectures. Reviews: *Rete* 2 (1975), 369–72; *English Historical Review* 91 (1976), 428–29; *History of Science* 14 (1976), 196–207.

331. Webster, Charles, ed. *The Intellectual Revolution of the Seventeenth Century.* London: Routledge and Kegan Paul, 1974. 445 pp.

Anthology of fifteen major articles and thirteen shorter comments—on political theory, science and theology published in *Past and Present* between 1953 and 1973. Pp. 435–45: Name and subject index. Reviews: *Revue de synthèse* 96 (1975), 158–61; *Medical History* 19 (1975), 202–9; *History of Science* 14 (1976), 196–207.

332. Büttner, Manfred. "Die Emanzipation der Geographie zu Beginn des 17. Jahrhunderts: ein Beitrag zur Geschichte der Naturwissenschaft in ihren Beziehungen zur Theologie." *Sudhoffs Archiv* 59 (1975), 148–64.

On the emancipation of the geographer from the task of using divine Providence to explain geographic processes.

333. Webster, Charles. *The Great Instauration: Science, Medicine and Reform, 1626–1660.* London: Duckworth, 1975. 630 pp.

Study of the scientific, medical, and social ideas of the English Puritans. Reviews: *TLS* 2 July 1976, pp. 810–12; *English Historical Review* 91 (1976), 853–59; *Isis* 68 (1977), 485–87.

334. Jacob, J. R., and M. C. Jacob. "Seventeenth Century Science and Religon: The State of the Argument." *History of Science* 14 (1976), 196–207.

Essay review of: Charles Webster, ed., *The Intellectual Revolution of the Seventeenth Century* (1974); Frank E. Manuel, *The Religion of Isaac Newton* (1974); Christopher Hill, *Change and Continuity in Seventeenth Century England* (1974).

335. Jacob, Margaret C. *The Newtonians and the English Revolution, 1689–1720.* Hassocks, Sussex: Harvester Press, 1976. 288 pp.

Explores Newtonian scientific and religious ideology in its social context. Reviews: *British Journal for the History of Science* 11 (1978), 164–71; *History of Science* 16 (1978), 143–51.

336. Klaaren, Eugene M. *Religious Origins of Modern Science. Belief in Creation in Seventeenth-Century Thought.* Grand Rapids, Mich.: Eerdmans, 1977. 244 pp.

"Conflict and reformation in Western theologies of creation made the rise of many natural sciences from the older natural philosophy a distinct and lively possibility; belief in divine creation was presupposed in the rise of modern natural science."

337. Holmes, Geoffrey. "Science, Reason, and Religion in the Age of Newton." *British Journal for the History of Science* 11 (1978), 164–70.

Essay review of M. C. Jacob, *The Newtonians and the English Revolution 1689–1720* (item 335).

338. Mandrou, R. *From Humanism to Science, 1480–1700.* Pelican History of European Thought, Vol. 3. Harmondsworth, Middx., 1978. 329 pp. [Translated from the 1973 French ed.]

Attempts to analyse and reconstruct the "complex relations which linked the intellectuals with widely differing social milieux." Reviewed in *Isis* 71 (1980), 345–46.

339. Tyacke, Nicholas. "Science and Religion at Oxford before the Civil War." *Puritans and Revolutionaries: Essays in Seventeenth-Century History Presented to Christopher Hill.* Edited by Donald Pennington and Keith Thomas. Oxford: Clarendon Press, 1978, pp. 73–93.

Surveys the events of the twenty-five years during which the Savilian chairs of astronomy and geometry, the Tomlinson chair of anatomy, and the botanical garden were established.

340. Consiglio d'Europa, Sedicesima Esposizione Europea di Arte, Scienza e Cultura. *Firenze e la Toscana dei Medici nell'Europa del Cinquecento. La Corte, il mare, i mercanti. La Rinascita della Scienza. Editoria e Società, Astrologia, magia e alchimia.* Florence: Electa Editrice, 1980. 438 pp.; illus.

Catalogue of an exhibition held in Florence, 1980. In four sections: (1) The Medici and Europe 1532–1609: the Court, the sea, the merchants (2) The renaissance of science (humanism and science; mathematics; technology and scientific instruments; musical theory; engineering works; physicians and anatomists; the debate in astronomy; renewed interest in natural history; geography, cartography, and the science of navigation; Galileo and Europe) (3) Society and the publishing industry (4) Astrology, magic, and alchemy in the Florentine and European Renaissance. Sections 2, 3, and 4 have bibliographies; section 3 has no catalogue of exhibits.

341. Jacob, James R., and Margaret C. Jacob. "The Anglican Origins of Modern Science: the Metaphysical Foundations of the Whig Constitution. " *Isis* 71 (1980), 251–67.

Attempts to "specify a precise linkage between the dynamics of the [English] Revolution and the philosophical origins of modern science."

342. Hunter, Michael. *Science and Society in Restoration England.* Cambridge: Cambridge Univ. Press, 1981. 233 pp.

Study of the social relations of Restoration science. A bibliographical essay (pp. 198–219) provides a critical introduction to existing scholarship. Reviewed in *Isis* 73 (1982), 314–15.

343. Redondi, Pietro. *Galileo eretico.* Microstorie, 7. Turin: Einaudi, 1983. 463 pp.

Argues that Galileo's atomism conflicted with the Tridentine doctrine on the Eucharist. Refutes the thesis that he was made to abjure the heliocentric theory because it appeared incompatible with Biblical statements. Urges a complete reconsideration and reinterpretation of the 1633 trial. Reviews: *Isis* 76 (1985) 379–80; *Rivista storica italiana* 97 (1985), 177–238; 934–68. English translation by Raymond Rosenthal (Princeton Univ. Press, 1987) reviewed in *Isis* 79 (1988), 348–50. See also items 347, 351.

344. Gwynn, Robin D. *Huguenot Heritage. The History and Contribution of the Huguenots in Britain.* London: Routledge and Kegan Paul, 1985. 220 pp.

Pp. 60–90, 202–4: Crafts and trades; Professions. Includes a bibliographical note. (The term "Huguenot" is used here in the widest possible sense to cover all French-speaking Protestants, including Walloons who came to England in Tudor times and refugees from the Principality of Orange who arrived a century later.)

345. Murdoch, Tessa, comp. *The Quiet Conquest. The Huguenots 1685 to 1985.* A Museum of London Exhibition in Association with the Huguenot Society of London, 15 May to 31 October 1985. London: Museum of London, 1985. 326 pp.; illus.

Catalogue of an exhibition commemorating the third centenary of the Revocation of the Edict of Nantes. 475 exhibits, out of which about a third relate to the sixteenth and seventeenth centuries.

346. Osler, Margaret J., and Paul Lawrence Farber, eds. *Religion, Science and Worldview; Essays in Honor of Richard S. Westfall.* Cambridge: Cambridge Univ. Press, 1985. 350 pp.

Thirteen essays reflecting Westfall's scholarly interests and activities (in three sections: Newtonian studies; Science and religion; Historiography and the social context of science)—all relating to sixteenth- and seventeenth-century science. Includes a bibliography of W.'s writings (mostly on the Scientific Revolution). Reviews: *Isis* 77 (1986), 520–21; *TLS* 1 Aug. 1986, p. 836.

347. Ferrone, Vincenzo, and Massimo Firpo. "From Inquisitors to Microhistorians: a Critique of Pietro Redondi's *Galileo eretico.*" *Journal of Modern History* 58 (1986), 485–524. [Originally published in *Rivista storica italiana*, 1985]

See item 343.

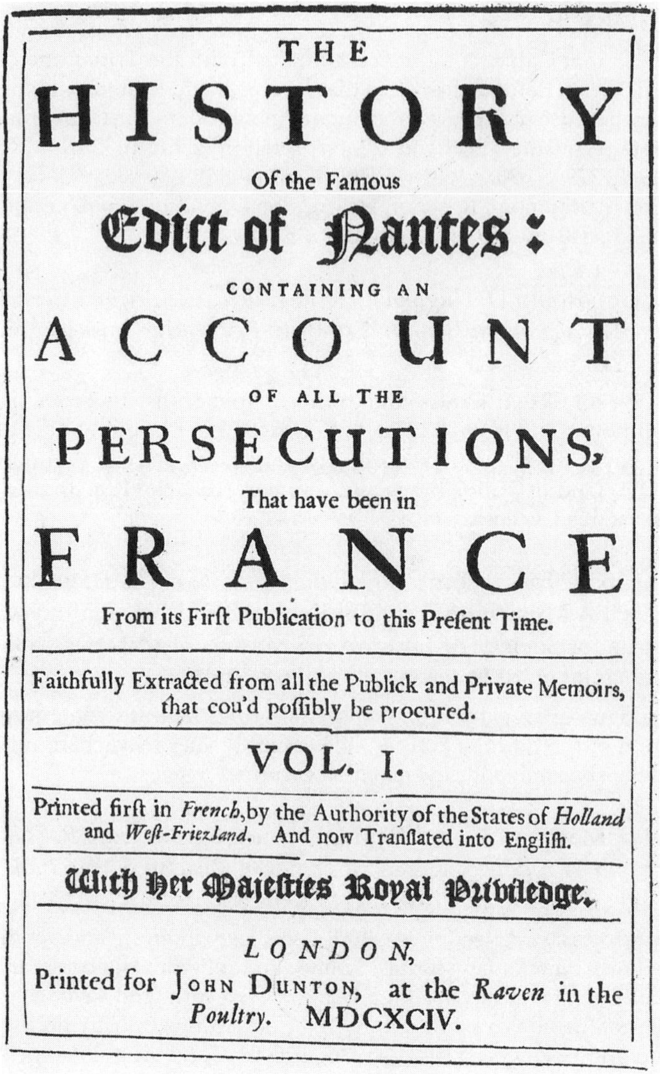

THE

HISTORY

Of the Famous

Edict of Nantes:

CONTAINING AN

ACCOUNT

OF ALL THE

PERSECUTIONS,

That have been in

FRANCE

From its Firſt Publication to this Preſent Time.

Faithfully Extracted from all the Publick and Private Memoirs, that cou'd poſſibly be procured.

VOL. I.

Printed firſt in *French*, by the Authority of the States of *Holland* and *Weſt-Friezland*. And now Tranſlated into Engliſh.

With Her Majeſties Royal Priviledge.

LONDON,
Printed for JOHN DUNTON, at the *Raven* in the *Poultry*. MDCXCIV.

Title page of the English translation of Elie Benoît's *Histoire de l'Edit de Nantes*, Delft, 1693–95 (University of London Library).

The gradual erosion of the privileges established by the Edict of Nantes (1598), culminating in their complete withdrawal by its Revocation (1685), resulted in the emigration of French Huguenots to Britain and elsewhere in increasing numbers. Among them were many skilled craftsmen, medical practitioners and scientists. ***Entries 248, 345, 1165.***

348. Funkenstein, Amos. *Theology and the Scientific Imagination from the Middle Ages to the Seventeenth Century.* Princeton: Princeton Univ. Press, 1986. 421 pp.

Explores the links between early modern science and medieval scholastic theology. Pp. 365–99: Bibliography. Pp. 401–21: Index. Reviewed in *Isis* 78 (1987), 664–65.

349. Lindberg, David C., and Ronald L. Numbers, eds. *God and Nature. Historical Essays on the Encounter between Christianity and Science.* Berkeley: University of California Press, 1986. 516 pp.

Collection based on an international conference held at the University of Wisconsin in April 1981. Eighteen papers of which seven relate to the sixteenth and seventeenth centuries (The Copernicans and the Churches; Galileo and the Church; Reformation theology and the mechanistic conception of nature; Puritanism, separatism, and science; Kepler, Descartes and Newton; Newtonian worldview; Catholicism and early modern science). Pp. 473–84: A guide to further reading. Pp. 489–516: Name and subject index.

350. "Symposium on the Fiftieth Anniversary of *Science, Technology and Society.*" *Isis* 79 (1988), 571–605.

Articles by three "commentators from very different backgrounds," written at the invitation of the editor of *Isis*, giving their personal and informal reactions to Robert K. Merton's *Science, Technology and Society in Seventeenth Century England* (see item 308): "The Publication of *Science, Technology and Society*: Circumstances and Consequences" (I. Bernard Cohen); "Distancing Science from Religion in Seventeenth-Century England" (Thomas F. Gieryn); "Understanding the Merton Thesis" (Stephen Shapin).

351. Westfall, Richard S. "*Galileo Heretic*: Problems, As They Appear to Me, with Redondi's Thesis." *History of Science* 26 (1988), 399–415.

Essay review of Redondi's *Galileo Heretic* (1987). See item 343.

352. Cohen, I. Bernard, ed. *Puritanism and the Rise of Modern Science. The Merton Thesis.* New Brunswick, N.J.: Rutgers Univ. Press, 1990. 402 pp.

Anthology of twenty-four articles by several authors intended to serve as an introduction to and a general survey of the Merton thesis and its consequences. Most of the articles were published during the fifty years or so following the publication of *Science, Technology and Society*. Includes an introduction (pp. 1–111) by I. B. Cohen, afterword by Merton himself (pp. 303–71), and an analytical synopsis by K. E. Dufferin and Stuart W. Strickland (pp. 373–89). Pp. 89–111: List of articles and books related to the thesis (with summaries). Pp. 391–402: Index of names and subjects. See item 308.

353. Brooke, John Hedley. *Science and Religion. Some Historical Perspectives.* Cambridge: Cambridge Univ. Press, 1991. 422 pp.

Critical study of the relationship between scientific thought and religious belief—from the heliocentric theory of Copernicus to *in vitro* fertilization. Pp. 52–151 (chapters 2, 3, 4) cover the sixteenth and seventeenth centuries. Pp. 348–403: Bibliographic essay. Pp. 408–22: Subject index.

354. Shapiro, Barbara. "Early Modern Intellectual Life: Humanism, Religion and Science in Seventeenth Century England." *History of Science* 29 (1991), 45–71.

Explores "the relationship between two *intellectual* movements not commonly connected in current scholarship: Erasmian humanism and science."

354A. Gamble, Richard C., ed. *Calvin and Science.* Articles on Calvin and Calvinism, Vol. 12. New York: Garland Publishing, 1992. 196 pp.

Eleven papers, exploring the relationship between Calvinism and the rise of modern science.

SCIENTIFIC SOCIETIES

355. Ornstein, Martha. *The Role of the Scientific Societies in the Seventeenth Century.* Chicago: University of Chicago Press, 1938. 308 pp. [Reprint of the 1913 ed.]

Topics: The role of individual scientists; The Academies (Lincei, Cimento, Royal Society, Académie des Sciences, German scientific societies); Scientific journals; Science in the universities. Reviews: *Isis* 12 (1929), 154–56; 31 (1939), 87–89.

356. Maylender, Michele. *Storia delle accademie d'Italia.* Bologna: Cappelli, 1926–30. 5 vols.

Posthumously published with an introduction and biographical note on Maylender by Luigi Rava.

Historical encyclopedia of Italian academies and societies (past and present). Alphabetically arranged, with notes of varying length and detail. The term academy is not defined; both public and private associations are included. Each volume has an index of cities but no subject index or classification. For a topographical index and bibliography, see Giuseppe

Gabrieli, "Repertorio alfabetico e bibliografico delle accademie d'Italia nell' opera di M. Maylender," *Accademie e biblioteche d'Italia* 10 (1936), (1), 71–99.

357. Brown, Harcourt. *Scientific Organizations in Seventeenth-Century France (1620–1680).* History of Science Society publications, n.s., 5. Baltimore, Md.: Williams and Wilkins, 1934. 306 pp.

Study of the history of private assemblies of the amateurs of natural science.

358. Yates, Frances A. *The French Academies of the Sixteenth Century.* London: Warburg Institute, 1947. 376 pp.

Based on lectures delivered at the Warburg Institute in 1940. Pp. 95–104: Natural philosophy in the academies. Pp. 275–316: Connections of the sixteenth-century French academies with those of the seventeenth century. Reviewed in *Speculum* 25 (1948), 737.

359. Fisch, Max H. "The Academy of the Investigators." *Science, Medicine and History. Essays in Honour of Charles Singer.* E. Ashworth Underwood, ed. London: Oxford Univ. Press, 1953. Vol. 1, pp. 521–63.

Describes one of the lesser-known Italian academies which led the fight for the moderns in the seventeenth century.

360. Daumas, Maurice. "Esquisse d'une histoire de la vie scientifique." *Histoire de la science.* Encyclopédie de la Pléiade. Volume publié sous la direction de Maurice Daumas. Paris: Librairie Gallimard, 1957, pp. 1–192.

Pp. 41–109: Towards experimentation; scientific *milieux* of the sixteenth and seventeenth centuries; scientific academies; the scientific journal. Reviewed in *Isis* 52 (1961), 106–7.

361. Wall, Cecil H., Charles Cameron, and E. Ashworth Underwood. *A History of the Worshipful Society of Apothecaries of London.* Vol. 1. *1617–1815.* Revised, annotated, and edited by E. Ashworth Underwood. London: Wellcome Historical Medical Museum, 1963. 450 pp.

Traces the history from the granting of the Charter to the Apothecaries Act. The Society's activities "have had an influence of central importance on the structure of medical practice" in England.

362. Clark, Sir George. *A History of the Royal College of Physicians of London.* Oxford: Clarendon Press, 1964–66. 2 vols. 800 pp.

Official history, from the charter of 1518 to the coming of the Medical Act, 1850–58. Pp. 1–479 take the history up to 1704. Reviewed in *Isis* 56 (1965), 220–21.

363. Webster, Charles. "The College of Physicians: 'Solomon's House' in Commonwealth England." *Bulletin of the History of Medicine* 41 (1967), 393–412.

On the College as a major growth point of science in the Commonwealth.

364. Hoppen, K. Theodore. *The Common Scientist in the Seventeenth Century. A Study of the Dublin Philosophical Society, 1683–1708.* London: Routledge and Kegan Paul, 1970. 297 pp.

Pp. 59–61: The provision of science books at Trinity College, Dublin, from the evidence of early seventeenth-century catalogues and loan books. Pp. 212–74: Bibliography and notes. Reviewed in *Isis* 62 (1971), 547–49.

365. Middleton, W. E. Knowles. *The Experimenters: a Study of the Accademia del Cimento.* Baltimore, Md.: Johns Hopkins Univ. Press, 1971. 415 pp.

Study of the first society founded for the sole purpose of making scientific experiments, set up in Florence in 1657 (dissolved 1667). Includes a translation of the *Saggi di naturali esperienze* (1657).

366. Webster, Charles. "New Light on the Invisible College: the Social Relations of English Science in the Mid-Seventeenth Century." *Transactions, Royal Historical Society* (5), 24 (1974), 19–42.

A new reconstruction of the Invisible College.

367. Evans, R. J. W. "Learned Societies in Germany in the Seventeenth Century." *European Studies Review* 7 (1977), 129–51.

Outlines the effect of the Thirty Years' War on German intellectual life.

368. Boehm, L., and E. Raimondi, eds. *Università, accademie e società scientifiche in Italia e in Germania dal Cinquecento al Settecento. Atti della Settimana di Studio, 15–20 settembre 1980.* Annali dell'Istituto Storico Italo-Germanico, Quaderno 9. Bologna: Il Mulino, 1981. 460 pp.

Fourteen papers presented at a conference held in Trent. "Provides useful direction and helps clarify the role played by intellectual communities in Italy and Germany." Reviewed in *Minerva* 21 (1983), 471–76.

369. Galluzzi, P., C. Poni, and M. Torrini, eds. "Accademie scientifiche del '600." *Quaderni storici* 48 (1981), 757–921.

Four essays dealing with the most important scientific academies of seventeenth-century Italy (Lincei, Cimento, Investiganti, Istituto di Bologna).

370. Righini Bonelli, M. L., and A. van Helden. "Divini and Campani: A Forgotten Chapter in the History of the Accademia del Cimento." *Annali dell'Istituto e Museo di Storia della Scienza di Firenze* 6 (1981), (1), 3–176.

Pp. 3–43: The struggle between Eustachio Divini (1610–85) and Giuseppe Campani (1635–1715) for supremacy in telescope making, 1663–1665. Pp. 44–176: Documents relating to the comparison of the telescopes of Divini and Campani with those of the Grand Duke of Tuscany (Ferdinand II).

371. Stroup, Alice. *A Company of Scientists: Botany, Patronage, and Community at the Seventeenth-Century Parisian Royal Academy of Sciences.* Berkeley, Calif.: University of California Press, 1990. 387 pp.

A study of botanical research carried out by the Academy in the seventeenth century (pp. 65–166), along with its finances and its relations with the rest of the scholarly community. Based on archival sources. Pp. 339–69: Bibliography; pp. 371–87: Index. Review: *Annals of Science* 49 (1992), 283.

The Royal Society of London

BIBLIOGRAPHIES

372. "Bibliography of Writings on the History of the Royal Society." *Notes and Records, Royal Society of London* 10 (1952–53)–33 (1978–79).

Current bibliography of recent books and articles dealing with the history of the Royal Society or its fellows.

MONOGRAPHS AND ARTICLES

373. Stimson, Dorothy. "Dr. Wilkins and the Royal Society." *Journal of Modern History* 3 (1931), 539–63.

Aims to provide evidence for Wilkins's leadership of the group which projected the Royal Society and to show the influence of the society on his own literary and scientific activity.

374. Stimson, Dorothy. *Scientists and Amateurs: A History of the Royal Society.* New York: Henry Schuman, 1948. 270 pp.

Pp. 1–115 cover the history up to 1700.

375. Syfret, R. H. "The Origins of the Royal Society." *Notes and Records, Royal Society of London* 5 (1948), 75–137.

Notes on the prehistory of the society (Theodore Haak, Samuel Hartlib, Comenius, and the meetings of 1645).

376. Hartley, Harold, ed. *The Royal Society: Its Origins and Founders.* London: Royal Society, 1960. 275 pp.

Tercentenary commemoration volume. Pp. 1–45: Origins of the society (Douglas McKie); pp. 47–264: Sketches of seventeenth-century fellows (22 essays by various authors). Reviewed in *Isis* 53 (1962), 410–11.

377. Hall, Marie Boas. "Sources for the History of the Royal Society in the Seventeenth Century." *History of Science* 5 (1966), 62–76.

Survey including a description of published and unpublished MSS.

378. Purver, Margery. *The Royal Society: Concept and Creation.* London: Routledge and Kegan Paul, 1967. 246 pp.

The society's main purpose was "to cut away the whole existing system of natural sciences, and deliberately to begin the process of creating new sciences in which an organized body of related inductive knowledge, capable of continuous, unlimited development, could be built up, for the long-term benefit of mankind" (p. 235). Reviewed in *Isis* 59 (1968), 211–13.

379. Webster, C. "Origins of the Royal Society." *History of Science* 6 (1967), 106–28.

Essay review of Margery Purver, *The Royal Society: Concept and Creation* (1967). See item 378.

380. Rattansi, P. M. "The Intellectual Origins of the Royal Society." *Notes and Records, Royal Society of London* 23 (1968), 129–43.

Discusses the influence of Hermeticism, neo-Aristotelianism, and the mechanical philosophy.

381. Scala, Gail Ewald. "An Index of Proper Names in Thomas Birch, *The History of the Royal Society* (London, 1756–1757)." *Notes and Records, Royal Society of London* 28 (1974), 263–329.

An index to some 2,500 names. Dates of birth and death given where known.

382. Hall, M. Boas. "The Royal Society's Role in the Diffusion of Information in the Seventeenth Century." *Notes and Records, Royal Society of London* 29 (1975), 173–92.

On correspondence and publication in the *Philosophical Transactions*.

383. Jacob, J. R. "Restoration, Reformation and the Origins of the Royal Society." *History of Science* 13 (1975), 155–76.

On the misleading views of the earliest chroniclers of the society, Thomas Sprat and John Wallis. Although the leadership of the society may have foresworn conventional theology and politics, their own ideology of science was neither apolitical nor unrelated to religion. "Instead it was an aggressive, acquisitive, mercantilistic ideology justified in the name of both Restoration and Reformation."

384. Hoppen, K. T. "The Nature of the Early Royal Society." *British Journal for the History of Science* 9 (1976), 1–24, 243–73.

Study of the wide range of interests of the active members of the society.

385. Hall, Marie Boas. "Salomon's House Emergent: the Early Royal Society and Cooperative Research." *The Analytic Spirit. Essays in the History of Science in Honor of Henry Guerlac.* Harry Woolf, ed. Ithaca: Cornell Univ. Press, 1981, pp. 177–94.

The society tried to establish a working cooperation among its highly individual members. It was no Salomon's House, with a hierarchy of investigators but "rather a loosely knit association of men of differing capabilities, but free and independent minds."

386. Hunter, Michael. *The Royal Society and Its Fellows, 1660–1700. The Morphology of an Early Scientific Institution.* BSHS monographs, 4. Chalfont St. Giles: British Society for the History of Science, 1985. 272 pp. [Reprinted with corrections. First published 1982]

On the early fellows and their activities. Deals with the society's "institutional life in London and not with its role as a centre of a network of scientific correspondence and journalism." Pp. 1–52: The fellows and their role; the society's composition; its changing fortunes; the nature of the early Royal Society. Pp. 53–110: Appendices. Pp. 111–21: Tables. Pp. 122–57: Notes. Pp. 159–253: Catalogue of fellows; Pp. 256–70: Name index. Reviewed in *Annals of Science* 41 (1984), 494 and *History of Education* 13 (1984), 239–41.

387. Hunter, Michael, and Paul B. Wood. "Toward Solomon's House: Rival Strategies for Reforming the Early Royal Society." *History of Science* 24 (1986), 49–108.

Considers "not so much what the early Royal Society was like, but what it might have been like." Study based on a series of documents (mostly surviving only in MS) which have hitherto been overlooked.

388. Hunter, Michael. *Establishing the New Science. The Experience of the Early Royal Society.* Woodbridge, Suffolk: The Boydell Press, 1989. 382 pp.

A fresh view of the society's formative years. Consists of "a series of detailed case-studies of key episodes in the society's early evolution, based on the extensive and illuminating documentation that survives." Some of the essays have already appeared elsewhere. Pp. 356–68 contain a bibliographical essay on recent studies of the early Royal Society and its milieu. Reviews: *Annals of Science* 47 (1990), 518–19; *Isis* 82 (1991), 375–76.

Philosophy

BIBLIOGRAPHIES

389. Schwab, Moise. *Bibliographie d'Aristote.* Burt Franklin, Bibliography and Reference Series, 235. New York: Burt Franklin, 1967. 380 pp. [Originally published in France, 1896]

List of 3,742 items (eds. and secondary literature), arranged in eleven chapters (including: logic, natural philosophy). See also items 394, 401, 403.

390. Riedl, John O., ed. *A Catalogue of Renaissance Philosophers (1350–1650).* Milwaukee: Marquette Univ. Press, 1940. 179 pp.

Biobibliography of some seven hundred "philosophers" classified under 102 headings (including "many scientists, ascetics, theologians, and rhetoricians"). Compiled by a class of undergraduate students under the editor's direction. With bibliography.

391. Kristeller, Paul Oskar, and John Herman Randall, Jr. "The Study of the Philosophies of the Renaissance." *Journal of the History of Ideas* 2 (1941), 449–96.

Bibliographical essay. "During the Renaissance not only theologians and scientists, but also politicians, physicians, classical philologists, literary men, and artists were driven to philosophic reflection. In consequence, the thought of the Renaissance has elicited the interest not only of historians of philosophy, but also of scholars in many other fields" including that of the history of science and of education.

392. Gibson, R. W. *Francis Bacon. A Bibliography of His Works and of Baconiana to the Year 1750.* Comfort: Scrivener Press, 1950. 369 pp. Supplement [typescript]. 1959. 20 pp.

Pp. 1–222: Bibliography of Bacon's works (items 1–256). Pp. 223–324: List of Baconiana and other related works (items 257–680). Pp. 340–69: General index.

393. Varet, Gilbert. *Manuel de bibliographie philosophique.* 1. *Les philosophies classiques.* 2. *Les sciences philosophiques.* Paris: Presses Universitaires de France, 1956. 2 vols.; 1,056 pp.

Select bibliography of some 20,000 items, minutely classified. Vol. 1, pp. 1–39: Encyclopedias, histories and regional bibliographies. Vol. 2, pp. 923–1048: Name index. Reviews: *Bulletin des bibliothèques de France*, 1957, pp. 164–70; *Revue d'histoire des sciences* 12 (1959), 75–76.

394. Riley, Lyman W. *Aristotle Texts and Commentaries to 1700 in the University of Pennsylvania Library. A Catalogue.* Philadelphia: Univ. of Pennsylvania Press, 1961. 109 pp.

Some five hundred books and forty MSS (texts and commentaries including forty printed eds. of the complete works). With indexes of: commentaries and paraphrases; translators, editors and others. Bibliographies noted in preface (p. 5). See also items 389, 401, 403.

395. Schüling, Hermann. *Bibliographie der im 17. Jahrhundert in Deutschland erschienenen logischen Schriften.* Berichte und Arbeiten aus der Universitätsbibliothek Giessen, 3. Giessen: Universitätsbibliothek, 1963. 143 pp.

An attempt to provide a complete bibliography of works of logic published in Germany in the seventeenth century.

396. Sebba, Gregor. *Bibliographia Cartesiana. A Critical Guide to the Descartes Literature, 1800–1960.* The Hague: Martinus Nijhoff, 1964. 510 pp.

Introduction to Descartes studies (items 1–562, arranged in twelve sections); Alphabetical bibliography (items 1001–3612). Indexes (systematic; analytical). Reviewed in *Isis* 56 (1965), 459.

397. Tonelli, Giorgio. *A Short-Title List of Subject Dictionaries of the Sixteenth, Seventeenth, and Eighteenth Centuries as Aids to the History of Ideas.* London: The Warburg Institute, 1971. 64 pp.

Includes about one hundred major dictionaries of the sixteenth and seventeenth centuries. With subject index.

398. Equipe Descartes. "Bulletin cartésien: bibliographie critique des études cartésiennes." *Archives de philosophie* 35 (1972), 263–319; 36 (1973), 431–95; 37 (1974), 453–97; 38 (1975), 253–309.

International critical bibliography of Cartesian studies, 1970–73.

399. Tobey, Jeremy A. *The History of Ideas. A Bibliographical Introduction.* Vol. 2: *Medieval and Early Modern Europe.* Santa Barbara, Calif.: Clio, 1977. 320 pp.

Introduction (pp. 1–13); Philosophy (pp. 14–93); Science (pp. 94–145); Religion (pp. 146–82); Aesthetics (pp. 183–244). Author-title index.

400. Totok, Wilhelm. *Handbuch der Geschichte der Philosophie.* Vol. 3. *Renaissance.* Vol. 4. *Frühe Neuzeit. 17. Jahrhundert.* Frankfurt-am-Main: Klostermann, 1980–81. 658 pp.; 612 pp.

Monumental bibliography of the history of philosophy, covering primary and secondary literature since 1920. Classified and arranged by topic and subtopic, with occasional introductions. Articles in journals included. No annotations, but classification is minute. Includes sections on Fracastoro, Cardano, Campanella, Bruno, Leonardo da Vinci, Galileo, Paracelsus, Van Helmont, Copernicus, Kepler, Gassendi, Bacon, Servetus, Descartes, Leibniz, Comenius, Locke, Newton.

Literature up to 1920 is covered in: Ueberweg, F., *Grundriss der Geschichte der Philosophie*, Vol. 3 (Berlin: Mittler, 1924).

401. Cranz, F. Edward. *A Bibliography of Aristotle Editions 1501–1600.* 2nd ed., with addenda and revisions by Charles B. Schmitt. Bibliotheca Bibliographica Aureliana, 38. Baden-Baden: Valentin Koerner, 1984. 247 pp.

Pp. 161–224: Systematic index. Pp. 225–41: Index of translators, commentators, and editors. Review: *The Library* (6), 7 (1985), 278–80.

402. Müller, Kurt, ed. *Leibniz-Bibliographie. Die Literatur über Leibniz bis 1980.* Begründet von Kurt Müller. Herausgegeben von Albert Heinekamp. Zweite Auflage. Veröffentlichungen des Leibniz-Archivs 10. Frankfurt-am-Main: Klostermann, 1984. 742 pp. [First published 1967]

Guide to the literature on Leibniz—contemporary to 1980. Bibliography continued in *Studia Leibnitiana* 1 (1969)– .

403. Lohr, Charles H. *Latin Aristotle Commentaries.* II. *Renaissance Authors.* Florence: Leo S. Olschki, 1988. 515 pp.

Survey of commentaries of the period 1500–1650. With biobibliographical notes on the authors. Originally published in *Studies in the Renaissance* 21 (1974) and *Renaissance Quarterly* 28 (1975)–35 (1982). See also items 389, 394, 401.

404. Blackwell, Constance T. "Towards a History of Renaissance Philosophy: A Bibliography of the Writings of C.B. Schmitt." *New Perspectives on Renaissance Thought.* John Henry and Sarah Hutton, eds. London: Duckworth, 1990, pp. 291–308.

A complete bibliography of books written, edited or jointly edited by C.B.S. (17), articles (117) and book reviews (152). See also item 462.

GENERAL WORKS

405. *Encyclopedia of Philosophy.* Paul Edwards, editor-in-chief. New York: Macmillan, 1967. 8 vols.

Aims to provide a truly encyclopedic presentation of philosophical theories and concepts. Contains some 1,500 articles most of which "are sufficiently explicit to be read with pleasure and profit by the intelligent nonspecialist." Vol. 8, pp. 387–544: Index of subjects (some 36,000 entries).

406. Wiener, Philip P., ed. *Dictionary of the History of Ideas. Studies of Selected Pivotal Ideas.* New York: Scribner's, 1973–74. 5 vols.

Some three hundred articles. Subjects include: alchemy; astrology; atomism; cosmology; matter; nature; Newton and the method of analysis; optics and vision; technology, time; time and measurement; Renaissance; witchcraft; Platonism; China in western thought and culture. With bibliographies. Index (Vol. 5).

407. Bietenholz, Peter G., ed. *Contemporaries of Erasmus. A Biographical Register of the Renaissance and Reformation.* Toronto: Univ. of Toronto Press, 1985–87. 3 vols.

Provides biographical information on persons (more than 1,900) mentioned in the correspondence and works of Erasmus, who died after 1450 and were thus approximately his contemporaries.

408. Schmitt, Charles B., Quentin Skinner, Eckhard Kessler, and Jill Kraye, eds. *The Cambridge History of Renaissance Philosophy.* Cambridge: Cambridge Univ. Press, 1988. 968 pp.

An introduction to the intellectual life of the Renaissance—a synthesis of the past century's research. With bio-bibliographies of some of the most important philosophers of the period and a bibliography of primary and secondary sources. The book is the subject of five essay reviews in *Bulletin of the Society for Renaissance Studies* 6 (1989), (2), 12–43. Also reviewed in *British Journal for the History of Science* 22 (1989), 377–96.

An earlier work, still useful, is: Jacob Brucker, *Historia critica philosophiae, a mundi incunabilis ad nostram usque aetatem deducta*, 2nd ed. (Leipzig, 1766–67), Vols. 4–5. Each volume has name and subject indexes.

MONOGRAPHS AND ARTICLES

409. Stones, G. B. "The Atomic View of Matter in the XVth, XVIth, and XVIIth Centuries." *Isis* 10 (1928), 445–65.

On the continuous atomic tradition from Nicholas of Cusa to Charleton.

410. Burtt, Edwin Arthur. *The Metaphysical Foundations of Modern Physical Science*. 2nd ed. London: Routledge and Kegan Paul, 1932. 349 pp.

Historical study of "the philosophy of early modern science, locating its key assumptions as they appear, and following them out to their classic formulation in the metaphysical paragraphs of Sir Isaac Newton."

411. Lenoble, Robert. *Mersenne ou la naissance du mécanisme*. Paris: Vrin, 1943. 633 pp.

Study of the life and work of the "Secrétaire de l'Europe savante." Pp. xi–lxiii: Bibliography of works written, read, and cited by Mersenne.

412. Collingwood, R. G. *The Idea of Nature*. Oxford: Clarendon Press, 1945. 183 pp.

Discusses Greek cosmology, the Renaissance view of nature, and the modern view of nature.

413. Anderson, Fulton Henry. *The Philosophy of Francis Bacon*. Chicago: Univ. of Chicago Press, 1948. 312 pp.

Study of Bacon's scheme for the reform of learning and the sciences.

414. Farrington, Benjamin. *Francis Bacon, Philosopher of Industrial Science*. London: Lawrence and Wishart, 1951. 198 pp.

Study of Bacon's ideas. Includes excerpts from his writings and an analysis of *The Great Instauration*. Reviewed in *Isis* 41 (1950), 215–16.

415. Boas, Marie. "The Establishment of the Mechanical Philosophy." *Osiris* 10 (1952), 412–541.

On the "development of a sophisticated mechanical philosophy and its struggle to overthrow the doctrine of substantial forms." A revision of the author's doctoral thesis: "Robert Boyle and the Corpuscular Philosophy. A

Study of Theories of Matter in the Seventeenth Century." Pp. 525–41: Bibliography.

416. Meyer, R.W. *Leibniz and the Seventeenth-Century Revolution.* Cambridge: Bowes and Bowes, 1952. 227 pp. [Translation of a revised version of the 1948 German ed.]

Pp. 169–227: Notes and index of names. Reviews: *TLS* 12 July 1950, p. 439 (German ed.); *TLS* 17 Oct. 1952, p. 681; *Isis* 46 (1955), 74–76.

417. Gilbert, Neal W. *Renaissance Concepts of Method.* New York: Columbia Univ. Press, 1960. 255 pp.

Sketch of the methodology of the Renaissance. Pp. 237–45: Selective bibliography. Reviewed in *Isis* 53 (1962), 406–8.

418. Geymonat, L. *Galileo Galilei. A Biography and Inquiry into His Philosophy of Science.* Text translated from the Italian with additional notes and appendix by Stillman Drake. New York: McGraw-Hill, 1965. 260 pp. [First published in Italian in 1957]

Reviewed in *Isis* 49 (1958), 370–71.

419. Kristeller, P. O. "Renaissance Aristotelianism." *Greek, Roman, and Byzantine Studies* 6 (1955), 157–74.

"Renaissance Aristotelianism was overthrown during the seventeenth century in the physical sciences and in metaphysics, but by that time it had fulfilled its historical function. For about three centuries, the Aristotelian school had carried on with partial success the main professional work in physics and logic, and thus it was able to bequeath to its successors, modern science and modern philosophy, their chief subject matter and some of their problems and concepts."

420. Nauert, Charles G., Jr. *Agrippa and the Crisis of Renaissance Thought.* Illinois Studies in the Social Sciences, 55. Urbana, Ill.: Univ. of Illinois Press, 1965. 374 pp.

Traces Agrippa's life and intellectual development and discusses a number of key themes developed in his writings (occultism, scepticism, and his views on religion). Pp. 116–56: Agrippa's background and reading. Pp. 334–55: Bibliography. Reviewed in *Journal of the History of Philosophy* 5 (1967), 86–88.

421. Kargon, Robert Hugh. *Atomism in England, from Harriot to Newton.* Oxford: Clarendon Press, 1966. 168 pp.

On the Northumberland and Newcastle groups of natural philosophers and their contribution to the atomic doctrine in the seventeenth century. Pp. 140–63: Select bibliography. Reviewed in *Isis* 60 (1969), 572–75.

422. Laudan, Laurens. "The Clock Metaphor and Probabilism: The Impact of Descartes on English Methodological Thought, 1650–65." *Annals of Science* 22 (1966), 73–104.

The author's conclusions are challenged by G. A. J. Rogers in "Descartes and the Method of English Science," *Annals of Science* 29 (1972), 237–55.

423. Crombie, A. C. "The Mechanistic Hypothesis and the Scientific Study of Vision: Some Optical Ideas as a Background to the Invention of the Microscope." *Proceedings, Royal Microscopical Society* 2 (1967), 3–112. [Reprinted in *Historical Aspects of Microscopy*, eds. S. Bradbury and G. L'E. Turner (Cambridge: Heffer, 1967)]

Paper read at a one-day conference held by the Royal Microscopical Society, at Oxford, 18 March 1966. Discusses the concepts of the mechanism of vision from antiquity to the seventeenth century. Pp. 87–107: References.

424. Smith, Cyril S., and John G. Burke. *Atoms, Blacksmiths, and Crystals. Practical and Theoretical Views of the Structure of Matter in the Seventeenth and Eighteenth Centuries. Papers Read at a Clark Library Seminar, November 26, 1966.* Los Angeles: William Andrews Clark Memorial Library, 1967. 59 pp.

Two papers: the texture of matter as viewed by artisan, philosopher, and scientist in the seventeenth and eighteenth centuries; snowflakes and the constitution of crystalline matter.

425. Crapulli, Giovanni. *Mathesis universalis: genesi di una idea nel XVI secolo.* Lessico Intellettuale Europeo 2. Rome: Edizioni dell'Ateneo, 1969. 285 pp.

"Brings to our attention the ideas of a number of sixteenth-century thinkers on the place of mathematics in the general scientific complex of the period." Reviewed in *Isis* 63 (1972), 277–78.

426. Grant, Edward. "Medieval and Seventeenth-Century Conceptions of an Infinite Void Space beyond the Cosmos." *Isis* 60 (1969), 39–60.

On conceptions of the "extracosmic void."

427. Schmitt, C. B. "Experience and Experiment: A Comparison of Zabarella's View with Galileo's in *De motu*." *Studies in the Renaissance* 16 (1969), 80–138.

Analyses the use of the notions "experience" and "experiment."

428. Schüling, Hermann. *Die Geschichte der axiomatischen Methode im 16. und 17. Jahrhundert*. Studien und Materialen zur Geschichte der Philosophie 13. Hildesheim: Ohms, 1969. 199 pp.

History of the axiomatic method. Pp. 168–99: Sources.

429. Rossi, Paolo. *Philosophy, Technology and the Arts in the Early Modern Era*. Benjamin Nelson, ed. New York: Harper and Row, 1970. 194 pp. [Trans. of *I Filosofi e le macchine* (1962)]

Essays on Francis Bacon and his influence. Originally published in Italian between 1955 and 1967.

430. Oldroyd, David R. "Robert Hooke's Methodology of Science as Exemplified in his *Discourse of Earthquakes*." *British Journal for the History of Science* 6 (1972), 109–30.

Argues that "Hooke had a much clearer idea than Bacon of the importance of so-called 'hypothetico-deductive' methods in science, and that in many ways his methodological views represent a significant advance on those of Bacon."

431. Ariotti, Piero E. "Toward Absolute Time: the Undermining and Refutation of the Aristotelian Conception of Time in the Sixteenth and Seventeenth Centuries." *Annals of Science* 30 (1973), 31–50.

The introduction of the concept of an absolute time independent of external motion was one of the major conceptual innovations of seventeenth-century science.

432. Heninger, S. K., Jr. *Touches of Sweet Harmony. Pythagorean Cosmology and Renaissance Poetics*. San Marino, Calif.: The Huntington Library, 1974. 446 pp.

Study of: Pythagoreanism in the Renaissance; Pythagorean doctrine; Poetics. Some 840 works cited (list on pp. 401–28). Reviews: *British Journal for the History of Science* 11 (1978), 78–79; *Shakespeare Studies* 10 (1977), 357–65.

433. Soppelsa, Marialaura. *Genesi del metodo galileiano e tramonto dell' aristotelismo nella scuola di Padova*. Centro per la storia della tradizione aristotelica nel Veneto, Università di Padova, Saggi e testi 13. Padua: Editrice Antenore, 1974. 225 pp.

Analysis of the writings on natural philosophy of the Paduan school from 1600 to 1750. Pp. 208–19: Bibliographical note. Reviewed in *Annals of Science* 32 (1975), 80–81.

434. XVIe Colloque international de Tours. *Platon et Aristote à la Renaissance.* De Pétrarque à Descartes 32, Centre d'Etudes Supérieures de la Renaissance. Paris: Vrin, 1976. 587 pp.

Pp. 93–154: Plato and Aristotle in the universities and colleges of the sixteenth century. Pp. 217–76: Plato and Aristotle in the scientific movement of the Renaissance. Pp. 277–402: Readings of Plato and Aristotle during the Renaissance.

435. Canguilhem, Georges. *La formation du concept de réflexe au XVIIe et XVIIIe siècles.* 2nd ed. Paris: Vrin, 1977. 208 pp. [First published 1955]

Pp. 9–78: The problem of muscular movement before Descartes; Cartesian theory of involuntary movement; Formation of the concept of reflex movement. Pp. 193–202: Bibliography.

436. Sloan, Phillip R. "Descartes, the Sceptics, and the Rejection of Vitalism in Seventeenth-Century Physiology." *Studies in the History and Philosophy of Science* 8 (1977), 1–28.

"A preliminary attempt to account for the suddenness and tenacity with which the 'mechanical philosophy' took hold in the middle of the seventeenth century."

437. Butts, Robert E., and Joseph C. Pitt, eds. *New Perspectives on Galileo. Papers Deriving from and Related to a Workshop on Galileo Held at Virginia Polytechnic Institute and State University, 1975.* Dordrecht: Reidel, 1978. 262 pp.

On Galileo's philosophy of science and his methodology. Reviews: *Isis* 70 (1979), 311–12; *British Journal for the Philosophy of Science* 31 (1980), 196–99.

438. Grant, Edward. "The Principle of the Impenetrability of Bodies in the History of Concepts of Separate Space from the Middle Ages to the Seventeenth Century." *Isis* 69 (1978), 551–71.

On the responses to the problems posed by Aristotle's criticisms against a separate space.

439. McMullin, Ernan, ed. *The Concept of Matter in Modern Philosophy.* Notre Dame, Ind.: Univ. of Notre Dame Press, 1978. 304 pp.

Collection of papers on the history and philosophy of the concept of matter from the seventeenth to the twentieth century. Pp. 59–75: Descartes' concept of matter (Richard J. Blackwell); pp. 76–103: Matter in seventeenth-century science (Marie Boas Hall). Reviews: *Annals of Science* 37 (1980), 110–11; *Isis* 71 (1980), 486.

440. Tocanne, Bernard. *L'idée de nature en France dans la seconde moitié du XVIIe siècle. Contributions à l'histoire de la pensée classique.* Bibliothèque française et romaine 67. Paris: Klincksieck, 1978. 501 pp.

An enquiry into all the aspects of the word "nature." Reviewed in *Modern Language Review* 75 (1980), 194–95.

441. Popkin, Richard H. *The History of Scepticism from Erasmus to Spinoza.* Berkeley: Univ. of California Press, 1979. 333 pp.

A revised and expanded ed. of *The History of Scepticism from Erasmus to Descartes* (1960, 1964, 1968). "An initial attempt to see the role of scepticism in modern thought." P. 249: List of reviews of earlier eds. Pp. 300–26: Bibliography. Reviewed in *TLS* 26 Sep. 1980, p. 1074.

442. Schmitt, Charles B. *Studies in Renaissance Philosophy and Science.* Variorum reprint CS 146. London: Variorum Reprints, 1981. 342 pp.

Collection of twelve papers written between 1966 and 1978 relating to: the history of Platonism during the Renaissance; the philosophical and scientific context of Galileo Galilei.

443. Wallace, William A. *Prelude to Galileo. Essays on the Medieval and Sixteenth-Century Sources of Galileo's Thought.* Boston Studies in the Philosophy of Science, Vol. 62. Dordrecht: Reidel, 1981. 369 pp.

Essays based on MS sources, written over some fifteen years. Reviewed in *Annals of Science* 90 (1983), 211–12.

444. Carter, Richard B. *Descartes' Medical Philosophy. The Organic Solution to the Mind-Body Problem.* Baltimore, Md.: Johns Hopkins Univ. Press, 1983. 301 pp.

Part 1. Thinking bodies and the bodies of mathematical physics: preparation for medicine and mechanics. Part 2. Compound bodies: living and non-living systems. Name and subject indexes. Reviewed in *Bulletin of the History of Medicine* 58 (1984), 594–96.

445. Olivieri, Luigi, ed. *Aristotelismo veneto e scienza moderna. Atti del 25° anno accademico del Centro per la Storia della Tradizione Aristotelica nel Veneto.* Padua: Antenore, 1983. 2 vols. 1,133 pp.

Fifty-two papers based on lectures given and seminars held at the Centre. Subjects covered include: Galileo, medicine, and mathematics.

446. Schmitt, Charles B. *Aristotle and the Renaissance.* Martin Classical Lectures, Oberlin College. Cambridge, Mass.: Harvard Univ. Press, 1983. 187 pp.

Four lectures on Renaissance Aristotelianism and its "geographical, chronological, and intellectual variations." With a concluding essay: "Towards a History of Aristotle in the Renaissance." Pp. 116–83: Appendices, notes, and bibliography. Reviewed in *Isis* 75 (1984), 228–29.

447. Shapiro, Barbara J. *Probability and Certainty in Seventeenth-Century England. A Study of the Relationships between Natural Science, Religion, History, Law, and Literature.* Princeton, N.J.: Princeton Univ. Press, 1983. 347 pp.

Reviews: *Social Studies of Science* 14 (1984), 137–52; *Minerva* 21 (1983), 459–65; *TLS* 20 May 1983, p. 518.

448. Entry deleted.

449. Alexander, Peter. *Ideas, Qualities and Corpuscles: Locke and Boyle on the External World.* Cambridge: Cambridge Univ. Press, 1985. 330 pp.

"Substantial and often radical reinterpretation of some of the central themes of Locke's thought." Reviewed in *British Journal for the History of Science* 19 (1986), 357–58.

450. Fattori, Marta, ed. *Francis Bacon: terminologia e fortuna nel XVII secolo. Seminario internazionale, Roma, 11–13 Marzo 1984.* Lessico intelettuale europeo, 33. Rome: Edizioni dell' Ateneo, 1985. 327 pp.

Sixteen papers representing three different traditions of Baconian scholarship (English, French and Italian). Reviewed in *Isis* 78 (1987), 129–30.

451. Finocchiaro, Maurice A. "Toward a Philosophical Reinterpretation of the Galileo Affair." *Nuncius* 1 (1986), (1), 189–202.

Critical review of recent writings in particular, Paul Poupard, ed., *Galileo . . . 350 anni di storia, 1633–1983* (Rome, 1984). See item 580.

452. Freudenthal, Gideon. *Atom and Individual in the Age of Newton. On the Genesis of the Mechanistic World View.* Boston Studies in the Philosophy of Science, 88. Dordrecht: Reidel, 1986. 276 pp. [Translation of the 1982 German ed.]

On the role of philosophical presuppositions in the formation of scientific concepts and the influence of social relations on the construction of such concepts. Reviews: *Annals of Science* 43 (1986), 498–502; *Archives internationales d'histoire des sciences* 37 (1987), 360–61.

453. Blumenberg, Hans. *The Genesis of the Copernican World. Studies in Contemporary German Social Thought.* Cambridge, Mass.: MIT Press, 1987. 772 pp. [Translated from the German]

The author challenges the view of the Copernican reform as "an event in the history of science that happened to have great repercussions outside science but had no essential extrascientific preconditions." He shows how Copernicus's reform of astronomy was made possible by processes that involved the whole range of European thought, religious, philosophical, and metaphorical as well as literal and scientific. Pp. 687–707: Translator's notes. Pp. 709–66: Author's notes. Reviews: *Isis* 79 (1988), 727–28; *Journal for the History of Astronomy* 19 (1988), 276–77.

454. Brundell, Barry. *Pierre Gassendi. From Aristotelianism to a New Natural Philosophy.* Synthese historical library 30. Dordrecht: Reidel, 1987. 251 pp.

Presents a "new interpretation of Gassendi's philosophical work with the aim of clarifying the goals and tactics of this pioneer architect of a mechanistic system, thereby hoping to throw some further light on the origins of mechanicism in the seventeenth century." Bibliography (pp. 228–45).

455. Dear, Peter. *Mersenne and the Learning of the Schools.* Ithaca, N.Y.: Cornell Univ. Press, 1988. 264 pp.

Study of Mersenne seen here not as the "Secretary of a learned Europe," but as a "natural philosopher wrestling with some of the philosophical problems raised by the New Learning and coming down quite soundly on the side of modernity." Pp. 239–59: Bibliography.

456. Giard, Luce. "Charles Schmitt, reconstructeur d'une histoire de la Renaissance savante." *Aristotelismus und Renaissance.* Eckhard Kessler, Charles H. Lohr, and Walter Sparn, eds. Wiesbaden: Harrassowitz, 1988, pp. 23–52.

For an English translation of this paper see item 462, pp. 264–90.

457. Joy, Lynne Sumida. *Gassendi the Atomist: Advocate of History in an Age of Science.* Ideas in context. Cambridge: Cambridge Univ. Press, 1988. 311 pp.

Describes the influence of humanism on Gassendi's development as a philosopher. Analyses his ability to "conjoin in a single historical discussion the arguments of the Greek atomists and the scientific and philosophical problems of his contemporaries."

458. Kessler, Eckhard, Charles H. Lohr, and Walter Sparn, eds. *Aristotelismus und Renaissance. In Memoriam Charles B. Schmitt.* Wolfenbütteler Forschungen 40. Wiesbaden: Harrassowitz, 1988. 237 pp.

Proceedings of a colloquium on Renaissance Aristotelianism held in the Herzog August Bibliothek, Wolfenbüttel, 23–25 October 1986. Ten papers, of which two relate to Charles Schmitt's contribution to the subject. Included is his introduction to the *Cambridge History of Renaissance Philosophy* (item 408). Pp. 217–32 contain a list of his writings.

459. Meinel, Christoph. "Early Seventeenth-Century Atomism: Theory, Epistemology and the Insufficiency of Experiment." *Isis* 79 (1988), 68–103.

"Historical typology and evaluation of the arguments presented in support of the corpuscular hypothesis during the first half of the seventeenth century."

460. Pérez-Ramos, Antonio. *Francis Bacon's Idea of Science and the Maker's Knowledge Tradition.* Oxford: Clarendon Press, 1988. 334 pp.

Mainly an enquiry into the key notions that make up Bacon's idea of science. Pp. 296–323: Bibliography. Reviewed in *British Journal for the History of Science* 23 (1990), 106–7.

461. Wollgast, Siegfried. *Philosophie in Deutschland zwischen Reformation und Aufklärung, 1550–1650.* Berlin: Akademie Verlag, 1988. 1,037 pp.

Pp. 65–127: From natural philosophy to modern science. Pp. 221–62: Johann Kepler. Pp. 423–70: Joachim Jungius. Pp. 909–88: Bibliography. Pp. 989–1037: Subject and name indexes.

462. Henry, John, and Sarah Hutton, eds. *New Perspectives on Renaissance Thought. Essays in the History of Science, Education and Philosophy, in Memory of Charles B. Schmitt* [1933–1986]. London: Duckworth, 1990. 324 pp.

Eighteen essays of which the majority relate to the period 1450–1700. Topics include: Gesner and Paracelsus; Egnazio Danti; Renaissance Aristotle; Copernicus and French universities; medicine; magnetic philosophy; Ramus. Concludes with an appreciation of Schmitt's work by Luce Giard (pp. 264–90) and a bibliography of his writings by Constance Blackwell (see item 404). Pp. 309–24: Name index.

Science

BIBLIOGRAPHIES

463. Hallam, Henry. *Introduction to the Literature of Europe in the Fifteenth, Sixteenth, and Seventeenth Centuries.* Seventh ed. London: John Murray, 1864. 4 vols. [First published 1837–39]

Monumental survey, by a reputed English historian, of the "progress of letters in every part of Europe," from theology, philosophy, poetry and drama to mathematics, science and medicine. Mathematics of the second half of the seventeenth century excluded because of the "slightness of my own acquaintance with subjects so momentous and difficult" (author). Each volume contains a separate chapter for scientific literature. Index to the four volumes in Vol. 4, pp. 370–424. Reviewed anonymously in *Quarterly Review* 58 (1837), 29–60; 65 (1840), 340–83.

464. Picatoste y Rodríguez, Felipe. *Apuntes para una biblioteca científica española del siglo XVI. Estudios biográficos y bibliográficos de ciencias de exactas físicas y naturales, y sus immediatas aplicaciones en dicho siglo.* Madrid: Manuel Tello, 1891. 417 pp.

Annotated list of some one thousand sixteenth-century Spanish books and MSS, with biographical notes on the authors. Alphabetically arranged by author. With indexes.

465. Second International Congress of the History of Science and Technology, London 1931. *The First Century of Science in England. Giordano Bruno to Isaac Newton, 1584–1687. Notes Descriptive of an Exhibit at the British Museum.* London, 1931. 28 pp.

Catalogue of seventy-six books (some twenty-five authors) ranging from those of Giordano Bruno "surreptitiously printed in London in 1584" to the *Principia* of Newton. With notes on the authors.

466. Johnson, Francis R., and Sanford V. Larkey. "Science [in the Renaissance]." *Modern Language Quarterly* 2 (1941), 363–401.

Bibliographical survey of the current state of scholarship in the field of Renaissance science (physical sciences, biology, botany, anatomy, and physiology). Some 250 references.

467. Marañón, G. "La literatura científica en los siglos XVI y XVII." *Historia general de las literaturas hispánicas*, 3. *Renacimiento y Barroco.* Edited by Guillermo Diaz-Plaja. Barcelona: Barna, 1953, pp. 931–66.

Surveys the literature of: medicine, natural history, agriculture, and mathematics in Spain.

468. Pighetti, Clelia. "Cinquant'anni di studi newtoniani (1909–1959)." *Rivista critica di storia della filosofia* 15 (1960), 181–203, 295–318.

Bibliography of studies on Newton during the fifty years since the 2nd ed. of G. J. Gray's *Bibliography* (1907). Chronologically arranged, with some annotations. See also bibliographical survey by I. B. Cohen and A. P. Youschkevich in *DSB*, Vol. 9 (1974), pp. 93–103 and the more recent work by Peter and Ruth Wallis (item 475).

469. Heninger, S. K., Jr. "Tudor Literature of the Physical Sciences." *Huntington Library Quarterly* 32 (1969), 101–33, 249–70.

Survey of scientific literature (mathematics; cosmography; astrology; astronomy; geography, chorography, and navigation; physics; magnetism; meteorology; divinatory sciences). Select bibliography of secondary material (125 items).

470. *Dictionary of Scientific Biography.* Charles C. Gillispie, editor-in-chief. New York: Charles Scribner's Sons, 1970–90. 18 vols.

See also item 54. Contains bio-bibliographical articles on some six hundred scientists who lived between the years 1450 and 1700.

471. Stringer, Gail Griffin. "A Bibliography of Works about Sir Christopher Wren." *Papers, The American Association of Architectural Bibliographers* 9 (1972), 29–50.

With the exception of Thieme-Becker (*Allgemeines Lexikon*), the bibliography is restricted to books and articles in English.

472. Dehergne, Joseph. *Répertoire des Jésuites de Chine de 1552 à 1800.* Rome: Institutum Historicum S.I., 1973. 430 pp.

Bio-bibliography of 920 Jesuit missionaries to China of whom about sixty (astronomers, mathematicians, cartographers, engineers, and physicians) lived before 1700.

473. Wellisch, Hans. "Conrad Gessner: a Bio-bibliography." *Journal, Society for the Bibliography of Natural History* 7 (1975), 151–247.

Outline of Gesner's life and work. Supplements Bay's account (item 124). With a complete bibliography of his works and a list of 79 works about him. Name index.

474. Marcan, Peter, comp. *Poetry Themes. A Bibliographical Index to Subject Anthologies and Related Criticism in the English Language, 1875–1975.* London: Clive Bingley, 1977. 301 pp.

Aims to provide a subject approach to poetry—by indexing "subject anthologies which bring together poetry on one subject or a group of related subjects." Annotated. Items 1545–1964 (Science and the natural world, pp. 224–81) include many works relevant to Renaissance and seventeenth-century science.

475. Wallis, Peter, and Ruth Wallis. *Newton and Newtoniana, 1672–1975: A Bibliography.* Project for Historical Bibliography (PHIBB), University of Newcastle-upon-Tyne. Folkestone, Eng.: Dawson, 1977. 362 pp.

A bibliography of Newton's works and of Newtonian studies. Classified under: Collected works, bibliography; *Principia*; Optics; Fluxions; *Arithmetica Universalis*; Minor mathematical works; Chronological, theological, and miscellaneous works; Reports on coinage; works edited by Newton; Biographies and general works. Includes articles on Newton in encyclopedias and biographical dictionaries, and sections dealing with him in historical studies. No annotations but book reviews noted. Reviewed in *TLS*, 10 Nov. 1978, p. 1315 and *History of Science* 17 (1979), 147–49.

476. Schmitt, C. B. "Recent Trends in the Study of Medieval and Renaissance Science." *Information Sources in the History of Science and Medicine.* Editors, Pietro Corsi and Paul Weindling. London: Butterworth Scientific, 1983, pp. 221–40.

See item 21. Covers research done during the period 1955 to 1981. With a select bibliography.

477. Ladendorf, Heinz. *Leonardo da Vinci und die Wissenschaften: Eine Literaturübersicht.* Kölner medizinhistorische Beiträge, Ed. 23/2. Cologne: Institut für Geschichte der Medizin der Universität, 1984. 190 pp.

Bibliographic index to the literature on Leonardo's science. 1,800 entries under 40 headings.

478. Kren, Claudia. *Medieval Science and Technology. A Selected, Annotated Bibliography.* Bibliographies of the History of Science and Technology, 11. New York: Garland Publishing, 1985. 369 pp.

Includes several works whose coverage extends beyond the year 1450. Reviews: *History and Philosophy of the Life Sciences* 10 (1988), 382–83; *Isis* 77 (1986), 337–38.

479. Costabel, Pierre, and Monette Martinet. *Quelques savants et amateurs de science au XVIIe siècle. Sept notices bio-bibliographiques caractéristiques.* Cahiers d'histoire et de philosophie des sciences, nouv. sér., 14. Paris: Société française d'histoire des sciences et des techniques, 1986. 136 pp.

Bio-bibliographical notes on Mersenne, Roberval, Huygens, Paradies, Du Hamel, Morin, Rohault.

480. Jensen, Michael. *Bibliographia Nicolai Stenonis.* Mørke: Impetus, 1986. 97 pp.

List of writings by and about Niels Stensen (1638–1686) from his own time up to 1986.

481. Schatzberg, Walter, Ronald A. Waite, and Jonathan K. Johnson, eds. *The Relations of Literature and Science. An Annotated Bibliography of Scholarship, 1880–1980.* New York: The Modern Language Association of America, 1987. 458 pp.

Pp. 1–28: Interaction of literature and science (174 entries). Pp. 63–149: Renaissance, seventeenth century (studies and surveys; individual authors. Items 384–963). Author and subject indexes.

482. Guerrini, Mauro. *Bibliotheca Leonardiana, 1493–1989.* Presentazioni di Augusto Marinoni, Carlo Pedretti. Milan: Editrice Bibliografica, 1990. 3 vols. 2,216 pp.

Catalogue of works by and about Leonardo da Vinci in the Leonardian libraries in Vinci and Milan, comprising what amounts to a comprehensive bibliography of Vinciana (complete up to ca. 1985). Vol. 1: Catalogue arranged chronologically in two sequences (works by and about Leonardo)—6,192 titles. Vol. 2: Indexes to Vol. 1 (authors, titles, and serials) and concordances. Vol. 3: Subject catalogue (pp. 1461–1533: Index of subject headings).

482A. Selin, Helaine. *Science across Cultures: An Annotated Bibliography of Books on Non-Western Science, Technology, and Medicine.* Garland Reference Library of the Humanities, Vol. 1597. New York: Garland Publishing, 1992. 431 pp.

836 items arranged in ten chapters by place: multicultural, Africa, The Americas, Asia, China, Japan, India, Islamic Science, The Middle East, The Pacific. Pp. 383–431: Indexes (author, title, subject). No chronological divisions.

MONOGRAPHS AND ARTICLES

483. Morley, Henry. *The Life of Girolamo Cardano of Milan, Physician.* London: Chapman and Hall, 1854. 2 vols.

Detailed, well-documented biography of Cardano, based on "a steady search among his extant works, and by collecting into a body statements and personal allusions which occur in some of them, assigning to each its due place."

Vol. 2, pp. 315–28: Name and subject index (including a chronology of leading events).

484. Caverni, Raffaello. *Storia del metodo sperimentale in Italia.* Florence: G. Civelli, 1891–1910. 6 vols. [Reprinted 1970 by Forni, Bologna, with new introduction by G. Tabarroni; and in 1972 by Johnson Reprint Corporation, New York (Sources of Science, 134)]

Monumental study, unfortunately incomplete, of the experimental method in Italy. Vol. 1, pp. 25–264: Introduction (Pp. 25–126: Man's knowledge of the external world, from the beginnings, through Plato and Aristotle, and the peripatetics, to the Renaissance, the Accademia de' Lincei and Bacon. Pp. 127–215: From Galileo to the Accademia del Cimento and other academies. Pp. 217–64: From Newton to the naturalists of the eighteenth century). Pp. 265–526: On the principal instruments employed in scientific experiments (includes: thermometer, pendulum clock, telescope, micrometer, barometer, areometer, pluviometer). Vol. 2: The physical sciences (including astronomy). Vol. 3: Anatomy, physiology, natural history. Vols. 4 and 5: Mechanics (Galileo and his successors). Vol. 6 [NOT SEEN]: Hydraulics (Galileo and Castelli).

It is said that the author's criticism of Galileo (described by some as malicious) offended many historians of science, and the book was virtually placed on a lay index of prohibited books. As a result the work remained for a long time outside the mainstream of Italian historiography of science. A series of articles on the author and his work by Aldo Mieli, Giovanni Giovannozzi, Carlo del Lungo and Antonio Favaro appeared in *Archivio di storia della scienza* 1 (1919–20), 264–96 and 2 (1921–22), 147–66. The book was reviewed by S. Timpanaro in *Dizionario letterario Bompiani delle opere* (Milan: Bompiani, 1949), Vol. 7, pp. 164–65. The saga of the writing of this book and the reactions it provoked have been described by G. Tabarroni in *Physis* 11 (1969), 564–70, and by V. Cappelletti and F. di Trocchio in *Dizionario biografico degli italiani*, Vol. 23 (Rome: Trecanni, 1979), pp. 85–88.

485. Olschki, Leonardo. *Geschichte der neusprachlichen wissenschaftlichen Literatur. 1. Die Literatur der Technik und der angewandten Wissenschaften vom Mittelalter bis zur Renaissance. 2. Bildung und Wissenschaft im Zeitalter der Renaissance in Italien. 3. Galileo und seine Zeit.* Heidelberg, Leipzig, and Halle, 1919–27. 3 vols. [Reprinted by Krause, 1965]

Survey of the scientific and technological literature of the Renaissance. The author, born and educated in Italy (later in Germany), was professor of Romance philology at Heidelberg.

486. Clark, Cumberland. *Shakespeare and Science. A Study of Shakespeare's Interest in, and Literary and Dramatic Use of, Natural Phenomena; with an Account of the Astronomy, Astrology, and Alchemy of His Day, and His Attitude towards These Sciences.* New York: Haskell House Publishers, 1970. 262 pp. [First published in 1929]

A detailed study of the "attitude and feeling of the master-poet" towards natural phenomena. Name and subject index but no index of citations.

487. Sarton, George. "Science in the Renaissance." *The Civilization of the Renaissance.* By James Westfall Thompson, George Rowley, Ferdinand Schevill, George Sarton. Mary Tuttle Bourbon Lectures, Mount Holyoke College, 1928–29. Chicago: Univ. of Chicago Press, 1929, pp. 75–95.

An "impolite piece" by Sarton, rejecting the claim of the Renaissance to a notable place in the history of science and exposing some of its darker sides. Recalls the cold reception of the humanists to the two greatest events of the age—the development of printing and the geographical discoveries. See Harcourt Brown in *Studies in the Renaissance* 7 (1960), 27–42.

488. Smith, Preserved. *A History of Modern Culture.* Vol. 1: *The Great Renewal, 1543–1687.* New York: Holt, 1930. 672 pp.

Surveys the intellectual progress of Western culture as a whole. Pp. 17–176: The sciences (astronomy; physics; mathematics; geography, biology, anatomy; the Scientific Revolution). Pp. 609–53: Bibliography.

489. Brugmans, Henri L. *Le séjour de Christian Huygens à Paris et ses relations avec les milieux scientifiques français suivi de son journal de voyage à Paris et à Londres.* Paris: Droz, 1935. 200 pp.

Study of Huygens's relations with French scientists (from his early friendship with Descartes and Mersenne to his election to the Academy of Sciences. Reviewed in *Annals of Science* 1 (1936), 235–37.

490. Nicolson, Marjorie. *A World in the Moon. A Study of the Changing Attitude toward the Moon in the Seventeenth and Eighteenth Centuries.* Smith College Studies in Modern Languages, Vol. 17 (2). Northampton, Mass. 1936. 72 pp.

Attempts to show how "poetic, satiric, religious, and philosophical imagination seized upon new proof for old ideas, and also developed ideas not possible before telescopic observation." Deals primarily with England. Covers the period from the *Siderius nuncius* (1610) of Galileo to the balloon ascents of the Montgolfiers (1784).

491. Coffin, Charles Monroe. *John Donne and the New Philosophy.* New York: The Humanities Press, 1958. 311 pp. [First published 1937]

Study of John Donne (1572–1631), his "knowledge of science and his indebtedness to that learning." Pp. 297–311: Index of names, subjects, and titles of books.

492. Houghton, Walter E. "The History of Trades: Its Relation to Seventeenth-Century Thought as Seen in Bacon, Petty, Evelyn, and Boyle." *Journal of the History of Ideas* 2 (1941), 33–60.

On the project for a history of trades, first sketched in *The Advancement of Learning.* It was "an idea closely associated with the progress of science, education, and society. And without recognizing its influence, we cannot fully appreciate the work of Bacon and the Royal Society, of Petty, Evelyn, and Boyle."

493. Houghton, Walter E. "The English Virtuoso in the Seventeenth Century." *Journal of the History of Ideas* 3 (1942), 51–73, 190–219.

An introduction to the role of the English virtuoso (amateur, dilettante, inquirer into nature) in the formation of modern culture.

494. Tillyard, E. M. W. *The Elizabethan World Picture.* London: Penguin, 1972. 125 pp. [First published 1943]

Extracts and expounds "the most ordinary beliefs about the constitution of the world as pictured in the Elizabethan age."

495. Mieli, Aldo, Desiderio Papp, and José Babini. *Panorama general de historia de la ciencia.* Buenos Aires: Espasa-Calpe Argentina, 1945–1961. 12 vols.

Well-documented survey, from antiquity to the nineteenth century, begun by Aldo Mieli (first editor of *Archivio di storia della scienza* and founder member of the International Academy of the History of Science) and completed by his friends and pupils in Argentina. Vol. 3: *Eclosíon del Renacimiento* (1951; 400 pp.). Vol. 4: *Lionardo da Vinci, sabio* (1951; 223 pp.). Vols. 5 and 6: *La ciencia del Renacimiento; mathemática y ciencias naturales. Astronomía, física y biología* (1952; 246 pp., 193 pp.). Vol. 7: *Las ciencias exactas en siglo XVII* (1954; 234 pp.). Vol. 9: *Biología y medicina en los siglos XVII y XVIII* (1959; 258 pp.). Vol. 12 contains chronological table, indexes and table of contents for the whole work. Vols. 3–12 reviewed in *Revue d'histoire des sciences* 5 (1952), 372–73; 9 (1956), 93–94 and 17 (1964), 166–67.

496. Sergescu, P. *Coups d'oeil sur les origines de la science exacte moderne.* Paris: Société d'Edition d'Enseignement Supérieur, 1951. 203 pp.

Pp. 1–101: Series of radio talks. Pp. 102–84: Name index with biographical notes. Review: *Rivista di storia della scienza* 45 (1952), 280–81.

497. Crombie, A. C. *Medieval and Early Modern Science.* Vol. 2: *Science in the Later Middle Ages and Early Modern Times, XIII–XVII Centuries.* Garden City, N.Y.: Doubleday, 1959. 380 pp. [First published in 1952 under the title: *Augustine to Galileo* by Falcon Press, London]

Pp. 121–333: The revolution in scientific thought in the sixteenth and seventeenth centuries. With bibliography. Reviewed in *Isis* 44 (1953), 398–403 and 51 (1960), 591–93.

498. Crombie, A. C. *Robert Grosseteste and the Origins of Experimental Science, 1100–1700.* Oxford: Clarendon Press, 1953. 369 pp.

Pp. 260–319: Experimental method and the transmission of thirteenth- and fourteenth-century writings on the rainbow, color, and light to the seventeenth century. The historical foundations of the modern theory of experimental science. Pp. 320–52: Bibliography. Reviewed in *Isis* 46 (1955), 66–69.

499. Dijksterhuis, E. J. "Christian Huygens." *Centaurus* 2 (1953), 265–82.

Sketches Huygens's role in the scientific life of the seventeenth century. Address delivered on the occasion of the completion of the publishing of his collected works.

500. Villoslada, Riccardo Garcia. *Storia del Collegio Romano dal suo inizio (1551) alla soppressione della Compagnia di Gesù (1773).* Analecta Gregoriana, Vol. 66, Sec. A (2). Rome, 1954. 352 pp.

Pp. 194–213: Galileo and the professors of the College. Pp. 321–36: List of rectors and professors (including professors of natural philosophy and of mathematics).

501. Sarton, George. *The Appreciation of Ancient and Medieval Science during the Renaissance (1450–1600).* Philadelphia: Univ. of Pennsylvania Press, 1955. 233 pp.

Study of the transmission of the scientific heritage of antiquity and the Middle Ages. Reviewed in *Isis* 48 (1957), 373–75.

502. Lenoble, Robert, ed. "Les sciences au XVIIe siècle." *XVIIe siècle,* 30 (1956), 1–143.

Seven articles, with preface. "La réprésentation du monde physique à l'époque classique" (R. Lenoble); "La biologie au XVIIe siècle" (Maurice Caullery); "La pharmacie au XVIIe siècle" (Charles Bedel); "L'enseignement des mathemátiques au XVIIe siècle" (F. de Dainville); "L'oeuvre astronomique de Kepler" (A. Koyré); "La vie scientifique au XVIIe siècle" (M. Daumas); "Remarques sur l'affaire Galilée" (B. Rochot).

503. Nicolson, Marjorie. *Science and Imagination.* Ithaca, N.Y.: Great Seal Books, Cornell Univ. Press, 1956. 238 pp.

Collection of essays, first published in periodicals (1935–37), studying the interplay of science and literature; 1. The telescope and imagination. 2. The "New Astronomy" and English imagination. 3. Kepler, the *Somnium,* and John Donne. 4. Milton and the telescope. 5. The scientific background of Swift's *Voyage to Laputa* (with Nora M. Mohler). 6. The microscope and English imagination.

504. Svendsen, Kester. *Milton and Science.* Cambridge, Mass.: Harvard Univ. Press, 1956. 304 pp.

A comprehensive study of natural science in Milton based on the vernacular medieval and Renaissance encyclopedias of science. Pp. 251–304: Notes and index. Reviewed in *Isis* 48 (1957), 494–95.

505. Sarton, George. *Six Wings: Men of Science in the Renaissance.* London: Bodley Head, 1957. 318 pp.

Series of six lectures (based on the work of some thirty persons): The frame of the Renaissance; Mathematics and astronomy; Physics, chemistry, technology; Natural history; Anatomy and medicine; Leonardo da Vinci— art and science. Although the book is outdated, Sarton's views are still of interest. Reviews: *Isis* 48 (1957), 375–77; *Bulletin of the History of Medicine* 32 (1958), 575–76.

506. Clagett, M., ed. *Critical Problems in the History of Science. Proceedings of the Institute for the History of Science at the University of Wisconsin, September 1–11, 1957.* Madison: University of Wisconsin Press, 1959. 555 pp.

Of the sixteen papers presented and discussed, six are relevant: The scholar and the craftsman in the Scientific Revolution (A. R. Hall); The role of art in the Scientific Renaissance (Giorgio de Santillana); The significance of medieval discussions of scientific method for the Scientific Revolution (A. C. Crombie); The origins of classical mechanics from Aristotle to Newton (E. J. Dijksterhuis); Contra-Copernicus, a critical re-estimation of the mathematical planetary theory of Ptolemy, Copernicus, and Kepler (Derek J. de S. Price); Structure of matter and chemical theory in the seventeenth and eighteenth centuries (M. Boas). Reviewed in *Isis* 53 (1962), 230–31 and retrospectively in *Isis* 72 (1981), 267–83.

507. Hooykaas, R. *Humanisme, science et réforme: Pierre de la Ramée (1515–1572).* Leiden: Brill, 1958. 133 pp. [First published in *Free University Quarterly* 5 (1958), 167–294]

Analyses Ramus's works and compares them with other writings of the sixteenth century. Stresses that a study of Ramus and of Ramism can throw light on (a) the transformation of medieval natural science into experimental science and (b) the influence of humanism, of the Reformation, and of the economic and social life on the rise of modern science.

508. Elia, Pasquale M. D'. *Galileo in China. Relations through the Roman College between Galileo and the Jesuit Scientist-Missionaries (1610–1640).* Cambridge, Mass.: Harvard Univ. Press, 1960. 115 pp. [Originally published in Italian in *Analecta Gregoriana* 37 (1947)]

Study of the impact of Galileo on the Far East. The "zeal with which the Jesuits made Galileo's discoveries known in China, Korea, and Japan, while the astronomer was yet alive, suggests that they might have followed him in his heliocentric conclusions had it not been for the injunction of 1616 and especially the sentence of 1633" (p. vii). Appendix contains the first European document on the Chinese calendar (by Father Sabatino de Ursis, S.J., August 1612).

509. Nicolson, Marjorie. *The Breaking of the Circle. Studies in the Effect of the "New Science" upon Seventeenth-Century Poetry.* Rev. ed. New York: Columbia Univ. Press, 1960. 216 pp.

Based on lectures delivered at Northwestern University. The theme of the lectures is that the cosmology of the Renaissance poets was most often interpreted in terms of the perfect circle. This Circle of Perfection, "from which man for so long deduced his ethics, his aesthetics, and his metaphysics, was broken during the seventeenth century."

510. Rouleau, Francis. Review of *Science and Civilization in China.* Vol. 3: *Mathematics and the Sciences of the Heavens and the Earth* (Cambridge, 1959), by Joseph Needham. *Archivum Historicum Societatis Jesu* 30 (1961), 299–303.

511. Metropolitan Museum of Art. *The Renaissance.* Six essays by Wallace K. Ferguson, Robert S. Lopez, George Sarton, Roland H. Bainton, Leicester Bradner, Erwin Panofsky. New York: Harper Torchbooks, 1962. 184 pp. [Originally published in 1953 under the title: *The Renaissance: A Symposium*]

Papers on the Renaissance as seen from different points of view: political; social and economic; scientific; religious; literary; artistic.

512. Forbes, R. J. *A History of Science and Technology.* Vol. 1: *Ancient Times to the Seventeenth Century.* Harmondsworth: Penguin Books, 1963. 294 pp.

A history of the physical sciences for the general reader. Reviewed in *Isis* 55 (1964), 101–2.

513. Manuel, Frank E. *Isaac Newton, Historian.* Cambridge: Cambridge Univ. Press, 1963. 328 pp.

Study of Newton's views on history and biblical chronology. Pp. 225–328: Bibliography, notes, and index. Reviewed in *Isis* 55 (1964), 119–20.

514. Hodgen, Margaret T. *Early Anthropology in the Sixteenth and Seventeenth Centuries.* Philadelphia: Univ. of Pennsylvania Press, 1964. 523 pp.

"Deals with the organizing ideas employed by students of man and culture in the sixteenth and seventeenth centuries." It was during this time that the scientific method in the study of culture and society emerged. Notes and bibliographies after each chapter. Reviewed in *Isis* 55 (1964), 454–55.

515. Taton, René, ed. *The Beginnings of Modern Science. From 1450 to 1800. A General History of the Sciences,* Vol. 2. London: Thames and Hudson, 1964. 665 pp. [Translated from the 1958 French ed.]

Pp. 1–177: The Renaissance (mathematics; the Copernican revolution; physics; geology; chemistry; human biology and medicine; zoology; botany). Pp. 179–391: The seventeenth century (scientific revolution; from symbolic algebra to infinitesimal calculus; the birth of a new science of mechanics; the golden age of observational astronomy; the birth of mathematical optics; magnetism and electricity; the chemistry of principles; human and animal biology; medicine; botany; the birth of geology). Pp. 631–65: Name and subject indexes. Bibliographies.

516. Wightman, William P. D. "Science and the Renaissance." *History of Science* 3 (1964), 1–19.

A further exposition of the ideas expressed in the book of the same title (Item 100)—written at the invitation of the editors of the journal.

517. Lach, Donald F., and Edwin J. Van Kley. *Asia in the Making of Europe.* Vol. 1: *The Century of Discovery.* Vol. 2: *A Century of Wonder (The Visual Arts. The Literary Arts. The Scholarly Disciplines).* Vol. 3: *A Century of Advance (Trade, Missions, Literature. South Asia. South-East Asia. East Asia).* Chicago: Univ. of Chicago Press, 1965–93. 3 vols. in 9. illus.

First three volumes of a projected 6-volume work on the interaction between Asia and Europe in the sixteenth, seventeenth, and eighteenth centuries. Vols. 1 and 2 deal with the sixteenth century. Vols. 3 and 4 deal with the seventeenth century. See item 595.

Vol. 1, pp. 1–86: Europe's knowledge of Asia, from ancient times to the opening of the Cape route. Pp. 87–331: Opening of cultural and geographical vistas (the spice trade; the printed word; the Christian mission). Pp. 332–821: European images of India, S.E. Asia, Japan, and China in the sixteenth century. Pp. 837–912: Bibliography. Pp. 913–65: Subject index.

Vol. 2, Book 1. Impact of Asian art objects, artifacts, flora, fauna and crafts upon the arts of Europe. Pp. 123–85: The iconography of Asian animals. Pp. 200–36: Bibliography. Pp. 237–57: Subject index.

Vol. 2, Books 2 and 3. Asia's influence upon Europe's learned world, national literatures and scholarly disciplines. Pp. 39–79: Books, libraries and reading. Pp. 397–489: Technology and the natural sciences; cartography and geography. Pp. 567–734: Bibliography. Pp. 735–64: Subject index.

Reviewed in: *English Historical Review* 82 (1967), 351 and 94 (1979), 593; *TLS* 26 Aug. 1965, p. 730; *TLS* 26 Nov. 1971, p. 1468; *TLS* 3 Nov. 1978, p. 1289; *NYTBR* 11 Apr 1965, p. 7; *NYTBR* 3 Jan. 1971, p. 8; *NYTBR* 25 June 1978, p. 9.

Vol. 3 (Books 1–4) deals with the seventeenth century in the same way that Vol. 1 covers the sixteenth century. Pp. 5–129: Empire and trade (Decline of the Iberian maritime empire; the Dutch East India Co. and the English East India Co.). Pp. 130–298: Christian missions. Pp. 301–597: European literature on Asia (Iberian, Italian, French, Dutch, German and Danish, English). Pp. 601–1888: European images of Asia (from India to Japan). Pp. 1889–1917: Epilogue: a composite picture. Pp. 1919–2077: General bibliography. Pp. xxxi–cxii: Cumulative index (each book of Vol. 3 has an index). Reviews: *NYTRB* 5 Sep. 1993, p. 16; *TLS* 31 Dec. 1993, p. 19.

Vol. 4 not published.

518. Hall, Marie Boas. *Robert Boyle on Natural Philosophy. An Essay with Selections from His Writings.* Bloomington, Ind.: Indiana Univ. Press, 1965. 406 pp.

Pp. 3–115: Life and work of Boyle. Pp. 117–391: Selections from his writings, with introductory notes. Reviewed in *TLS* 23 Dec. 1965, p. 1190.

519. Herivel, John. *The Background to Newton's Principia. A Study of Isaac Newton's Dynamical Researches in the Years 1664–84.* Oxford: Clarendon Press, 1965. 337 pp.

Study based on Newton's MSS. Reviewed in *Isis* 57 (1966), 403–4.

520. Koyré, Alexandre. *Newtonian Studies.* London: Chapman and Hall, 1965. 288 pp.

Seven essays (written between 1950 and 1961) "each illustrating a different aspect of Newton's thought." Reviews: *Annals of Science* 21 (1965), 204–5; *British Journal for the History of Science* 3 (1966), 84–85.

521. *Le soleil à la Renaissance. Science et mythe. Colloque international tenu en avril 1963.* Travaux de l'Institut pour l'Etude de la Renaissance et de l'Humanisme, Université Libre de Bruxelles, 2. Brussels: Presses Universitaires de Bruxelles, 1965. 584 pp.

Twenty-five papers. Subjects treated include: Copernicus, heliocentrism, Renaissance astronomy, Galileo, navigation, optics, scientific instruments, Leonardo da Vinci, Cabbalists, Fludd, Vesalius, Nicolas of Cusa, Bruno, Thomas More, Ronsard. *Debus I think.*

522. Nicolson, Marjorie Hope. *Pepys' Diary and the New Science.* Charlottesville: Univ. Press of Virginia, 1965. 198 pp.

Three essays based on lectures given at the University of Virginia: Samuel Pepys, amateur of science; the first blood transfusions; "Mad Madge" [Margaret Cavendish, Duchess of Newcastle] and "the Wits." The material for these essays is drawn from Pepys's *Diary*, 1660–69. Pepys's interest in science (begun many years before his election to the Royal Society) was "symptomatic of a remarkable period in history, the first age that was as 'science conscious' as ours has become." Elected to the Royal Society in 1665, he was installed as President in 1684.

523. McGuire, J. E., and P. M. Rattansi. "Newton and the Pipes of Pan." *Notes and Records, Royal Society of London* 21 (1966), 108–43.

A study of Newton's "classical" scholia.

524. Fischer, Hans, ed. *Conrad Gesner, 1516–1565: Universalgelehrter, Naturforscher, Arzt.* Mit Beiträgen von Hans Fischer, Georges Petit, Joachim Staedtke, Rudolf Steiger, Heinrich Zoller. Zurich: Orell Füssli, 1967. 234 pp.

Commemorative publication. Seven essays on different aspects of Gesner's life and work. With annotated bibliography.

525. Maccagni, Carlo, ed. *Atti del primo convegno internazionale di ricognizione delle fonti per la storia della scienza italiana: i secoli XIV–XVI. Pisa, Domus Galilaeana, 14–16 settembre 1966.* Pubblicazioni di Storia della Scienza, Sezione V, Atti di Convegni, Vol. 1. Florence: Barbèra, 1967. 322 pp.

On the sources of the history of Italian science during the Renaissance. Five papers (cultural context; mathematics; physics and astronomy; medicine and biology; natural history) followed by a discussion and conclusion. Index of MSS.

526. McMullin, Ernan, ed. *Galileo: Man of Science.* New York: Basic Books, 1967. 455 + cii pp.

Twenty-three essays on various aspects of Galileo's work. Includes a supplement to the *Bibliografia Galileiana* of Carli-Favaro for 1564–1895 and of Boffito for 1896–1940. Reviewed in *Isis* 59 (1968), 451–52.

527. Singleton, Charles S., ed. *Art, Science, and History in the Renaissance.* Baltimore, Md.: Johns Hopkins Press, 1967. 446 pp.

Proceedings of a seminar sponsored by the Humanities Center of Johns Hopkins University. In three parts: Art and music; Science; History. Subjects covered in part 2 include: navigation; Platonism; Hermetic tradition; Galileo; Empiricism and the Scientific Revolution.

528. Manuel, F. *A Portrait of Isaac Newton.* Cambridge, Mass.: Harvard Univ. Press, 1968. 478 pp.

Study of "Newton the man—his person, his world outlook, and his style of life." Reviewed in *Isis* 61 (1970), 141–42.

529. Zoubov, V. P. *Leonardo da Vinci.* Cambridge, Mass.: Harvard Univ. Press, 1968. 335 pp. Tr. from the Russian.

Biographical sketch of Leonardo and an outline of the main themes of his work and his "relationship to different aspects of nature and human activity." Reviewed in *British Journal for the History of Science* 5 (1970), 203–4.

530. O'Malley, C. D., ed. *Leonardo's Legacy. An International Symposium.* UCLA Center for Medieval and Renaissance Studies, Publication no. 2. Berkeley and Los Angeles: Univ. of California Press, 1969. 225 pp.

Nine papers. Subjects include: physiology, technology, automation, architecture, movement of water.

531. Gould, Heywood. *Sir Christopher Wren, Renaissance Architect, Philosopher, and Scientist.* London: Franklin Watts, 1970. 216 pp.

Study of the life and work of one of the creative spirits that guided the seventeenth-century world over "the threshold of the modern scientific age."

532. Palter, Robert, ed. *The* Annus Mirabilis *of Sir Isaac Newton, 1666–1966.* Cambridge, Mass.: MIT Press, 1970. 351 pp. [First published in *The Texas Quarterly* 10 (1967), (3) 1–287]

Sixteen papers on Newton's life and society, his scientific achievements (and philosophical analysis thereof), and his influence, read at a Symposium by outstanding scholars. Present ed. includes critical comments on the major papers.

533. Hessen, Boris. "The Social and Economic Roots of Newton's *Principia.*" *Science at the Cross Roads. Papers Presented to the International Congress of the History of Science and Technology, London, from June 29–July 3, 1931, by the Delegates of the USSR.* 2nd ed. London: Frank Cass, 1971, pp. 149–212.

A Marxist analysis of the genesis and development of Newton's work.

534. Shea, William R. *Galileo's Intellectual Revolution.* London: Macmillan, 1972. 204 pp.

An investigation of Galileo's "middle period"—from 1610 (return to Florence) to 1632 (publication of the *Dialogo*)—during which he "worked out the methodology of his intellectual revolution." Covers: the debate on floating bodies; the controversies over the sunspots (1612) and the comets (1618); the end of the Aristotelian cosmos; the *Dialogo*; the ill-fated theory of the tides. Reviews: *Rivista critica di storia della filosofia* 29 (1974), 328–34; *British Journal for the Philosophy of Science* 26 (1975), 81–82; *Renaissance Quarterly* 28 (1975), 363–66; *Historia Mathematica* 3 (1976), 103–9.

535. Roger, Jacques, ed. *Sciences de la Renaissance. VIIIe Congrès international de Tours.* Paris: Vrin, 1973. 308 pp.

Twenty-four papers. Subjects covered: the intellectual climate; mathematics; astronomy; anatomy; medicine; life sciences; chemistry; alchemy; pharmacy. Index of names. Reviewed in *Revue de synthèse* 96 (1975), 371–77.

536. Cipolla, Carlo M., ed. *The Fontana Economic History of Europe: the Sixteenth and Seventeenth Centuries.* London: Collins/Fontana Books, 1974. 640 pp.

Comprehensive survey. Topics covered: Population in Europe; Patterns and structure of demand; technology in the age of the Scientific Revolution; Rural Europe; European industries; European trade; The emergence of modern finance in Europe.

537. Shirley, John, ed. *Thomas Harriot, Renaissance Scientist.* Oxford: Clarendon Press, 1974. 181 pp.

Collection of studies, based on papers read at a symposium in April 1971. Pp. 166–74: Bibliography. Reviewed in *Renaissance Quarterly* 29 (1976), 94–96.

538. Wilson, Dudley, ed. *French Renaissance Scientific Poetry.* London: Athlone Press, 1974. 185 pp.

"The poetry in this anthology will illustrate abundantly—that the science of this age is an amalgam of science, philosophy and magic and that the scientific poet is best seen as the Magus who interprets a personal vision of these aspects of the universe" (Introduction). Thirty-eight texts arranged under: Cosmos and microcosmos; fish, plants, and medicine; the microcosm—man, his anatomy, senses and feelings; meteorology and astronomy; mathematics and music; the poet as magus (his vision and interpretation of the universe). Selected general bibliography. Bio-bibliographical notes on individual poets.

539. Wisan, W. "The New Science of Motion: A Study of Galileo's *De motu locali.*" *Archive for History of Exact Sciences* 13 (1974), 103–306.

Study of the development of Galileo's science of motion. (The *De motu locali* constitutes the third and fourth *giornate* of the *Discorsi* [1638].) Summary (pp. 296–97) and bibliography (pp. 299–306). Review: *Annali, Istituto e Museo di Storia della Scienza di Firenze* 1 (1976), (2), 89–97.

540. Crosland, M. P., ed. *The Emergence of Science in Western Europe.* London: Macmillan, 1975. 201 pp.

Six (out of 11) papers deal with Italy, England, and the Netherlands during the sixteenth and seventeenth centuries. Reviewed in *British Journal for the History of Science* 9 (1976), 320–23.

541. *Leonardo nella scienza e nella tecnica. Atti del simposio internazionale di storia della scienza, Firenze-Vinci 23–26 giugno 1969.* Florence: Giunti-Barbèra, 1975. 302 pp.

Twenty-six papers on various aspects of Leonardo da Vinci's work from cosmology and mechanics to anatomy and technology, including one on the discovery of the Madrid Codices in April 1965.

542. Rowse, A. L. *Oxford in the History of the Nation.* London: Weidenfeld and Nicolson, 1985. 256 pp. [First published 1975]

Popular study by a leading Oxford historian of the part played by the town and University in the history of the nation. Pp. 55–141 cover the sixteenth and seventeenth centuries. With index but no documentation.

543. Whiteside, D. T. "In Search of Thomas Harriot. Essay Review [of]: *Thomas Harriot: Renaissance Scientist.* Edited by J. W. Shirley (Clarendon Press, Oxford, 1974)." *History of Science* 13 (1975), 61–70.

See item 537.

544. Brown, Harcourt. *Science and the Human Comedy. Natural Philosophy in French Literature, from Rabelais to Maupertuis.* Toronto: Univ. of Toronto Press, 1976. 221 pp.

Pp. 1–125 deal with Rabelais, Pascal, Molière, and the *Journal des sçavans.*

545. Cochrane, Eric. "Science and Humanism in the Italian Renaissance." *American Historical Review,* 81 (1976), 1039–57.

On Renaissance humanism and the birth of modern science. Views of Thorndike and Sarton reconsidered.

546. Redwood, John, ed. *European Science in the Seventeenth Century.* Newton Abbot: David and Charles, 1977. 208 pp.

Source book (forty-one extracts in twenty sections). Aims at showing the "diversity of intellectual endeavour and the range of intellectual achievement" of seventeenth-century science. Select bibliography and index. With notes.

547. López Piñero, José Maria. *Ciencia y técnica en la sociedad española de los siglos XVI y XVII.* Barcelona: Editorial Labor, 1979. 511 pp.

General survey of scientific activity in Spain during the sixteenth century. Seventeenth-century Spain and the Scientific Revolution. Pp. 457–507: Classified bibliography. Reviewed in *Technology and Culture* 23 (1982),112–14 and *Journal of the History of Biology* 16 (1983), 433–40.

548. Shapiro, Barbara, and Robert G. Frank, Jr. *English Scientific Virtuosi in the Sixteenth and Seventeenth Centuries. Papers Read at a Clark Library Seminar 5 February 1977.* Los Angeles: Univ. of California, William Andrews Clark Memorial Library, 1979. 123 pp.

Two papers 1. History and natural history in sixteenth and seventeenth-century England: an essay on the relationship between humanism and science. 2. The physician as virtuoso in seventeenth-century England. Documented.

549. Bos, H. J. M., M. J. S. Rudwick, H. A. M. Snelders, and R. P. W. Visser, eds. *Studies on Christian Huygens. Invited Papers from the Symposium on the Life and Work of Christian Huygens, Amsterdam, 22–25 August 1979.* Lisse: Swets and Zeitlinger, 1980. 321 pp.

Topics covered include: Cartesianism; concept of matter; mathematics; astronomy; mechanics; optics; instruments; *Horologium,* music. With notes and chronology. Reviewed in *Isis* 73 (1982), 137–38.

550. Cohen, I. Bernard. *Album of Science. From Leonardo to Lavoisier, 1450–1800.* New York: Charles Scribner's Sons, 1980. 306 pp.

Provides "a pictorial record of the growth of the scientific enterprise, an attempt to show in images what science was like in the distant and recent past and to convey a sense of the perception of science by men and women, both scientists and non-scientists." The 368 illustrations (annotated) relate to the major discoveries of the period. In seven parts (The new world of science; The astronomical universe; The exact sciences; The experimental physical sciences; The earth sciences; The life sciences; Science and society), subdivided into 28 chapters, each with an introduction. Pp. 278–87: Bibliography. Reviewed in *Isis* 72 (1981), 648–49.

551. Durbin, P. T., ed. *A Guide to the Culture of Science, Technology, and Medicine.* New York: Free Press, 1980. 723 pp.

"Set of state-of-the-field surveys of nine academic disciplines that take as their object science, technology and medicine" (History of science; History of technology; History of medicine; Philosophy of science; Philosophy of technology; Philosophy of medicine; Sociology of science and technology; Medical sociology and science and technology in medicine; Science policy studies). Bibliographies, subject and author index. Review: *Isis* 73 (1982), 567–71.

552. Schmitz, Rudolf, and Fritz Krafft, eds. *Humanismus und Naturwissenschaften.* Boppard: Harald Boldt, 1980. 210 pp.

Eleven papers read at a Symposium in Nuremberg of the Deutsche Forschungsgemeinschaft (12–14 October 1977). Topics covered include: physical science; astrology of Manilius; Regiomontanus and Peuerbach; Gesner; botany, alchemy, and geography. Reviewed in *Isis* 73 (1982), 312–13 and *Sudhoffs Archiv* 66 (1982), 195–96.

553. Westfall, R. S. *Never at Rest. A Biography of Isaac Newton.* Cambridge: Cambridge Univ. Press, 1980. 908 pp.

Detailed scientific biography by a leading Newton scholar, based on recent scholarship. Pp. 875–84: Bibliographical essay, including a brief guide to the location of Newton MSS. Reviewed in: *British Journal for the Philosophy of Science* 33 (1982), 305–15; *Isis* 73 (1982), 100–7; and *British Journal for the History of Science* 16 (1983), 101–5.

554. Bynum, W. F., E. J. Browne, and R. Porter, eds. *Dictionary of the History of Science.* London: Macmillan Press, 1981. 494 pp.

Consists of some seven hundred articles dealing with leading fields of science, and ideas and topics within them, suitably cross referenced. Prelims contain an analytical table of contents arranged under: astronomy, biology, chemistry, earth sciences, historiography and sociology of science, human sciences, mathematics, medicine, miscellaneous, philosophy of science, physics. Pp. 452–94: Biographical index of scientists.

555. Castillejo, David. *The Expanding Force in Newton's Cosmos, as Shown in His Unpublished Papers.* Madrid: Ediciones de Arte y Bibliofilia, 1981. 125 pp.

Stresses the need to study Newton's MSS on alchemy, prophesy, and chronology in conjunction with his better-known works on mechanics and optics. Reviewed in *British Journal for the History of Science* 17 (1984), 112–13.

556. Harig, Gerhardt. *Physik und Renaissance. Zwei Arbeiten zum Entstehen der klassischen Naturwissenschaften in Europa.* Ostwalds Klassiker der exakten Wissenschaften, Vol. 260. Leipzig: Akademische Verlagsgesellschaft Geest und Portig, 1981. 89 pp.

Two essays from a Marxist viewpoint: 1. Walter Hermann Ryff and Tartaglia and the history of dynamics. 2. The emergence of classical science in Europe.

557. Shirley, John W., ed. *A Source Book for the Study of Thomas Harriot*. The Development of Science: Sources for the History of Science. New York: Arno Press, 1981. Not paginated.

Collection of twenty-five articles ("more original and significant studies") by twelve authors on Harriot—the earliest dated 1833. With bibliography and introductory note.

558. Struik, Dirk J. *The Land of Stevin and Huygens. A Sketch of Science and Technology in the Dutch Republic during the Golden Century*. Dordrecht: Reidel, 1981. 162 pp. [First published in Dutch in 1958]

Survey of the period 1580/90–1700, with bibliography.

559. Bechler, Zev, ed. *Contemporary Newtonian Research*. Studies in the History of Modern Science, Vol. 9. Dordrecht: Reidel, 1982. 241 pp.

Collection of essays by three philosophers and four historians. Topics include: the "roots of universal gravity"; Newton's mathematical education; Newton's theological MSS. Reviewed in *Isis* 74 (1983), 609–10 and in *International Studies in Philosophy* 18 (1986), (1), 65–66.

560. Slaughter, M. M. *Universal Languages and Scientific Taxonomy in the Seventeenth Century*. Cambridge: Cambridge Univ. Press, 1982. 277 pp.

Survey of the scientific demand for clear, unambiguous, universal languages.

561. Burke, John G., ed. *The Uses of Science in the Age of Newton*. Berkeley: Univ. of California Press, 1983. 204 pp.

Seven essays on various aspects of science and technology, providing interpretations of Newtonian science, some of which challenge the contextualist approach. Reviews: *Isis* 76 (1985), 274–75; *TLS* 25 Jan. 1985, p. 89.

562. Palter, Robert. "The Newton Myths: Some Reflections on Westfall's *Never at Rest*." *Archives internationales d'histoire des sciences* 33 (1983), 344–53.

See item 553.

563. Shirley, John W. *Thomas Harriot: a Biography*. Oxford: Clarendon Press, 1983. 508 pp.

Detailed study of Harriot's life by one of the leading non-scientist "Harrioteers" of our time. Pp. 476–90: Bibliography. Pp. 490–508: Name and subject index. Reviewed in: *Annals of Science* 41 (1984), 289–91 ("As so often in Harriot studies, the mathematician vanishes among his own manuscripts"—A. R. Hall); *Isis* 75 (1984), 759–60 ("Harriot the scientist remains strangely elusive"—John Henry); *TLS* 13 Jan. 1984, p. 34.

564. Snelders, H. A. M. "Science in the Low Countries during the Sixteenth Century. A Survey." *Janus* 70 (1983), 213–27.

Surveys the state of science before and after the revolt against Spain.

565. Christianson, Gale E. *In the Presence of the Creator: Isaac Newton and His Times.* New York: Free Press, 1984. 623 pp.

Popular biography of Newton. Reviewed in *Science* 226 (1984), 39–40.

566. Pagano, Sergio M., ed. *I documenti del processo Galileo Galilei.* Pontificiae Academiae Scientiarum Scripta Varia, 55. Vatican City: Archivio Segreto Vaticano, 1984. 280 pp.

Basically a critical edition of the Galileo file (first compiled after the 1633 trial and now kept in the Vatican Archives) and of the relevant documents of the Inquisition. Both sets of documents are essentially the same as those published in Vol. IX of Antonio Favaro's ed. of Galileo's *Opere.* Includes a criticism of Redondi's thesis (item 343). For critical reviews see *Isis* 76 (1985), 380–81 (Finocchiaro) and *Journal for the History of Astronomy* 17 (1986), 66–67 (Shea).

567. Wallace, William A. *Galileo and His Sources. The Heritage of the Collegio Romano in Galileo's Science.* Princeton: Princeton Univ. Press, 1984. 371 pp.

On the influence of the Jesuits on Galileo. A detailed study of his Latin notebooks and their sources. Review: *TLS* 3 Jan. 1986, p. 13.

568. Coyne, G. V., M. Seller, and J. Życiński, eds. *The Galileo Affair: A Meeting of Faith and Science. Proceedings of the Cracow Conference, 24 to 27 May 1984.* Vatican City: Specola Vaticana, 1985. 179 pp.

Reappraisals of Galileo by a group of historians of science, philosophers, and theologians (Galileo's concept of science; the rhetoric of proof in his writings; the problem of annual parallax; his religion; the young Bellarmine; Galileo and the development of science; cultural repercussions of the Galileo affair). Reviewed in *Archives de philosophie* (Paris) 49 (1986), 333–36.

569. D'Elia, Alfonsina. *Christiaan Huygens: una biografia intellettuale.* Filosofia e scienza nel cinquecento e nel seicento, serie I, studi 27. Milan: Angeli, 1985. 360 pp.

84   *Index Libror. Prohib.*
cuius Generalis permittuntur; Ea verò,quæ funt
ex editione Pauli Manutij permittuntur .
F.   Dethmanuus à Bolman , *uide , Symbolum Militare.*
Deuota admonitio . *uide , Amica .*
Deuotiſſimæ preces ad Beatiſſimam Virginem. *vide*
preces deuotiſſima ,
Deus, & Rex *ſic inſcribitur quidam liber .*
Diaboli Bulla .
De Diabolicis incantationibus . *vide, Magica . ſeu Mi-*
rabilium .
Dialectica legalis . *etiam cum nomine Auctoris ,*
Dialectica Reſolutio . *vide Analyſis reſolutio .*
Dialecticarum partitionum . *uide, Rodulphi Goelenÿ.*
Dialacticæ Exercitationes . *vide, Ioannis Scholÿ .*
Dialexeon Academiarum . *vide , Adami Theodo-*
ri Siberi ,
Dialogi aduerſus Ioannem Eckium .
Dialogi Iſtorici ouero Compendio hiſtorico dell'
Italia, e dello ſtato preſente de'Prencipi , e Re-
publiche Italiane dell'Accademico Incognito ,
Parte prima, ſeconda, e terza .
Dialogi Luciani . *vide, Luciani Samoſatenſis ,*
Dialogi di Mercurio, e Caronte .
Dialogi Olympiæ Fuluiæ Moratæ .
Dialogi Politici. ouero la Politica , che vſano in
queſti tempi i Prencipi , e le Republiche Italia-
ne, per conſeruare i loro ſtati .
Dialogi Sacri *ſinc nomine Auctoris , qui tamen ſunt*
Sebaſtiani Caſtalionis Haretici .
Dialogi, & ſcripta contra Ioannem XXII. *vide ,*
Gulielmi Ochami .
Dialogo di Chriſtofaro Bronzini della dignità , e
nobiltà delle donne, donec corrigatur .
Dialogo di Giacopo Riccamati Oſſaneſe , Inter-
locutori il Riccamati, e'l Mutio .
Dialogo di Galileo Galilei , doue ne i congreſſi di
quattro giornate ſi diſcorre ſopra i duoi Maſſi-
mi Siſtermi del mõdo Tolemaico,e Copernicano.
Dialogo per inſtruire i fanciulli . *Vide , Catechiſmo*
ſiue Formulario ,

Dia-

Index librorum prohibitorum, Rome, 1670, p. 84 with entry for Galileo's *Dialogo*
(1632). (Bodleian Library, Oxford, Mason AA.227). **Entry 566.**

Biography covering all aspects of Huygens's work. Documented, with bibliography and index of names. Reviewed in *Nuncius* 1 (1986), (2), 190–91 and in *History of European Ideas* 8 (1987), 750–51.

570. Impey, Oliver, and Arthur MacGregor, eds. *The Origins of Museums: The Cabinet of Curiosities in Sixteenth- and Seventeenth-Century Europe.* Oxford: Clarendon Press, 1985. 335 pp.; pls.

Thirty-three articles based on papers read at a symposium, "The Cabinet of Curiosities," held at Oxford to commemorate the tercentenary of the Ashmolean Museum. Pp. 281–312: Bibliography. Pp. 313–35: Name and subject index. Reviews: *TLS* 25 July 1986, p. 808; *Isis* 79 (1988), 452–67.

571. Ochs, Kathleen H. "The Royal Society of London's History of Trades Programme: An Early Episode in Applied Science." *Notes and Records, Royal Society of London* 39 (1985), 129–58.

Points out some of the shortcomings of Houghton's study (item 492).

572. Riondato, Ezio, ed. *Trattati scientifici nel Veneto fra il XV e XVI secolo.* Vicenza: Pozza, 1985. 156 pp.

Seven essays on the scientific-publishing industry in Venetia of the late fifteenth and sixteenth centuries. Subjects covered: Aristotelianism, mathematics, medicine, Biringuccio's *Pyrotecnia*, architecture, culinary arts.

573. Schlosser, Louis. *La vie de Michel Nostradamus.* Paris: Pierre Belfond, 1985. 285 pp.

Study of Nostradamus (1503–1566)—physician, humanist, astrologer, and prophet—and his times.

574. Trevor-Roper, Hugh. *Renaissance Essays.* London: Secker and Warburg, 1985. 312 pp.

Collection of thirteen essays on topics of European history (between the Renaissance and the Thirty Years' War) of which the following are published for the first time: The Paracelsian movement (pp. 149–99); Robert Burton and *The Anatomy of Melancholy* (pp. 239–74).

575. Bottazzi, Filippo. *Leonardo scienziato.* A cura di Leonardo Donatelli, Francesco Ghiretti, e Andrea Russo. Naples: Giannini, 1986. 448 pp.

Collection of eighteen studies by Bottazzi (1867–1941), distinguished physiologist, written between 1902 and 1941 on various aspects of the scientific work of Leonardo da Vinci (natural philosophy, physiology, biology, anatomy, natural history, nutrition, physics, chemistry). Pp. 13–20: List of Leonardo's MSS. Pp. 439–45: Bio-bibliographical note on Bottazzi.

576. Gjertsen, Derek. *The Newton Handbook.* London: Routledge and Kegan Paul, 1986. 665 pp.

Some five hundred entries (in seven categories: works of Newton; personalities; scientific entries; other intellectual interests; institutions; private life; legends), varying in length from a few lines to forty-seven pages (*Principia*), arranged in one alphabetical sequence. With bibliography (pp. 626–40) and index. "Despite its bulk and appearance of thoroughness, I found this to be an unsatisfactory work that will probably be of slight value to any historian"—Alan M. Shapiro in *Isis* 78 (1987), 490. "A genuinely informative *vade-mecum* which (do I let the secret out?) even those solidly established scholars who are given potted biographies inside keep to hand along with their well-worn *Concise O.E.D.s* and other trusty reference books"—D. T. Whiteside in *Notes and Records, Royal Society of London* 44 (1990), 111.

577. Negri, Lionello, Nicoletta Morello, and Paolo Galluzzi. *Niccolò Stenone e la scienza in Toscana alla fine del '600: mostra documentaria ed iconografica, Firenze, 23 settembre–6 dicembre 1986. Catalogo.* Florence: Biblioteca Medicea Laurenziana, 1986. 172 pp.

Catalogue of an exhibition held to commemorate the third centenary of Steno's death. Detailed notes on 192 exhibits, mostly books and MSS. Pp. 11–16: Biographical notice. Pp. 31–48: Steno, the anatomist. Pp. 67–88: Steno and inorganic nature. Pp. 113–29: The scientific debate inTuscany.

578. Manno, Antonio, ed. *Cultura, scienze e tecniche nella Venezia del Cinquecento. Atti del Convegno internazionale di studio Giovan Battista Benedetti e il suo tempo.* Venice: Istituto Veneto di Scienze, Lettere ed Arti, 1987. 503 pp.

Proceedings of conference held 3–4 October 1985 dedicated to the life, times, and work of Benedetti (1530–1590). Thirty-two papers in five sections. Topics include: The Academy of Venice; Daniele Barbaro; Collegio Romano; philosophy and science, mechanics; arithmetic; Tartaglia; perspective; musical science; dialling; naval architecture; Arsenal of Venice; hydraulics. Two exhibitions were held (of books, manuscripts, and archival documents) to coincide with the conference by the Archivio di Stato, Venice, and the Biblioteca Marciana, respectively, *Ambiente scientifico veneziano tra cinque e seicento* and *La scienza a Venezia tra quattro e cinquecento.* Catalogues (NOT SEEN) published 1985.

579. Marinoni, Augusto. "La biblioteca di Leonardo." *Raccolta vinciana* 22 (1987), 291–342.

Essay on the library of Leonardo da Vinci, including a summary of earlier studies. For catalogue see 184.

580. Poupard, Paul, Cardinal, ed. *Galileo Galilei: Toward a Resolution of 350 Years of Debate—1633–1983*. Pittsburgh: Duquesne Univ. Press, 1987. 208 pp. [First published in Italian and French in 1984]

These papers are the result of collaboration by a group of theologians, scientists and historians, who, at the request of Pope John Paul II, re-examined the Galileo affair in the light of the 350 years that have lapsed since his condemnation by the Holy Office. Topics: Galileo Galilei, 350 years of history; G.'s predecessors (Copernicus, the Jesuit professors of the Collegio Romano); G. and the culture of his time (philosophy, theology); G. in the eighteenth century; G. and modern science. Conclusion (by Pope John Paul II). Reviewed by Maurice A. Finocchiaro in *Isis* 78 (1987), 634–35 ("falls far short of what Catholic scholarship is capable of accomplishing, and of what the seriousness of the issue demands.")

581. Shapin, Steven. "Essay Review of *The Correspondence of Henry Oldenburg*." Edited by A. Rupert Hall and Marie Boas Hall. Madison: Univ. of Wisconsin Press, 1955–73, Vols. 1–9; London: Mansell, 1975–77, Vols. 10–11; London: Taylor and Francis, 1986, Vols. 12–13. *Isis* 78 (1987), 417–24.

Oldenburg was Secretary of the Royal Society of London from its inception (1662) till his death (1677).

582. Vickers, Brian, ed. *English Science, Bacon to Newton*. Cambridge English prose texts. Cambridge: Cambridge Univ. Press, 1987. 244 pp.

Anthology. Highlights: development of the experimental method from being purely empirical to being more theoretical in its approach; discussions concerning the proper nature of scientific language. With bibliography.

583. Coyne, G. V., M. Hellers, and J. Zýciński, eds. *Newton and the New Direction in Science. Proceedings of the Cracow Conference, 25 to 28 May 1987*. Vatican City: Specola Vaticana, 1988. 269 pp.

Multi-faceted analyses, by seventeen specialists, of Newton's work, that of his followers and opponents, and of Newtonian themes in contemporary science and philosophy. Pp. 21–134 relate to the years before 1700. (Pp. 109–14: Newton's controversy with Leibniz over the invention of the calculus.)

584. Goodman, David C. *Power and Penury. Government, Technology and Science in Philip II's Spain.* Cambridge: Cambridge Univ. Press, 1988. 275 pp.

The theme of the book is Spain's involvement under Philip II (1527–98) with technology and natural science—during a period of war and expansion. Topics: 1. The occult sciences: the crown's support and controls (Philip II and the occult. The Inquisition's censure and other clerical opinion. Lay opinion and popular beliefs. *Conversos,* Moriscos and the occult). 2. Cosmography and the crown (Criticism of Aristotelian cosmology and its repercussions. Longitude and politics. Royal projects to survey Spain and the Indies. Provisions for the perfection of navigation). 3. Technology for war (Philip II and shipbuilding. Artillery and munitions. Gunners and engineers, Secret inventions for war). 4. Producing the king's silver (The rise and fall of Guadalcanal. Silver from the Indies. A new role for the mine at Almadén). 5. The crown's interest in medicine (Royal hospitals for the poor, the sick, and the workers. The king's control of medical practice. The king's medical establishment and the search for medical plants. Medical services for the military). Pp. 265–70: Bibliography (primary and secondary sources).

585. Henry, John. "The Origins of Modern Science: Henry Oldenburg's Contribution." Essay Review of: *The Correspondence of Henry Oldenburg,* edited and translated by A. Rupert Hall and Marie Boas Hall, 13 vols., Madison and London, 1965–1986. *British Journal for the History of Science* 21 (1988), 103–10.

See item 581.

586. Marcorini, Edgardo, ed. *The History of Science and Technology. A Narrative Chronology,* Vol. 1: *Prehistory–1900.* Vol. 2: *1900–1970.* New York: Facts on File, 1988. 889 pp. [Translated from the 1975 Italian ed.]

Vol. 1, pp. 134–218 cover the years 1450–1700. Vol. 2, pp. 841–89: Indexes. No bibliography, but primary sources are cited in the text.

587. Scheurer, P. B., and G. Debrock, eds. *Newton's Scientific and Philosophical Legacy.* International archives of the history of ideas, 123. Dordrecht: Kluwer, 1988. 382 pp.

Proceedings of colloquium held in Nijmegen, 9–12 June 1987. Twenty-five papers arranged in four groups, of which the first, Newton's science (pp. 25–131), is relevant.

588. Hunter, Michael, and Simon Schaffer, eds. *Robert Hooke: New Studies.* Woodbridge, Suffolk: Boydell Press, 1989. 310 pp.

Revised version of nine papers read at a conference on Hooke held at the Royal Society, London, 19–21 July 1988. Topics: Instruments for astronomy and navigation; Practical optics; Marine timekeeper for determining longitude; *Micrographia;* Mechanism and magic; Motion in a curved path; Geological theories; Hooke's Diary; Hooke's complex and ambiguous identity. Pp. 295–304: Bibliography of secondary works on Hooke published this century (some 200 items).

589. Maffioli, C. S., and L. C. Palm, eds. *Italian Scientists in the Low Countries in the XVIIth and XVIIth Centuries. Invited Papers from the Congress held in Utrecht on 25–27 May 1988 to Commemorate the 350th Anniversary of the Publication of Galileo Galilei's 'Discorsi e dimostrazioni matematiche intorno a due nuove scienze' (Leyden, 1638).* Nieuwe Nederlandse Bijdragen tot de Geschiedenis der Geneeskunde en der Naturwetenschappen, 34. Amsterdam: Rodopi, 1989. 334 pp.

Sixteen papers of which nine are relevant (Galileo and the Scientific Revolution; Italian and Dutch universities; science and religion in the Northern Netherlands; mechanics and hydrostatics in the late Renaissance; Galileo in Holland before the *Discorsi;* the problem of longitude; Giovanni Francesco Buonamico and the fossils; Italian influences on Leeuwenhoek; editorial fortune of Bolognese scientists in Holland).

590. North, J. D. *The Universal Frame. Historical Essays in Astronomy, Natural Philosophy, and Scientific Method.* London: Hambledon Press, 1989. 384 pp.

Reprint of essays concerning "themes in the history and methodology of the exact sciences" published between 1969 and 1987, of which eleven relate to the sixteenth and seventeenth centuries. Pp. 373–84: Index.

591. Schiebinger, Londa. *The Mind Has No Sex: Women in the Origins of Modern Science.* Cambridge, Mass.: Harvard Univ. Press, 1989. 355 pp.

History of women's contributions to the development of modern science. Examines the "persistent effort to distance science from women and the feminine" and inquires why women scientists have been so rare. Selected bibliography. Reviews: *Nature* 342 (1989), 309–10; *Annals of Science* 47 (1990), 314; *TLS* 9 Mar. 1990, pp. 243–44.

592. Grafton, Anthony. "Humanism, Magic and Science." *The Impact of Humanism on Western Europe.* [Essays in Honour of Denys Hay.] Edited by Anthony Goodman and Angus Mackay. London: Longman, 1990, pp. 99–117.

One of eleven survey articles on the impact of humanism on European life and culture.

593. Hall, A. Rupert. *Henry More: Magic, Religion and Experiment.* Oxford: Basil Blackwell, 1990. 304 pp.

Biographical study restricted to the scientific aspects of More's thought and his relationship to Newton in particular.

594. Hallyn, Fernand. *The Poetic Structure of the World: Copernicus and Kepler.* New York: Zone Books, 1990. 367 pp. [Tr. from the French ed. of 1987]

Analysis of the theories of Copernicus and Kepler in the light of the poetics of the age. Review: *Isis* 83 (1992), 657–58.

595. Lach, Donald F., and Edwin J. Van Kley. "Asia in the Eyes of Europe: the Seventeenth Century." *The Seventeenth Century* 5 (1990), 93–109.

Notice of the forthcoming Vol. 3 of *Asia in the Making of Europe* by the present authors. See item 517.

595A. Laeven, Augustinus Hubertus. *The "Acta Eruditorum" under the Editorship of Otto Mencke (1644–1707). The History of an International Learned Journal between 1682 and 1707.* Translated from the Dutch by Lynne Richards. Mit einer Zusammenfassung in deutscher Sprache. Amsterdam and Maarssen: APA-Holland and University Press, 1990. 431 pp.

Examines the history of the *Acta Eruditorum* (Leipzig), one of the most important periodicals of the Enlightenment, founded by Otto Mencke, professor of philosophy at the University of Leipzig, during the first twenty-five years of its existence. The journal provided scientists, among them Leibniz, with an international platform on which they could present their discoveries and ideas. Includes: a list of all contributors (hitherto anonymous), an inventory of the editor's correspondence, a bibliography, and an index of names.

596. Olby, R.C., G.N. Cantor, J.R.R. Christie, and M.J.S. Hodge, eds. *Companion to the History of Modern Science.* London: Routledge, 1990. 1,081 pp.

"Descriptive and analytical guide to the development of western science from 1500, and to the diversity and course of that development first in Europe and later across the world." Sixty-seven essays including: "The Copernican revolution" (J. R. Ravetz, pp. 201–16); "The Scientific Revolution" (John A. Schuster, pp. 217–42); "Newton and Natural Philosophy" (Alan Gabbey, pp. 243–63); "The Heart and Blood from Vesalius to Harvey"

(Andrew Wear, pp. 568–82); "Magic and Science in the Sixteenth and Seventeenth Centuries" (John Henry, pp. 583–96); "Atomism and the Mechanical Philosophy" (Martin Tamny, pp. 597–609).

597. Serrai, Alfredo. *Conrad Gesner.* A cura di Maria Cochetti (con una bibliografia delle opere allestita da Marco Menato). Il Bibliotecario, 5. Rome: Bulzoni, 1990. 430 pp.

Analysis of Gesner's works with a biographical note and an intellectual profile.

598. Whiteside, D. T. "The latest on Newton . . . . " *Notes and Records, Royal Society of London* 44 (1990), 111–17.

Review of fifteen recently published books on Newton, including five works written in commemoration of the tercentenary of the *Principia.*

599. Pumfrey, Stephen, Paolo L. Rossi, and Maurice Slawinski, eds. *Science, Culture and Popular Belief in Renaissance Europe.* Manchester: Manchester Univ. Press, 1991. 331 pp.

Twelve commissioned essays, with notes and a select bibliography (pp. 293–317). Topics: The new learning; Natural philosophy and the science of history; Natural philosophy and rhetoric; Natural philosophy and its public concerns; The role of the church; The dissemination of knowledge; Artisans, instruments and practical mathematics; Medicine and popular healing; Witchcraft; Astrology; The decline of astrological beliefs. Pp. 319–31. Index.

600. Goodman, David, and Colin Russell, eds. *The Rise of Scientific Europe, 1500–1800.* Sevenoaks, Kent: Hodder and Stoughton, for the Open University, 1991. 437 pp.

Written for Open University students. "A history of scientific Europe, of its institutions as well as its achievements, of public attitudes as well as private attainment." In sixteen chapters (with conclusion), the first nine relating to the period 1500–1700. Bibliographies at the end of each chapter. Index.

601. Hall, A. Rupert. *Isaac Newton: Adventurer in Thought.* Oxford: Blackwell, 1993. 468 pp.

Biography of Newton "the mathematician and philosopher" by the editor of his correspondence. Pp. 399–457: Notes and bibliography. Pp. 459–68: Index. Review: *Annals of Science* 51 (1994), 544–45.

601A. Meli, Domenico Bertoloni. *Equivalence and Priority: Newton versus Leibniz. Including Leibniz's Unpublished Manuscripts on the* Principia. Oxford: Clarendon Press, 1993. 318 pp.

Examines the "competing world systems put forward by Newton and Leibniz in the late 1680s and their reception up to the beginning of the eighteenth century." Based on the *Principia* and Leibniz's *Tentamen de motuum coelestium causis* (1689).

601B. Kessler, Eckhard, ed. *Girolamo Cardano: Philosoph, Naturforscher, Arzt.* Wolfenbütteler Abhandlungen zur Renaissanceforschung, Bd. 15. Wiesbaden: Harrassowitz Verlag, 1994. 376 pp.

Fifteen articles on various aspects of Cardano's life and work. Based on papers read at a symposium held in Wolfenbüttel, 8–12 October 1989.

Occult Science

BIBLIOGRAPHIES

602. Grässe, Johann Georg Theodor. *Bibliotheca Magica et Pneumatica oder Wissenschaftlich geordnete Bibliographie der wichtigsten in das Gebiet des zauber-, wunder-, geister- und sonstigen Aberglaubens vorzüglich älterer Zeit einschlagenden Werke. Mit Angabe der aus Wissenschaften auf der Königl. Sächs. Oeff. Bibliothek zu Dresden befindlichen Schriften. Ein Beitrag zur sittengeschichtlichen Literatur.* Leipzig: Engelmann, 1843. 176 pp.

Some three thousand titles, arranged in twenty-four sections. No annotations, but there are subject and author indexes. Pp. 118–42: Contents of: Eberhard David Hauber, *Bibliotheca Acta et Scripta Magica* (1739–45); Georg Conrad Horst, *Zauberbibliothek* (1821).

603. *Bibliotheca Magica et Pneumatica. Geheime Wissenschaften, Sciences Occultes, Occult Sciences, Folklore. Kataloge 31–35.* Munich: Jacques Rosenthal [1895?]. 680 pp.

Bookseller's catalogue. 8,875 titles classified in 47 sections with occasional annotations. Topics include: Alchemy and Rosicrucians; prophesies; astrology, comets; witchcraft; Index of Prohibited Books; curiosa, jocosa, satirica; wives, love, marriage; gastronomy, cookery; hunting, fowling, fishing; agriculture.

604. Gardner, F. Leigh. *A Catalogue Raisonné of Works on the Occult Sciences. Vol. 1: Rosicrucian Books.* With an introduction by Dr. William Wynn Westcott. London: privately printed, 1903. 82 pp.

List of 604 titles, arranged alphabetically by author, with occasional annotations.

605. Gardner, F. Leigh. *A Catalogue Raisonné of Works on the Occult Sciences. Vol. 2: Astrological Books.* With a sketch of the history of astrology by Dr. William Wynn Westcott (Supreme Magus of the Rosicrucians of England). London: privately printed, 1911. 164 pp.

List of 1,340 titles (of which some eight hundred are pre-1700), with occasional annotations.

606. Caillet, Albert L. *Manuel bibliographique des sciences psychiques ou occultes. Sciences des mages, Hermétique, Astrologie, Kabbale, Franc-Maçonnerie, Médecine ancienne, Mesmérisme, Sorcellerie, Singularités, Aberrations de tout ordre, Curiosités. Sources bibliographiques et documentaires sur ces sujets.* Paris: Lucien Dorbon, 1912. 3 vols.

11,648 titles partly annotated and arranged alphabetically by author. Pp. x–lxvii: Subject index.

607. Robbins, Rossell Hope. *The Encyclopedia of Witchcraft and Demonology.* Ninth printing. New York: Crown Publishers, 1972. 571 pp. [First published 1959]

The work consists of a series of articles (from a few lines to several pages) on the history of the witch-craze "from its beginnings in the fifteenth century, through its peak about 1600, to its ending in the eighteenth century." It attempts to show what witchcraft meant to the men and women accused as witches and to the judges who condemned them. Witchcraft is treated here *not* as "a department of anthropology, folklore, mythology or legend" but as a form of religion, a Christian heresy. To the author, it was "an intellectual aberration devised by inquisitors with the exceptional powers of torture and confiscation, and soon taken over and shared by civil authorities." Pp. 558–71: Select bibliography of 1,140 titles (with subject index, pp. 558–61). Reviews: *Renaissance News* 13 (1960), 178–80; *Modern Philology* 59 (1961), 128–30; *American Reference Books Annual* 11 (1980), 672.

608. Midelfort, H. C. E. "Recent Witch Hunting Research, or, Where Do We Go from Here?" *Papers, Bibliographical Society of America* 62 (1968), 373–420.

Classified bibliography of the period 1940–1967 (509 items) with index of authors. Includes a selective list of useful books (17) published before 1940, and a 12-page introductory note. No annotations.

609. Yale University Library. *Alchemy and the Occult. A Catalogue of Books and Manuscripts from the Collection of Paul and Mary Mellon.* Compiled by Ian Macphail. New Haven, Conn.: Yale Univ. Press, 1968. 2 vols. 581 pp.

Richly illustrated catalogue of a collection of 160 books (of which 145 are pre-1700). Full bibliographical descriptions with notes on the author and contents of each book. Chronologically arranged, with name and title index. Reviews: *Ambix* 17 (1970), 58–59; *Renaissance Quarterly* 24 (1971), 241–42. The collection was inspired by the work of C. G. Jung and based on his collection of books and MSS. For an account of the collection, see the compiler's article in *Ambix* 14 (1967), 198–202.

610. Yale University Library. *Alchemy and the Occult. A Catalogue of Books and Manuscripts from the Collection of Paul and Mary Mellon Given to the Yale University Library.* Vol. 3: *Manuscripts 1225–1671.* Vol. 4: *Manuscripts 1675–1922.* Compiled by Laurence C. Witten II and Richard Pachella. With an introduction by Pearl Kibre and additional notes by William McGuire. New Haven, Conn.: Yale Univ. Library, 1977. 2 vols. xciv + 853 pp.; illus.

Catalogue of 149 MSS of which nos. 10–79 are of the period 1450–1700. Gives detailed listing of contents, with bibliographical references, photographic reproductions and summary. Vol. 4 contains an index of incipits and an index of names and subjects. Reviews: *Papers, Bibliographical Society of America* 73 (1979), 375; *Isis* 71 (1980), 333–34.

611. *Bibliotheca esoterica. Catalogue annoté et illustré de 6,707 ouvrages anciens et modernes qui traitent des sciences occultes (alchimie, astrologie, cartomancie, chiromancie, démonologie, grimoires, hypnotisme, kabbale, magie, magnétisme, médecine spagirique, mysticisme, prophétie, recettes et secrets, sorcellerie, spiritisme, théosophie, etc.) comme aussi des sociétés secrètes (franc-maçonnerie, rose-croix, templiers, compagnonnage, illuminés, hérésies, etc.)* Brueil-en-Vexin: Editions du Vexin Français, 1975. 656 pp.

Paris bookseller's catalogue (Librairie Dorbon-Ainé), arranged alphabetically by author. Detailed annotations, no subject index.

612. Crowe, Martha J., ed. *Witchcraft: Catalogue of the Witchcraft Collection in Cornell University Library.* Milkwood, N.Y.: KTO Press, 1977. 644 + 9 pp.

Catalogue of the research collection of historical and scholarly materials pertaining to the phenomenon of witchcraft in Europe and North America, "conceived and laboriously acquired by Cornell's first president." In general, alchemy, astrology, cabbala, magic, the occult, prophesies, and superstition are *not* included. An author/title list of some three thousand works with a separate subject index. Pp. ix–xxcviii: Introduction by Rossell Hope Robbins. Reviewed in *American Reference Books Annual* 9 (1978), 703.

613. Schuler, Robert M. *English Magical and Scientific Poems to 1700. An Annotated Bibliography.* New York: Garland Publishing, 1979. 120 pp.

Alphabetically arranged list of 624 poems (including pseudo-science and verse charms). Some pre-1500 material. Subject and name indexes. Reviewed in *Ambix* 27 (1980), 212.

614. Pritchard, Alan. *Alchemy: A Bibliography of English Language Writings.* London: Routledge and Kegan Paul, 1980. 439 pp.

Attempts a "comprehensive coverage of English language writings in all forms of material" up to 1975. 3,188 entries in three sequences: Alchemical texts (arranged by country); Works about alchemy (by country); Works about alchemy (by subject). Pp. 407–39: Name and subject index. Reviewers: *Isis* 72 (1981), 512; *Ambix* 28 (1981), 56–57; *Annals of Science* 39 (1982), 524–26.

615. *Bibliotheca magica: dalle opere a stampa della Biblioteca Casanatense di Roma (secc. XV–XVIII).* Florence: Leo S. Olschki Editore, 1985. 225 pp.

Alphabetical catalogue of 1,266 titles (1,150 pre-1700) with references to standard bibliographies. Indexes: secondary authors; illustrators; places of publication; chronological; subjects (19 groups). Reviewed in *Ambix* 33 (1986), 45.

616. Hogart, R. Charles, ed. *Alchemy. A Comprehensive Bibliography of the Manly P. Hall Collection of Books and Manuscripts, including Related Material on Rosicrucianism and the Writings of Jacob Boehme.* Los Angeles: The Philosophical Research Society, 1986. 314 pp.

Catalogue of 164 books and 245 MSS, copiously annotated. Includes: complete transcription of title; collation; notes on illustrations; description of binding; description of contents; discussion of text, and biographical note on author/translator.

617. Kies, Cosette N. *The Occult in the Western World. An Annotated Bibliography.* London: Mansell, 1986. 233 pp.

List of 890 books (in English, with a few exceptions), arranged in thirteen chapters (with subdivisions): General works on the occult; Traditional witchcraft and satanism; Modern witchcraft and satanism; Magic and the hermetic arts; Secret societies, exotic religions and mysticism; Psychics and psychical research; Ghosts, poltergeists, and hauntings; Primitive magic and beliefs; Myths, legends, and folklore; Ancient astronauts, disappearances, and UFOs; Prophesy and fortune-telling; Astrology; Skeptics and debunkers. Glossary; name and title indexes.

618. Verginelli, Vinci. *Bibliotheca hermetica: catalogo alquanto ragionato della raccolta Verginelli-Rota di antichi testi ermetici (secoli XV–VIII).* Florence: Nardini, 1986. 382 pp.

Catalogue of some four hundred books (of which 238 are pre-1700) and 57 MSS on alchemy collected by Vinci Verginelli and donated to the library of the Accademia Nazionale dei Lincei. Some entries are detailed, with references to bibliographies and other relevant works.

619. Kren, Claudia. *Alchemy in Europe. A Guide to Research.* Garland Reference Library of the Humanities, 692. New York: Garland Publishing, 1990. 144 pp.

Bibliography of 520 items—eds. of alchemical treatises and secondary materials (books, journal articles and essays)—some annotated. Classification is thematic and chronological. Items 1–33: Research aids (mostly bibliographies and catalogues). 34–94: General works. 95–253: Early and medieval alchemy. 254–317: Alchemy in the Renaissance. 318–401: Alchemy in early modern Europe. 402–36: Alchemy and the arts. 437–56: Alchemy and society. 457–88: Alchemy and Newton. Pp. 117–30: Index of authors and anonyma.

MONOGRAPHS AND ARTICLES

620. Notestein, Wallace. *A History of Witchcraft in England from 1558 to 1718.* Washington, D.C.: American Historical Association, 1911. 442 pp.

"It is quite impossible to grasp the social conditions, it is impossible to understand the opinions, fears, and hopes of the men and women who lived in Elizabethan and Stuart England, without some knowledge of the part played in that age by witchcraft" (p. 1). Pp. 345–83: Survey of pamphlet literature. Pp. 384–419: List of cases of witchcraft (1558–1718), with references to sources and literature.

Based on Ph.D. dissertation, Yale University, 1908. This work was awarded the Herbert Baxter Adam Prize in European History.

621. Evans, Joan. *Magical Jewels of the Middle Ages and the Renaissance, Particularly in England.* Oxford: Clarendon Press, 1922. 264 pp.

Study of the belief in the virtues of gems and jewels. Pp. 140–94 cover the sixteenth and seventeenth centuries.

622. Thorndike, Lynn. *A History of Magic and Experimental Science.* New York: Columbia Univ. Press, 1923–58. 8 vols.

Monumental study by a distinguished medievalist of the relations between magic and science from the time of Pliny the Elder to Newton. Vols. 4–8 cover the period 1450–1700. General indexes and indexes of MSS. Reviews: *Isis* 23 (1935), 471–75 [Vols. 3–4]; 33 (1942), 691–712 [Vols. 5–6]; 49 (1958), 453–55 [Vols. 7–8]. See item 626.

623. Prior, Moody E. "Joseph Glanvill, Witchcraft, and Seventeenth-Century Science." *Modern Philology* 30 (1932), 167–93.

Although the *Sadducismus Triumphatus* (1681) of Glanvill has been seen by others as a contradiction of the rest of his work, he thought it "fitted in quite harmoniously with his others." On Glanvill (1636–80, F.R.S.) see *DSB*.

624. Zilboorg, Gregory. *The Medical Man and the Witch in the Renaissance.* The Hideyo Noguchi Lectures. Publications, Institute of the History of Medicine, Johns Hopkins University, Ser. 3, Vol. 2. Baltimore, 1935. 215 pp.

Three essays on the problem of witchcraft by a psychiatrist. 1. "The Physiological and Psychological Aspects of the *Malleus Maleficarum* (1484), the 'Hammer of Witches.'" 2. "Medicine and the Witch in the Sixteenth Century." 3. "Johann Weyer (1515/16–1588), the Founder of Modern Psychiatry."

625. Howland, Arthur C., ed. *Materials toward a History of Witchcraft Collected by Henry Charles Lea.* With an introduction by George Lincoln Burr. Philadelphia: Univ. of Pennsylvania Press, 1939. 3 vols.; 1548 pp.

"Immense, ordered, scholarly compendium of notes and comments on a variety of sources," edited posthumously. Covers the three centuries ending in the Age of Reason. As the author died in 1902, many subjects now considered relevant to the theme are left untouched. Reviewed in *TLS* 8 Jan. 1960, p. 22 (1957 ed).

In three parts. Part 1: The older beliefs (the powers of evil; magic and sorcery; popular beliefs accepted by churchmen). Part 2: How the witch theory developed (assimilation of sorcery to heresy; witch trials to the middle of the sixteenth century; treatises on witchcraft to 1550; view of the Protestant reformers; mysticism and witchcraft). Part 3: The delusion at its height (its promoters and critics; witchcraft as viewed by the secular law; witchcraft literature of the Roman Inquisition; demoniacal posession; witchcraft by regions). Part 4: The decline of witchcraft (witchcraft and the philosophers; witchcraft and the moral theologians; The final controversies (skeptics and believers, witchcraft and disease, survivals into later times). Most of the books used are now in the Lea Library, University of Pennsylvania.

626. Durand, Dana B. "Magic and Experimental Science: The Achievement of Lynn Thorndike." *Isis* 33 (1942), 691–712.

Article includes a review of vols. 5–6 of Thorndike's *History* (item 622).

627. Walker, D. P. *Spiritual and Demonic Magic from Ficino to Campanella.* Studies of the Warburg Institute, 22. London, 1958. 244 pp.

Study of Ficino's magic and its influence through the sixteenth and early seventeenth centuries. Reviews: *Revue d'histoire des sciences* 12 (1959), 271–74; *Isis* 53 (1962), 254–55; *Renaissance News* 15 (1962), 299–303.

628. Teall, John L. "Witchcraft and Calvinism in Elizabethan England: Divine Power and Human Agency." *Journal of the History of Ideas* 23 (1962), 21–36.

Examines the thesis that "English Calvinism was especially responsible for inducing the hysteria [against witchcraft] in the latter part of the sixteenth century."

629. Wilkinson, Ronald Sterne. "George Starkey [1627–1665], Physician and Alchemist." *Ambix* 11 (1963), 121–52.

Starkey was an "early advocate of the experimental philosophy, a friend and associate of Robert Boyle, and the first American alchemist to receive wide attention in England."

630. Secret, F. *Les Kabbalistes chrétiens de la Renaissance.* Paris: Dunod, 1964. 372 pp.

On the Christian Cabbala and its influence on the history of ideas in Europe. Bibliographical notes to each chapter. Index of names. Reviewed in *Ambix* 13 (1965), 55–60.

631. Yates, Frances A. *Giordano Bruno and the Hermetic Tradition.* London: Routledge and Kegan Paul, 1964. 466 pp.

Includes a discussion of hermeticism and natural magic in Renaissance thought from the time of Ficino to the early seventeenth century. Reviewed in *Isis* 55 (1964), 389–91.

632. Debus, A. G. *The English Paracelsians.* London: Oldbourne, 1965. 222 pp.

Study of the Paracelsian influence on chemistry, medicine, and general natural philosophy in England prior to 1640. Reviews: *History of Science* 5 (1966), 100–4; *Science* 154 (1966), 758; *Ambix* 14 (1967), 63–64; *Journal of the History of Medicine* 22 (1967), 332.

633. Dumas, Francois Ribadeau. *Histoire de magie.* Paris: Les Productions, 1965. 621 pp.

Study of magic from antiquity to the present. Illustrated. No bibliography. Pp. 1–67: Prague in 1510, center of magic. Pp. 148–433: From Columbus to Louis XIV.

634. Deacon, Richard. *John Dee: Scientist, Geographer, Astrologer and Secret Agent to Elizabeth I.* London: Frederick Fuller, 1968. 309 pp.

Brief survey of Dee's career, based on MS sources. Reviewed in *TLS* 12 Sep. 1968, p. 1033.

635. Rossi, Paolo. *Francis Bacon: From Magic to Science.* London: Routledge and Kegan Paul, 1968. 280 pp. [Translation of the 1957 Italian ed.]

On the cultural environment that influenced, and was in turn influenced by Bacon's philosophy. Review of Italian ed. in *Isis* 68 (1977), 323–24.

636. Monter, E. William. *European Witchcraft.* New York: John Wiley and Sons, 1969. 177 pp.

Says that European witchcraft is a lost historical reality which has to be recaptured by a fresh effort of the imagination and that the literature on the subject seems to lie under a special curse that condemns its specialists to become slightly infected by the forms of lunacy they set out to describe. The author, a scholar-editor, has approached the subject through "a deft interweaving of primary sources and secondary analysis," with extracts from sixteen texts. Topics: Origins of organized witchcraft; The great witchcraft debate, 1560–1580; The apex of witchcraft trials, 1580–1650; The decline of witchcraft, 1650–1700; Recent perspectives on witchcraft. Pp. 173–77: Guide to further reading.

637. Yates, Frances A. *Theatre of the World.* Chicago: Chicago Univ. Press, 1969. 218 pp.

"Primarily centred on John Dee and Robert Fludd as representatives of Renaissance philosophy in England, with particular reference to the evidence in their works of the influence of the Renaissance revival of Vitruvius." Studies the English public theatre movement as "one of the products of the Vitruvian influences stemming from Dee."

638. Hutin, Serge. *Robert Fludd (1574–1637), alchimiste et philosophe rosicrucien.* Paris: Les Editions de l'Omnium Littéraire, 1971. 174 pp.

Study of: the Rosicrucian Brotherhood; the life, work, and religious philosophy of Fludd. With a bibliography of works by and about him.

639. Rowse, A. L. *The Elizabethan Renaissance: The Life of the Society.* London: Macmillan, 1971. 294 pp.

Pp. 227–72: Mentality and belief; witchcraft and astrology.

640. Thomas, Keith. *Religion and the Decline of Magic. Studies in Popular Beliefs in Sixteenth-and Seventeenth-Century England.* London: Weidenfeld and Nicolson, 1971. 716 pp.

Seeks to explain why some of the systems of belief, rightly disdained by intelligent persons today, were taken seriously by equally intelligent persons in the past. Covers: magic, astrology, witchcraft, and allied beliefs. With bibliographical notes. Pp. 669–716: Name and subject index. Reviewed in *Renaissance Quarterly* 26 (1973), 70–72.

641. French, Peter J. *John Dee: The World of an Elizabethan Magus.* London: Routledge and Kegan Paul, 1972. 243 pp.

Series of essays on various aspects of Dee's career, based on the author's Ph.D. thesis. Pp. 210–29: Bibliography of primary and secondary sources. Reviews: *American Historical Review* 78 (1973), 97; *TLS* 19 May 1972, p. 579.

642. Monter, E. William. "The Historiography of European Witchcraft: Progress and Prospects." *Journal of Interdisciplinary History* 2 (1972), 435–51.

Research note. Concludes by pointing out some of the more important gaps in our present knowledge.

643. Rowse, A. L. *The Elizabethan Renaissance: The Cultural Achievement.* London: Macmillan, 1972. 386 pp.

Attempts at giving a "representative and significant portrait of the Age and its activities." (Ch. 4: Architecture and sculpture. Ch. 7–9: Science and society; Nature and medicine; Mind and spirit.)

644. Shumaker, Wayne. *The Occult Sciences in the Renaissance. A Study in Intellectual Patterns.* Berkeley: Univ. of California Press, 1972. 284 pp.

In five chapters: astrology, witchcraft, white magic, alchemy, Hermes Trismegistus. Reviews: *History of Science* 13 (1975), 300–1; *English Language* 12 (1974), 137–40.

645. Yates, Frances A. *The Rosicrucian Enlightenment.* London: Routledge and Kegan Paul, 1972. 269 pp.

PTo

Historical study of the Rosicrucian manifestos and their impact on seventeenth-century Europe. Reviews: *History of Science* 11 (1973), 306–10; *British Journal for the History of Science* 6 (1973), 442–44.

646. Kors, Alan C. and Edward Peters, eds. *Witchcraft in Europe, 1100–1700: A Documentary History.* London: Dent, 1973. 382 pp.

Study illustrated by 44 texts. Pp. 193–377: The witch persecutions of the sixteenth century; Witchcraft in the seventeenth century; Scepticism, doubt, and disbelief in the sixteenth and seventeenth centuries.

647. Dobbs, B. J. Teeter. *The Foundations of Newton's Alchemy; "The Hunting of the Greene Lyon."* Cambridge: University Press, 1975. 300 pp.

Study of Newton's chemical interests and researches during the years 1660–90. Reviewed in *Isis* 68 (1977), 116–21. See also 650.

648. Anglo, Sydney, ed. *The Damned Art: Essays in the Literature of Witchcraft.* London: Routledge and Kegan Paul, 1977. 258 pp.

Collection of essays by different scholars on eleven texts, from the *Malleus Maleficarum* (1484) to *The Tryal of Witchcraft* (1705). Review: *TLS* 18 Nov. 1977, p. 1342.

649. Clulee, Nicholas H. "Astrology Magic, and Optics: Facets of John Dee's Early Natural Philosophy." *Renaissance Quarterly* 30 (1977), 632–80.

Investigates "the earliest stage of John Dee's natural philosophy, culminating in the publication of the *Propaedeumata aphoristica* in 1588, without regard for its connections with Dee's later works or any presumed general philosophic outlook."

650. Figala, Karin. "Newton as Alchemist." *History of Science* 15 (1977), 102–37.

Essay review of 647 above.

651. Westman, Robert S., and J. E. McGuire. *Hermeticism and the Scientific Revolution. Papers Read at a Clark Library Seminar, March 9, 1974.* [1. Magical Reform and Astronomical Reform: the Yates Thesis Reconsidered. 2. Neoplatonism and Active Principles: Newton and the *Corpus Hermeticum*]. Los Angeles: William Andrews Clark Memorial Library, Univ. of California, 1977. 150 pp.

On (1) Yates's *Giordano Bruno and the Hermetic Tradition* and (2) the relation of hermeticism to the thought of Newton. With bibliographical notes. For review, see 1618. 2 1616

652. Copenhaver, Brian P. *Symphorien Champier and the Reception of the Occultist Tradition in Renaissance France.* The Hague: Mouton, 1978. 368 pp.

Biography and "analysis of Champier's published thought on magic, astrology, alchemy, and related occultist beliefs and practices." Pp. 331–51: Bibliography. Reviews: *Isis* 71 (1980), 331–32; *History of European Ideas* 3 (1982), 126–28.

653. Hansen, Bert. "Science and Magic." *Science in the Middle Ages.* Edited by David C. Lindberg. Chicago: Univ. of Chicago Press, 1978, pp. 483–506.

Sets forth "the basic outlines of the magical world-view of the Middle Ages, comparing and contrasting magic with the contemporary natural philosophy, and briefly relating this to the interactions of magic and science in the sixteenth and seventeenth centuries." Pp. 498–506: Notes and references.

654. Godwin, Joscelyn. *Robert Fludd, Hermetic Philosopher and Surveyor of Two Worlds.* London: Thames and Hudson, 1979. 96 pp.

Collection of 121 illustrations from Fludd's works, annotated in detail with an introduction to his life and thought. Bibliography.

√have!

655. Yates, Frances A. *The Occult Philosophy in the Elizabethan Age.* London: Routledge and Kegan Paul, 1979. 217 pp.

Author's theme is that although we think of the Renaissance as an age of enlightenment, occult and mystical thoughts occupied a central place in the world of Renaissance man. This can be seen in the iconography of such significant figures as Spenser, Shakespeare, Dürer, Rembrandt, Pico della Mirandola, Marlowe, Chapman, and John Dee. Reviewed in *The Sunday Times*, 30 Dec. 1979, p. 44.

656. Easlea, Brian. *Witch Hunting, Magic and the New Philosophy: an Introduction to the Debates of the Scientific Revolution 1450–1750.* Brighton, Sussex: Harvester Press, 1980. 283 pp.

On the transition from the traditional world of magic to the modern world of science and technology. Reviewed in *Isis* 73 (1982), 602–3.

657. Nicholl, Charles. *The Chemical Theatre.* London: Routledge and Kegan Paul, 1980. 292 pp.

A study of alchemical symbols and themes in the works of Shakespeare and his contemporaries. Pp. 240–84: notes and bibliography. With illustrations.

658. Westfall, Richard S. "The Influence of Alchemy on Newton." *Science, Pseudo-science and Society.* Edited by Marsha P. Hanen, Margaret J. Osler, and Robert G. Weyant. The Calgary Institute for the Humanities. Ottawa: Wilfrid Laurier Univ. Press, 1980, pp. 145–69.

Discusses Newton's interest in alchemy and his turning away from it around 1700. "Like the tide Newton's concern with alchemy rose and fell, and like the tide it deposited some of its burden, which remained on the shore altering its contour. We can understand the phenomenon only if we follow it through the entire cycle" (p. 145).

659. Kubrin, David. "Newton's Inside Out! Magic, Class Struggle, and the Rise of Mechanism in the West." *The Analytic Spirit. Essays in the History of Science in Honor of Henry Guerlac.* Edited by Harry Woolf. Ithaca, N.Y.: Cornell Univ. Press, 1981, pp. 96–121.

On the role played by hermetic or magical ideas in Newton's natural philosophy and why his interest in such matters has been a well-kept secret until recent times.

660. Andreski, Stanislav. "The Syphilitic Shock. A New Explanation of the 'Great Witch Craze' of the Sixteenth and Seventeenth Centuries in the Light of Medicine and Psychiatry." *Encounter* 58 (1982), (5), 7–26.

In advancing his thesis the author (a sociologist) examines and rejects the previous explanations of the great witch-hunt. A revised and expanded version of this article can be found in the author's *Syphilis, Puritanism and Witch Hunts. Historical Explanations in the Light of Medicine and Psychoanalysis with a Forecast about AIDS* (London: Macmillan, 1989), pp. 1–84. (Reviewed in *TLS* 5 Oct. 1990, pp. 1051–52.)

661. Hutchinson, Keith. "What Happened to Occult Qualities in the Scientific Revolution?" *Isis* 73 (1982), 233–53.

Seeks to "re-evaluate current conceptions of the role of occult qualities in the Scientific Revolution."

662. Shumaker, Wayne. *Renaissance Curiosa. John Dee's Conversations with Angels. Girolamo Cardano's Horoscope of Christ. Johannes Trithemius and Cryptography. George Dalgarno's Universal Language.* Medieval and Renaissance Texts and Studies, Vol. 8.

Binghamton, N. Y.: Center for Medieval and Early Renaissance Studies, 1982. 208 pp.

Detailed discussion of four texts. The work is intended to complement the author's *The Occult Sciences in the Renaissance* (item 644). Translations or paraphrases of the Latin excerpts have been provided. Reviews: *Bibliothèque d'humanisme et renaissance* 47 (1985), 257–59; *Modern Philology* 82 (1985), 417–19.

663. Webster, Charles. *From Paracelsus to Newton: Magic and the Making of Modern Science.* The Eddington Memorial Lectures. Cambridge: Cambridge Univ. Press, 1982. 107 pp.

Argues that "there were remarkable elements of continuity sufficient to indicate an important degree of contiguity between the world views of the early sixteenth and late seventeenth centuries." Explores: prophesy, spiritual magic, and witchcraft. Reviewed in *Medical History* 27 (1983), 438 and *Annals of Science* 41 (1984), 310. See also 666.

664. Trevor-Roper, Hugh. "The European Witch-Craze of the Sixteenth and Seventeenth Centuries." *Religion, the Reformation and Social Change, and Other Essays.* Revised 3rd ed. London: Secker and Warburg, 1984, pp. 90–192.

Essay on the growth and decline of the witch-craze, the establishment of witchcraft as a doctrine, and the persecution of witches.

665. Vickers, Brian, ed. *Occult and Scientific Mentalities in the Renaissance.* Cambridge: Cambridge Univ. Press, 1984. 452 pp.

Inquires to what extent the occult "sciences" (alchemy, astrology, numerology, and natural magic) contributed to the Scientific Revolution. Includes studies of: Kepler, Bacon, Mersenne, Newton, Dee, Fludd, Cardano. Reviewed in *Isis* 76 (1985), 393–96. See also item 666.

666. Curry, P. "Revisions of Science and Magic." *History of Science* 23 (1985), 299–325.

Essay review of: *From Paracelsus to Newton*, by Charles Webster (item 663) and *Occult and Scientific Mentalities in the Renaissance*, edited by Brian Vickers (item 665).

667. Troncarelli, Fabio, ed. *La città dei segreti. Magia, astrologia e cultura esoterica a Roma (XV–XVIII).* Milan: Franco Angeli, 1985. 380 pp.

Twenty-six papers presented at a symposium, *Roma Ermetica*, held in Rome, 24–27 October 1983. "Questo volume a più voci si propone di gettare luce sull'universo oscuro di Roma ermetica, cercando di decifrarne i lineamenti poco noti o addiritura sconosciuti: l'altra Roma, contrapposta e complementare rispetto alla Roma trionfale ed. effimera di Cesare e di Pietro." (The papers presented here are intended to throw light on that dark world of hermetic Rome—by trying to unravel its little known, or altogether unknown, features—that other Rome, opposed and complementary to the triumphal and ephemeral Rome of Caesar and St Peter.) Pp. 367–80: Name index.

668. Van Lennep, Jacques. *Alchimie. Contributions à l'histoire de l'art alchimique.* Brussels: Crédit Communal de Belgique, 1985. 447 pp.

Illustrations (823) from MSS and printed books, arranged in six chapters, with commentary, notes and bibliography. Indexes: subjects; sources; names and titles.

669. Henry, John. "Occult Qualities and the Experimental Philosophy: Active Principles in Pre-Newtonian Matter Theory." *History of Science* 24 (1986), 335–81.

"Only by recognizing that Newton's use of occult qualities and active principles is the culmination of seventeenth-century English mechanical philosophy, will we fully appreciate the significance of thinkers like Charlton, Power, Boyle, Petty, and Hooke, to say nothing of lesser thinkers."

670. Meinel, Christoph, ed. *Die Alchemie in der europäischen Kultur- und Wissenschaftsgeschichte.* Wolfenbütteler Forschungen, 32. Wolfenbüttel: Herzog August Bibliothek, 1986. 356 pp.

Collection of seventeen papers presented at a symposium held in Wolfenbüttel, 2–5 April 1984. Seeks to reconcile various approaches to the historical study of alchemy. Reviewed in *Isis* 78 (1987), 482–83.

671. Suster, Gerald, ed. *John Dee: Essential Readings.* Wellinborough, Northants.: Crucible, 1986. 157 pp.

Seventeen extracts of which sixteen are from Dee's writings and letters. With introduction, prefatory notes, and select bibliography.

672. Levack, Brian P. *The Witch-Hunt of Early Modern Europe.* London: Longman, 1987. 267 pp.

"Attempts to explain why the great European witch-hunt took place in the period between 1450 and 1750, why it reached its peak in the late sixteenth and seventeenth centuries, why it was more severe in some countries than in others and why it came to an end." Pp. 238–56: Bibliographical note and bibliography. Reviewed in *Manuscripta* 32 (1988), 67–68 and *American Historical Review* 93 (1988), 1037–38.

673. Scarre, Geoffrey. *Witchcraft and Magic in Sixteenth and Seventeenth-Century Europe.* Studies in European History. Basingstoke, Hants: Macmillan Education, 1987. 75 pp.

Essay based on recent research and literature. With annotated bibliography (112 titles) and index.

674. Clulee, Nicholas H. *John Dee's Natural Philosophy. Between Science and Religion.* London: Routledge, 1988. 347 pp.

Dee's place and the role of the occult in sixteenth-century intellectual history. The result of intensive study of Dee's writings, the available biographical material, and careful consideration of his sources as reflected in his extensive library and, more importantly, numerous surviving annotated volumes from it. Pp. 302–35: Bibliography (includes list of Dee's extant writings). Reviewed in *Annals of Science* 47 (1990), 100–2.

675. Huffman, William H. *Robert Fludd and the End of the Renaissance.* London: Routledge, 1988. 252 pp.

Study of Fludd (1574–1637) and his Neoplatonist ideas against the background of the late Renaissance. Bibliography.

676. Klossowski de Rola, Stanislas. *The Golden Game. Alchemical Engravings of the Seventeenth Century.* With 533 illustrations. London: Thames and Hudson, 1988. 320 pp.

Engravings from thirty-eight alchemical works. With an introduction (pp. 8–22), bibliographical information, brief biographical notes, and commentaries on the illustrations. Select bibliography (thirty-two items) and name index.

677. Merkel, Ingrid, and Allen G. Debus, eds. *Hermeticism and the Renaissance. Intellectual History and the Occult in Early Modern Europe.* Washington, D.C.: Folger Shakespeare Library, 1988. 438 pp.

Twenty essays, based on papers read at a conference in March 1982, held at the Institute for Renaissance and Eighteenth Century Studies, Washington. In three parts: Background to the Renaissance (1–3); Magic, philosophy, and science (4–12); Literature and art (13–20).

678. Ruderman, David B. *Kabbalah, Magic, and Science: the Cultural Universe of a Sixteenth-Century Jewish Physician.* Cambridge, Mass.: Harvard Univ. Press, 1988. 232 pp.

Study of the intellectual and social world of Abraham ben Hananiah Yagel (1553–ca. 1623), an Italian Jewish physician, cabbalist, magician, and naturalist. Pp. 169–232: Notes and index.

679. Stevenson, David. *The Origins of Freemasonry: Scotland's Century, 1590–1710.* Cambridge: Cambridge Univ. Press, 1988. 246 pp.

The author argues that the real origins of freemasonry can be detected in early seventeenth-century Scotland, when "stonemasons first created a system of organised lodges, blending medieval mythology with late-Renaissance intellectual influences in a series of mysterious rituals, to be copied first in England and then throughout Europe." Pp. 236–46: Bibliographical note and index.

680. Clarke, Desmond M. *Occult Powers and Hypotheses. Cartesian Natural Philosophy under Louis XIV.* Oxford: Clarendon Press, 1989. 265 pp.

"Analysis of the concept of science developed by French disciples of Descartes in the period 1660–1700. Pp. 245–60: Bibliography.

681. Martin, Ruth. *Witchcraft and the Inquisition in Venice 1550–1650.* Oxford: Basil Blackwell, 1989. 282 pp.

Gives a vivid picture of popular and official beliefs in Counter-Reformation Italy and attempts to explain why that country escaped "the witch crazes which engulfed so much else of early modern Europe." Bibliography.

682. Roberts, Julian, and Andrew G. Watson, eds. *John Dee's Library Catalogue.* London: The Bibliographical Society, 1990. 253 pp.; plates.

Facsimile of the MS catalogue (1583) of Dee's library, with introduction and notes. Pp. 3–21: Biographical background to Dee's collecting from his undergraduate days at Cambridge up to 1583. Pp. 22–46: Physical features, contents, and use of the library. Pp. 47–73: Subsequent history of the library. Pp. 75–78: Annals of Dee's life. Pp. 155–87: List of books and MSS known to have formed part of the library but which do not appear in the catalogue or lists. Pp. 207–53: Indexes (authors, owners, subjects, MSS). Reviews: *Isis* 83 (1992), 533–34; *Renaissance Quarterly* 45 (1992), 871–73.

682A. Patai, Raphael. *The Jewish Alchemists. A History and Source Book.* Princeton, N.J.: Princeton Univ. Press, 1994. 617 pp.

A "prolegomenon" to the history of Jewish work in alchemy and the role of Jews in contributing to its theory and practice. Pp. 263–449 cover the fifteenth, sixteenth and seventeenth centuries.

Astrology

BIBLIOGRAPHIES

683. Gardner, F. Leigh. *A Catalogue Raisonné of Works on the Occult Sciences. Vol. 2. Astrological Books.* With a sketch of the history of astrology by Dr. William Wynn Westcott (Supreme Magus of the Rosicrucians of England). London, privately printed, 1911. 164 pp.

List of 1,340 titles, of which some eight hundred are pre-1700, with occasional annotations. Pp. ix–xx: The history of astrology.

684. Bosanquet, Eustace F. *English Printed Almanacks and Prognostications: A Bibliographical History to the Year 1600.* London: Bibliographical Society, 1917. 204 pp.

Two hundred items, annotated. Supplement: *The Library* (4), 8 (1928), 456–77.

685. Bosanquet, Eustace F. "Notes on Further Addenda to English Printed Almanacks and Prognostications to 1600." *The Library* (4), 18 (1938), 39–66.

See item 684.

686. Carmody, Francis J. *Arabic Astronomical and Astrological Sciences in Latin Translation. A Critical Bibliography.* Berkeley and Los Angeles: Univ. of California Press, 1956. 193 pp.

List of two thousand titles (texts of forty authors). Contains a brief description of each work with lists of eds. (mostly pre-1600) and MSS. Reviews: *Isis* 49 (1958), 457–58; *Speculum* 32 (1957), 339–41.

687. Houzeau, J. C., and A. Lancaster. "Astrologie." *Bibliographie générale de l'astronomie jusqu'en 1880.* London: Holland Press, 1964. Vol. 1, pp. 681–858.

2,319 entries, arranged under: antiquity; Arab and Oriental; Renaissance (before and after Gutenberg—569 items); conjunctions, eclipses and comets; medical astrology. See item 833.

688. Winder, Marianne. "A Bibliography of German Astrological Works Printed between 1465 and 1690 with Locations of Those Extant in London Libraries." *Annals of Science* 22 (1966), 191–220.

243 titles, arranged in four sections: German texts; Latin texts; Translations into German; Anti-astrological literature (42 items).

689. Capp, Bernard. *Astrology and the Popular Press: English Almanacs 1500–1800.* London: Faber and Faber, 1979. 452 pp.

General survey of the major interests and ideas of English almanac-makers. Includes chapters on: "The Development of the Almanac" (pp. 23–66); "Astrology, Science, and Medicine" (pp. 180–214). Pp. 293–340: Biographical notes. Pp. 347–86: Bibliography of English almanacs to 1700. Reviewed in *The Library* (6), 3 (1981), 253–54.

MONOGRAPHS AND ARTICLES

690. Percopo, E. "Luca Gaurico, ultimo degli astrologi. Notizie biografiche e bibliografiche." *Atti R. Accademia di Archeologia, Lettere, e Belle Arti* (Napoli), 17 (2), (1896), (1), 1–49.

Bio-bibliographical note on Gaurico (1476–1558), astrologer and mathematician. See also item 187.

691. Strauss, Heinz Artur. *Der astrologische Gedanke in der deutschen Vergangenheit mit 93 Abbildungen aus der altdeutschen Buchillustration.* Munich: Oldenbourg, 1926. 104 pp.

Introduction to astrological thought in late medieval and Renaissance Germany. Pp. 94–100: List of illustrations with explanatory notes. Pp. 101–2: Bibliography.

692. Bosanquet, Eustace F. "English Seventeenth-Century Almanacs. 1. Their History. 2. As Books of Reference. 3. As a Medium for Advertisement. 4. The MS Notes in Them by Contemporary Owners." *The Library* (4), 10 (1930), 361–97.

693. Camden, Carroll, Jr. "Elizabethan Almanacs and Prognostications." *The Library* (4), 12 (1931–32), 83–108, 194–207.

A study: almanacs and their contents; burlesques of almanacs; the series of 1583 (books published for the year 1583 in which there was predicted a conjunction of Saturn and Jupiter) and the 1588 predictions of calamities; comets and eclipses.

694. Camden, Carroll, Jr. "Astrology in Shakespeare's Day." *Isis* 19 (1933), 26–73.

"The sixteenth and seventeenth centuries were a golden age for astrology just as they were for English literature." Discusses the attitudes of the opponents and advocates of astrology. Includes quotations (and citations) of astrological passages from Shakespeare's works.

695. Allen, Don Cameron. *The Star-Crossed Renaissance: the Quarrel about Astrology and Its Influence in England.* Durham, N. C.: Duke Univ. Press, 1941. 280 pp.

Topics: Astrology in fifteenth-century Italy; Later Continental views; English astrologers and their opponents; Astrologers and English literature; Elizabethan and Jacobean satires on the almanac and prognostication; Some astrological physicians and their works. Bibliography. Reviewed in *Isis* 34 (1943), 377–78.

696. Parr, Johnstone. *Tamburlaine's Malady, and Other Essays on Astrology in Elizabethan Drama.* University, Ala.: Univ. of Alabama Press, 1953. 158 pp.

Study of literary allusions to astrology in Elizabethan drama in the light of Renaissance astrological doctrine. Pp. 112–50: Sources of the Renaissance Englishman's knowledge of astrology; a bibliographical survey and a bibliography. Reviewed in *Isis* 45 (1954), 398–99.

697. Parker, Derek. *Familiar to All. William Lilly [1602–1681] and Astrology in the Seventeenth Century.* London: Jonathan Cape, 1975. 272 pp.   last

Life and work of "the leading astrologer of the great age of astrology, when it was still possible for an astrologer to make a very good living indeed by practising his art (or science)."

698. Simon, Gérard. *Kepler, astronome, astrologue.* Bibliothèque des sciences humaines. Paris: Gallimard, 1979. 488 pp.

Points out the need to "understand both Kepler's astrology and his astronomy as integral and necessary parts of the same whole." Reviewed in *Isis* 71 (1980), 514–15.

699. Garin, Eugenio. *Astrology in the Renaissance. The Zodiac of Life.* London: Routledge and Kegan Paul, 1983. 145 pp. [Revised translation of the 1976 Italian ed.]

Text of four lectures given at the Collège de France in 1975. Discusses the role of astrology and astrological ideas during the years leading up to the Scientific Revolution. The emancipation of astronomy from astrology was a long and drawn-out process, and astrological themes played a crucial role in the thinking of Renaissance "astronomers." Reviews: *Annals of Science* 40 (1983), 675–76; *Renaissance Quarterly* 36 (1983), 577–80; *Speculum* 59 (1984), 650–51.

700. Eade, J. C. *The Forgotten Sky. A Guide to Astrology in English Literature.* Oxford: Clarendon Press, 1984. 230 pp.

In three parts. I. Astronomical background necessary for an understanding of the operations of astrology. Calendrical computation. II. Basic grammar of astrology. III. Representative selection of astrological and astronomical references from literary works (Chaucer to Laurence Sterne). Index of terms and subjects. Reviewed in *Bulletin of the Society for Renaissance Studies* 4 (1986), 67–68.

701. Field, J. V. "A Lutheran Astrologer: Johannes Kepler." *Archive for History of Exact Sciences* 31 (1984), 189–272.

A study of Kepler's astrology which had little in common with magic and witchcraft but belonged to the "high" astrology as taught in the *Quadrivium*. With a translation of Kepler's *De fundamentis astrologiae certioribus*.

702. Müller-Jahncke, Wolf Dieter. *Astrologisch-magische Theorie und Praxis in der Heilkunde der frühen Neuzeit.* Sudhoffs Archiv, Beiheft 25. Stuttgart: Steiner, 1985. 328 pp.

Study of the role of astrology and magic in the development of early modern medicine. "Excellent source book for the history of astromedicine of the fifteenth to seventeenth centuries." Reviews: *Gesnerus* 43 (1986), 157–58; *Ambix* 32 (1985), 153–54; *Medical History* 30 (1986), 110–11.

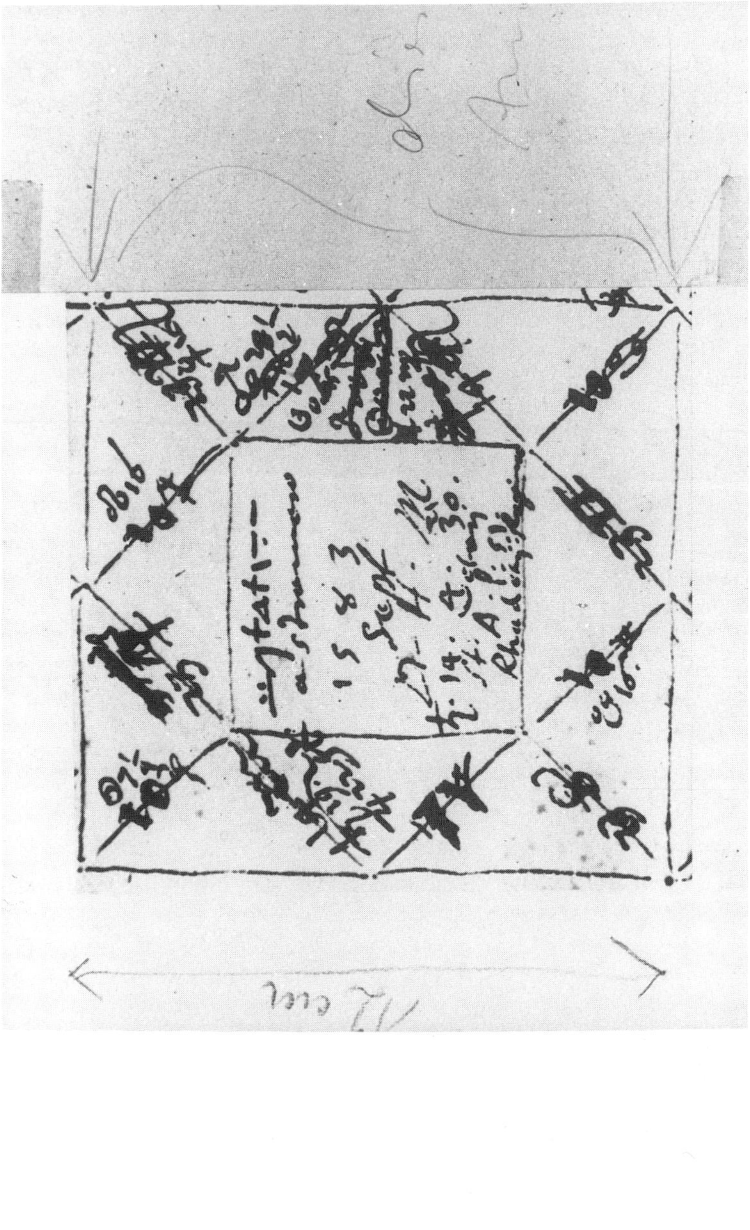

Horoscope cast by Kepler in 1608 for Albrecht Wenzel von Wallenstein (1583–1634). (Archives of the Kepler Commission, Bayerische Akademie der Wissenschaften). See item 858, pp. 338–45. *Entry 698.*

703. Zambelli, Paola, ed. *"Astrologi hallucinati": Stars and the End of the World in Luther's Time*. Berlin: Walter de Gruyter, 1986. 293 pp.

Papers read at a conference held in May 1984 in Berlin, on the role of astrological prevision—in particular, the prognostications of the 1524 flood. Reviewed in *British Journal for the History of Science* 20 (1987), 479–81.

704. Curry, Patrick, ed. *Astrology, Science and Society: Historical Essays*. Woodbridge, Suffolk: Boydell Press, 1987. 302 pp.

Based on a conference, held at the Warburg Institute, London, in March 1984, on the history of medieval and Renaissance astrology in Europe. Seven essays (pp. 95–300) relate to the Renaissance: Iconology of the Copernican revolution (Keith Hutchinson); Kepler's cosmology (J. V. Field); Metoposcopy (Angus G. Clarke); French astrological culture and its influence in England (Jacques Halbronn); Newton's comets (Simon Schaffer); Restoration England (P. Curry); An unpublished polemic by Flamsteed (M. Hunter). Bibliographies.

705. Knappich, Wilhelm. *Geschichte der Astrologie*. Zweite, ergänzte Auflage, mit einer Vorbemerkung zur Neuauflage und Ergänzung der Bibliographie von Bernwald Thiel. Frankfurt-am-Main: Klostermann, 1988. 400 pp. [First published 1967]

NOT SEEN. Pp. 185–291 of the 1st ed. relate to the period 1450–1700. With bibliography.

705A. Dunn, Richard. *Astrology in Harriot's Time*. The Durham Thomas Harriot Seminar, Occasional paper no. 14. Durham, 1993. 39 pp.

Looks at the status of astrology in Harriot's England. Argues that astrological concepts and practices were an important element of Elizabethan culture.

NB also:- S.J. Tester
    A History of Western Astrology
    [~ 1987]
    cf review by J.Henry TLS 27/11 – 3/12,

    – Geneva, Ann
    1995 book on Lilly!

Mathematics

BIBLIOGRAPHIES

706. Moore, Jonas. *Bibliotheca mathematica optimis libris diversarum linguarum refertissima: una cum variis philologicis historicis, et geographicis adornata: Honoratissimi Equitis Jonae Mori (Supervisoris Generalis Instrumentorum Bellicorum Regis Angliae) nuper defuncti.* London, 1684. 41 pp.

Sale catalogue of the library (some 1,600 items of which about one thousand are mathematical) of Jonas Moore (1617–79), mathematician and Surveyor-General of the Ordnance. His *A New Systeme of the Mathematicks* (1681) was one of the earliest English textbooks to present the Copernican hypothesis. See also item 826.

707. De Morgan, Augustus. *Arithmetical Books from the Invention of Printing to the Present Time Being Brief Notices of a Large Number of Works Drawn up from Actual Inspection.* London: Taylor and Walton, 1847. 124 pp.

Annotated bibliography with an introductory essay. Pp. 1–62: Works published from 1491 to 1700 (187 titles). 1966 reprint reviewed in *Annals of Science* 23 (1967), 311.

708. Riccardi, Pietro. *Biblioteca matematica italiana dalle origini della stampa ai primi anni del secolo XIX.* Milan: Società Tipografica Modenese, 1873–1928. [Reprinted 1952 by Görlich, Milan]. 7 pts.

Bibliography of some ten thousand books by Italians, on mathematics and its applications. Arranged in two sequences: (a) alphabetically by author, with annotations; (b) short-title catalogue, chronologically, under 21 main headings (262 subheadings). With list of some 800 works cited.

709. Bierens de Haan, David. *Bibliographie néerlandaise historique-scientifique des ouvrages importants dont les auteurs sont nés aux 16e,*

17e, et 18e siècles, sur les sciences mathématiques et physiques, avec leurs applications. Nieuwkoop: B. de Graaf, 1960. 424 pp. [First published in 1883 in *Bullettino di bibliografia e di storia delle scienze matematiche e fisiche*]

Bibliography of some 5,700 works published in the Netherlands, arranged alphabetically by author in one sequence. Pp. 334–82: subject index in 16 sections (116 subsections) including: mathematics (2–6), physics (7), chemistry (8), mechanics (9), architecture (10), astronomy (11), navigation (12), music (13), design (14), military science (15), philosophy (16). Pp. 383–414: Index of printers and publishers.

710. Riccardi, Pietro. "Saggio d'una bibliografia euclidea." *Memorie della Reale Accademia delle Scienze dell'Istituto di Bologna*, (4), 8 (1887), 401–523; 9 (1888), 321–43; (5), 1 (1890), 27–84; 3 (1892), 639–94. [Also published separately, Bologna 1887–1893. Reprinted by Georg Olms, Hildesheim in 1974]

Complete bibliography of Euclid's works (MS and printed) with indexes.

711. Smith, David Eugene. *Rara Arithmetica. A Catalogue of the Arithmetics Written Before the Year MDCI with a Description of Those in the Library of George Arthur Plimpton of New York.* Boston: Ginn, 1908. 507 pp.

Catalogue of some 550 books in the Plimpton Collection (now at the Columbia University Library, New York). With descriptions of their content and notes on other eds. In all some 1,200 titles are mentioned. Reviewed in *Bibliotheca Mathematica* (3), 9 (1909), 267–80.

712. Henderson, James. *Bibliotheca Tabularum Mathematicarum, Being a Descriptive Catalogue of Mathematical Tables. Part 1. Logarithmic Tables.* (A. *Logarithms of Numbers*). Cambridge: at the University Press, 1926. 208 pp.

Pp. 1–21: Introduction and historical account. Pp. 22–192: Bibliography (Naperian logarithms, Briggsian or common logarithms, antilogarithms, hyperbolic logarithms); Sixty pre-1700 titles with detailed annotations.

713. Thomas-Stanford, Charles. *Early Editions of Euclid's Elements.* Illustrated monographs, 20. London: Bibliographical Society, 1926. 64 pp. Reprinted in 1977 by Alan Wofsy Fine Arts, San Francisco.

Bibliography on 85 eds. (of which 38 are fragmentary) of the period 1482–1600. With an 18-page introductory note.

714. Kokomoor, F. W. "The Teaching of Elementary Geometry in the Seventeenth Century." *Isis* 10 (1928), 21–32.

A checklist of some 85 seventeenth-century manuals, with an introduction by L. C. Karpinski.

715. Rome, A. "Le R.P. Henri Bosmans, S.J. (1852–1928). Notice biographique et index analytique de ses travaux historiques." *Isis* 12 (1929), 88–112.

Pp. 94–110: Annotated bibliography of Bosmans' studies of the work of sixteenth- and seventeenth-century mathematicians including several Jesuit mathematicians and astronomers. See item 719.

716. Smith, David Eugene. "The Influence of the Mathematical Works of the Fifteenth Century upon Those of Later Times." *Papers, Bibliographical Society of America* 26 (1932), 143–71.

Brief bibliographical survey. Topics covered: arithmetic, algorism, abacus arithmetics [not *libri d'abaco*], commercial books, almanacs, astronomy, calendar reform, algebra, geometry.

717. Karpinski, Louis C. "Algebraical Works to 1700." *Scripta Mathematica* 10 (1944), 149–69.

Bibliography of sixteenth- and seventeenth-century published texts based on the collections at Michigan and Columbia University libraries.

718. Karpinski, L. C. "Bibliographical Checklist of All Works on Trigonometry Published up to 1700 A.D." *Scripta Mathematica* 12 (1946), 263–83; 18 (1952), 94.

A short-title list of some 320 works by 180 authors. With an additional list of four titles by A. W. Richeson.

719. Bernard-Maitre, H. "Un historien des mathématiques en Europe et en Chine: Le Père Henri Bosmans, S. J. (1852–1928)." *Archives internationales d'histoire des sciences* 3 (1950), 619–56.

Pp. 629–56: Bibliography of 241 articles and 278 book reviews (with index), mostly relating to sixteenth- and seventeenth-century mathematics. An annotated bibliography of 101 articles appeared in *Isis* 12 (1929), 88–112. See item 715.

720. Lacoarret, Marie. "Les traductions françaises des oeuvres d'Euclide." *Revue d'histoire des sciences* 10 (1957), 38–58.

List of French translations of the works of Euclid (forty-three titles, of which twenty-seven were published before 1700) with introductory note and biographical sketches of the translators.

721. Deubner, Fritz, and Hildegard Deubner. "Adam Ries, der Rechenmeister des deutschen Volkes." *NTM* 1 (1960), (3), 11–44; 7 (1970), (1), 1–22; 7 (1970), (2), 99–114; 8 (1971), (1), 58–69.

Complete bio-bibliography of Ries from the sixteenth century to 1970. Author and subject index. Most famous of all German *Rechenmeister* (teachers of practical arithmetic), Ries (1492–1559) wrote several books on arithmetic. His *Rechnung auff der linihen und federn* (1522) went through more than 108 eds. by 1656. For biography see *DSB*.

722. Smeur, A. J. E. M. *Die zestiende-eeuwse Nederlandse Rekenboeken.* [The Sixteenth-Century Arithmetics Printed in the Netherlands.] The Hague: Tijhoff, 1960. 167 pp.

Bibliography of some one hundred books with short descriptions of their contents and biographical notes on the authors (Part 1) and a systematic discussion of the subject matter (Part 2). Supplement in *Scientiarum Historia* 8 (1966), 153–59.

723. Simpkins, Diana M. "Early Editions of Euclid in England." *Annals of Science* 22 (1966), 225–49.

Survey of English eds. up to the end of the seventeenth century, by which time "Euclid was established as part of the liberal education of a gentleman, and as a basic discipline for navigators and military scientists." With bio-bibliographical notes on the editors (Recorde, Candish, Dee, Billingsley, Briggs, Fournier, Barrow, Mercator, De Chales).

724. May, Kenneth O. *Bibliography and Research Manual of the History of Mathematics.* Toronto: Univ. of Toronto Press, 1973. 818 pp.

A subject index to the secondary literature of the history of mathematics. Covers the period 1868–1965. Arranged alphabetically in five sequences: Biographies (from Abakanowicz to Zuse); Mathematical topics (from Addition to Zorn lemma); Epi-mathematical topics [i.e., external aspects of mathematics] (from Abacus to Zeno paradoxes); Historical classifications (time periods, countries, cities, organizations); Information retrieval (bibliographies, libraries, historiography, museums, etc.). Reviewed in *Isis* 65 (1974), 524–26 and *Historia Mathematica* 1 (1974), 192–94. The guides of George Sarton, *The Study of the History of Mathematics* (Cambridge: Harvard, 1936) and Gino Loria, *Guida allo studio della storia delle matematiche* (Milan, 1946) are now outdated (for details see item 20, p. 7). The journal *Historia Mathematica* (1974–) contains a section devoted to abstracts of current literature in the history of mathematics.

725. Van Egmond, Warren. *Practical Mathematics in the Italian Renaissance. A Catalog of Italian Abbacus Manuscripts and Printed Books to 1600.* Florence: L'Istituto e Museo di Storia delle Scienze, 1980. 442 pp.

Catalogue of some 450 *libri d'abaco*—manuscript and printed—of which about 350 relate to the period 1450 to 1600. Reviewed in *Isis* 74 (1983), 285.

726. Steck, Max. *Bibliographia Euclideana. Die Geisteslinien der Tradition in den Editionen der "Elemente" (ΣΤΟΙΧΕΙΑ) des Euklid (um 365–300). Handschriften—Inkunabeln—Frühdrucke (16. Jahrhundert). Textkritische Editionen des 17.–20. Jahrhunderts. Editionen der Opera minora (16.–20. Jahrhundert).* Mit einem wissenschaftlichen Nachbericht und mit faksimilierten Titelblättern, hauptsächlich der Erstausgaben und wichtiger Editionen. Hrsg. von Menso Folkerts. Arbor scientiarum, Reihe C, Vol. 1. Hildesheim: Gerstenberg Verlag, 1981. 444 pp.

Editions from 1482 to 1700 are given on pp. 39–112 and 218–371 (title pages). Reviewed in *Isis* 74 (1983), 283–84.

727. Rider, Robin E. *A Bibliography of Early Modern Algebra, 1500–1800.* Berkeley Papers in the History of Science 7. Berkeley: Univ. of California, Office for History of Science and Technology, 1982. 171 pp.

Chronological list of primary sources. No annotations. Pp. 19–63: publications from 1494 to 1700. Reviewed in *Annals of Science* 41 (1984), 296.

728. Jayawardene, S. A. "Mathematical Sciences." *Information Sources in the History of Science and Medicine.* Editors, Pietro Corsi and Paul Weindling. London: Butterworth Scientific, 1983, pp. 259–84.

Essay on the historiography of post-medieval mathematics. Select bibliography.

729. Dauben, Joseph W., ed. *The History of Mathematics from Antiquity to the Present. A Selective Bibliography.* Bibliographies of the History of Science and Technology, Vol. 6. New York: Garland, 1985. 467 pp.

Annotated bibliography compiled by forty-nine specialists. Classification is chronological, ethnographical, and systematic (2,384 entries). Author and subject indexes. Pp. 125–61: Renaissance and seventeenth-century mathematics (includes studies of individual mathematicians); Pp. 216–387: Sub-disciplines (include: algebra; geometry; logic; mathematical physics;

mathematics and navigation; numbers and number theory; numerical analysis; mathematical optics; potential theory; probability and statistics). Arrangement is alphabetical, by author, within each category. For a comprehensive bibliography of studies by Italian authors during the period 1961–1990, see Francesco Barbieri and Luigi Pepe, "Bibliografia italiana di storia delle matematiche, 1961–1990," *Bollettino di storia delle scienze matematiche* 12 (1), (1992), 1–181. (With chronological classification and index of names; some 800 entries for the Renaissance and seventeenth century.)

MONOGRAPHS AND ARTICLES

730. Cossali, Pietro. *Origine, trasporto in Italia, primi progressi in essa dell'algebra; storia critica di nuove disquisizioni analitiche e metafisiche arrichita.* Parma: Reale Tipografia, 1797–1799. 2 vols.

History of algebra in Italy, from Leonardo of Pisa to Bombelli. Vol. 2, pp. 96–492 deal with sixteenth-century developments (detailed analysis of the work of Cardano). Still useful. Needs to be updated in the light of recent studies, especially those of the *trattati d'abaco.*

731. Hutton, Charles. *Tracts on Mathematical and Physical Subjects.* London: F. C. and J. Rivington, 1812. 3 vols.

Collection of thirty-eight tracts. Vol. l, pp. 278–454: History of trigonometrical tables; History of logarithms; The construction of logarithms (tracts 19, 20, 21). Vol. 2, pp. 143–305: History of algebra (tract 33); pp. 201–302 cover the period from Pacioli to Ozanam.

732. Todhunter, I. *A History of the Theory of Probability from the Time of Pascal to That of Laplace.* Cambridge: Macmillan, 1865. 624 pp.

Pp. 1–77: Survey of literature—from a 1477 commentary on Dante's *Purgatorio* to James Bernoulli's *Ars Conjectandi* (1713).

733. Bierens de Haan, D. "Notice sur des tables logarithmiques hollandaises." *Bullettino di bibliografia e di storia delle scienze matematiche e fisiche* 6 (1873), 203–38.

On the work of Adriaan Vlacq (1600–ca. 1667) and his contemporaries on logarithms.

734. Bierens de Haan, D. "Notice sur quelques quadrateurs du cercle dans les Pays-Bas." *Bullettino di bibliografia e di storia delle scienze matematiche e fisiche* 7 (1874), 99–140.

Material for a history of squaring the circle during the sixteenth and seventeenth centuries.

735. Treutlein, P. "Das Rechnen im 16. Jahrhundert." *Abhandlungen zur Geschichte der Mathematik* 1 (1877), 1–100.

Survey of sixteenth-century arithmetic, based on the texts of Pacioli, De la Roche, Stevin, and the major German *Rechenmeister*.

736. Treutlein, P. "Die deutsche Coss." *Abhandlungen zur Geschichte der Mathematik* 2 (1879), 1–124.

Survey of the German contribution to Renaissance algebra (cossic terms and notation, the cossic algorithm, irrational quantities, solving equations, cossic texts). Needs to be updated with later studies of primary sources: see bibliography in Tropfke's *Geschichte der Elementarmathematik*, 4th ed., Vol. 1 (Berlin: De Gruyter, 1980).

737. Unger, Friedrich. *Die Methodik der praktischen Arithmetik in historischer Entwicklung, vom Ausgange des Mittelalters bis auf die Gegenwart nach den Originalquellen bearbeitet.* Leipzig: Teubner, 1888. 240 pp.

Pp. 1–111: Fifteenth and sixteenth centuries (Teaching of arithmetic in the West; Textbook writers from Peuerbach to Clavius and Stevin; The content of arithmetical works). Pp. 112–36: The seventeenth century. Reviewed in *Zeitschrift für Mathematik und Physik, hist.-lit. Abt.*, 34 (1889), 70–73.

738. Ball, W. W. Rouse. *A History of the Study of Mathematics at Cambridge.* Cambridge: University Press, 1889. 264 pp.

Pp. 9–83: Mathematics from John Hodgkins to Isaac Newton. Also relevant are Chapters 8 (The organization and subjects of education) and 11 (Outline of the history of the university). Reviewed in *Bibliotheca Mathematica*, N.F., 4 (1890), 25–26.

739. Ritter, Frederic. "François Viète, inventeur de l'algèbre moderne, 1540–1603. Essai sur sa vie et son oeuvre." *Revue occidentale philosophique, sociale et politique*, (2), 10 (1895), 234–74, 354–415.

Includes an analysis of Viète's algebraic works.

740. Cantor, Moritz. *Vorlesungen über Geschichte der Mathematik.* 2nd ed. Vols. 2–3. Leipzig: Teubner, 1900–1908.

Monumental survey of the history of mathematics, based on lectures given at the University of Heidelberg. Still useful though outdated. Contains extensive studies of primary sources. Vol. 2, pp. 197–844 and Vol. 3, pp. 1–251 cover the period 1450–1700. A series of additions and corrections was issued in *Bibliotheca Mathematica*, ser. 3, vols. 1–14. Reviewed in *Bibliotheca Mathematica* (N.F.), 13 (1899), 49–57; (3), 1 (1900), 276–78, 518–19.

741. Braunmühl, Anton von. *Vorlesungen über Geschichte der Trigonometrie.* Leipzig: Teubner, 1900–1903. 2 vols. [Reprinted 1971 by Martin Sändig, Wiesbaden]

Vols. 1, pp. 118–248 and 2, pp. 1–68 cover the period from Regiomontanus to Jakob Bernoulli. Each volume has a name and subject index. The work is based on lectures given at the Technische Hochschule in Munich. Reviewed in *Bibliotheca Mathematica* (3), 1 (1900), 280–84; 5 (1904), 74–78 and *Revue des questions scientifiques* (2), 19 (1901), 294–301; (3), 4 (1903), 277–80.

742. Grosse, Hugo. *Historische Rechenbücher des 16. und 17. Jahrhunderts und die Entwicklung ihrer Grundgedanken bis zur Neuzeit. Ein Beitrag zur Geschichte der Methodik des Rechenunterichts mit 5 Titelabbildungen.* Wiesbaden: Dr Martin Sändig oHG, 1965. 183 pp. [Reprint of the 1901 ed.]

Study of the teaching of arithmetic in German schools of the sixteenth and seventeenth centuries. The syllabus in different regions. Biographical notes on some *Rechenmeister* (Suevus, Adam Cüraus, Lucas Lossius, Matthäus Nefe, Joan Segherwitz, Caspar Jeigler, Georg Meichsner, Neudörffer, Schwenter, Hemeling, Horchen), analysis of their texts with examples of problems. Evaluation of the ethical and educational value of the arithmetic lesson. Reviewed in *Zeitschrift für mathematischen Unterricht* 33 (1902), 69–70.

743. Zeuthen, H. G. *Geschichte der Mathematik im XVI. und XVII. Jahrhundert.* Abhandlungen zur Geschichte der mathematischen Wissenschaften, 17. Leipzig: Teubner, 1903. 434 pp. [Reprinted 1966 by Johnson Reprint Corp., New York]

Useful survey though outdated. Pp. 1–80: Historical and biographical summary. Pp. 81–233: Finite analysis (from third- and fourth-degree equations to Descartes). Pp. 234–426: Origin and development of infinitesimal analysis. Reviewed in *Bibliotheca Mathematica*, (3), 5 (1904), 211–20.

744. Jackson, Lambert Lincoln. *The Educational Significance of Sixteenth Century Arithmetic. From the Point of View of the Present Time.* Teachers College, Columbia University, Contributions to Education, no. 8. New York, 1906. 232 pp.

Study based on some sixty texts published in Europe (from Barcelona to Vienna, from Rome to London) during the period 1478 to 1600. Ch. 1 (pp. 23–184): The essential features; an analysis of the texts, showing they were written by the leading scholars and teachers of the time, who were influenced by the "demands of the commercial and industrial world, the educational ideals and practices, the restraints of traditional customs, and the demands of science." Ch. 2 (pp. 185–228): The educational significance of the works from the point of view of the present time.

745. Knott, Cargill Gilson, ed. *Napier Tercentenary Memorial Volume.* London: Longmans, Green and Co., for the Royal Society of Edinburgh, 1915. 441 pp.

Addresses and essays communicated to the International Congress held to commemorate the tercentenary of the publication of Napier's *Mirifici logarithmorum canonis descriptio.* Pp. 1–242 relate to the life and work of Napier (includes a bibliography of books exhibited).

746. Glaisher, J. W. L. "On the Early History of the Signs + and − and on the Early German Arithmeticians." *Messenger of Mathematics* 51 (1921–22), 1–148.

Deals with the history of the signs (+ and −) and other questions relating to the arithmetic and algebra of the fifteenth and sixteenth centuries. Includes a study of the algebras of Ries, Rudolff and Stiefel, the German and Latin algebras in the Dresden Codex 80, and Widman's *Rechnung.*

747. Rey Pastor, J. *Los matemáticos españoles del siglo XVI.* Toledo: Imp. de A. Medina, 1926. 164 pp.

Brief notices on: Pedro Sanchez Ciruelo; Juan Martinez Siliceo; Fr. Juan de Ortega; Alvaro Tomas; Marco Aurel; El Bachiller Pérez de Moya; Antich Rocha; Pedro Nuñez; Molina Cano; Jaime Falcó. Pp. 155–62: Bibliography (40 titles). Review of 1913 ed. in *Bibliotheca Mathematica* (3), 14 (1913–14), 85–90.

748. Cajori, Florian. *A History of Mathematical Notations.* Chicago: Open Court Publishing Co., 1928–29. 2 vols.

Vol. 1, pp. 100–229: Symbols in the arithmetic and algebraical works of the sixteenth and seventeenth centuries. Vol. 2 deals with the history of notations in higher mathematics (arranged by topic).

749. Kokomoor, F.W. "The Distinctive Features of Seventeenth Century Geometry." *Isis* 10 (1928), 367–415.

Points out that the general revival of learning in the seventeenth century was followed by a renewal of interest in geometry. Includes an analysis of the contents of the books listed in item 714.

750. Johnson, Francis R., and Sanford V. Larkey. "Robert Recorde's Mathematical Teaching and the Anti-Aristotelian Movement." *Huntington Library Bulletin* 7 (1935), 59–87.

Recorde was a pioneer in the teaching of mathematics in the English language. The size and superiority of the school of practical mathematics in England can be attributed to the movement initiated by him.

751. Yeldham, Florence A. *The Teaching of Arithmetic through Four Hundred Years (1535–1935)*. London: Harrap, 1936. 143 pp.

Study of the history of teaching of arithmetic in England, based on textbooks in the library of the Mathematical Association at Reading University and in Dulwich College. Pp. 19–94 cover the period from Cuthbert Tunstall's *De arte supputandi* (1522) to John Ward's *Young Mathematician's Guide* (1707). Reviewed in *Isis* 27 (1937), 92–94.

752. Boyer, Carl B. *The Concepts of the Calculus: A Critical and Historical Discussion of the Derivative and the Integral*. New York: Columbia Univ. Press, 1939. 346 pp.

An outline of the development of the basic concepts. Pp. 96–223 deal with the sixteenth and seventeenth centuries. Reviewed in *Isis* 32 (1940), 205–10.

753. Coolidge, J. B. *A History of Geometrical Methods*. Oxford: Oxford Univ. Press, 1940. 451 pp.

Includes surveys of the contribution of Desargues and Pascal (projective geometry) and of Fermat, Descartes, Wallis, Barrow and Newton (algebraic geometry). Bibliography.

754. Zeller, Mary Claudia. "The Development of Trigonometry from Regiomontanus to Pitiscus." A dissertation for the degree of Doctor of Philosophy in the University of Michigan, 1944. Ann Arbor, Mich.: Edwards Bros., 1946. 119 pp.

Covers the development from Regiomontanus to Magini (1561–1617). With bibliography and synoptic table.

755. Bortolotti, Ettore. *La storia della matematica nella Università di Bologna*. Bologna: Zanichelli, 1947. 226 pp.

Pp. 35–144 give a general account of the history of mathematics in Italy during the sixteenth and seventeenth centuries. The study is not confined to the University of Bologna.

756. Sergescu, Pierre. "La littérature mathématique dans la première période (1665–1701) du 'Journal des Savants.'" *Archives internationales d'histoire des sciences* 1 (1947), 60–99.

A survey of mathematical reviews published in the first scientific journal.

757. Scott, J. F. *The Scientific Work of René Descartes* (1596–1650). London: Taylor and Francis, 1952. 211 pp.

Pp. 84–157 contain an analysis of *La géométrie*. Appendix contains biographical notes on some (40) of the less important persons mentioned in the text. Reviewed in *Isis* 44 (1953), 285–86.

758. Hofmann, J. E. *Geschichte der Mathematik.* Sammlung Göschen 226, 875, 882. Berlin: De Gruyter, 1953–57. 3 vols.

Bibliographically rich *aide-mémoire.* Vol. 1, pp. 89–155, Vol. 2, and Vol. 3, pp. 1–24 deal with the period 1450–1700. Reviewed in *Isis* 46 (1955), 57–59 and 49 (1958), 350–52.

759. Boyer, Carl B. *History of Analytic Geometry.* New York: Scripta Mathematica, 1956. 296 pp.

Pp. 54–191 cover the years from Pacioli and Chuquet to Euler. Documented with bibliography and index. Review: *Isis* 50 (1959), 489–92.

760. Russo, François. "La constitution de l'algèbre au XVIe siècle: étude de la structure d'une évolution." *Revue d'histoire des sciences* 12 (1959), 193–208.

Analyses the development of algebra from Pacioli to Descartes (notations; symbols for unknown and known quantities; the algebraic method).

761. Whiteside, D. T. "Patterns of Mathematical Thought in the Later Seventeenth Century." *Archive for History of Exact Sciences* 1 (1960–62), 179–388.

Contains an analysis of the concepts of function, geometry, and calculus. Emphasizes the work of British mathematicians. Select bibliography of primary sources.

762. David, F. N. *Games, Gods and Gambling: the Origins and History of Probability and Statistical Ideas from the Earliest Times to the Newtonian Era.* London: Charles Griffin and Co., 1962. 275 pp.

Pp. 36–178: From Luca Pacioli to Abraham de Moivre.

763. Fladt, Kuno. *Geschichte und Theorie der Kegelschnitte und der Flächen zweiten Grades.* Stuttgart: Elett, 1965. 374 pp.

Pp. 5–83 cover the history of conic sections and second-degree surfaces up to Euler. Review: *Janus* 53 (1966), 151.

764. Naux, C. *Histoire des logarithmes de Napier à Euler.* Paris: Blanchard, 1966–71. 2 vols.

Surveys the history from the beginnings (1614) to the middle of the eighteenth century. Reviewed in *Revue d'histoire des sciences* 20 (1967), 302; 26 (1973), 267–68.

765. Hofmann, J. E. "Leibniz als Vollender der Renaissance-Mathematik." *Sudhoffs Archiv* 50 (1966), 375–91.

Traces Leibniz's development as a mathematician and sees in him the culmination of Renaissance mathematics.

766. Scriba, Christoph J. *Studien zur Mathematik des John Wallis (1616–1703). Winkelteilungen, Kombinationslehre und zahlentheoretische Probleme. Im Anhang die Bücher und Handschriften von Wallis.* Wiesbaden: Steiner, 1966. 144 pp.

Wallis was Savilian Professor of Geometry (1649–1703) at Oxford and a founder member of the Royal Society. The study includes (a) an analysis of unpublished papers on number theory and (b) a list of MSS and books owned by him (now in the Bodleian Library).

767. Whiteside, D. T., ed. *The Mathematical Papers of Isaac Newton.* Cambridge: Cambridge Univ. Press, 1967–81. 8 vols.

This work is included, despite its being a primary source, because of its exceptional character. It is an edition of the extant corpus of Newton's mathematical writings, assembled in twenty groups and arranged chronologically, with historical introductions (some 500 pages), technical footnotes and English paraphrases. Each volume has an analytical table of contents and an index of names. Separate volume of cumulative indexes is in preparation. "Whiteside has built [with an extraordinary level of expertise joined to an extraordinary level of devotion] what every true scholar dreams of constructing, a monument which will endure through the centuries as part of mankind's precious legacy of learning." (Westfall in *Nature*, 1982, *295*: 265–66.)

768. Hofmann, Joseph E. *Michael Stifel, 1487?–1567. Leben, Wirken und Bedeutung für die Mathematik seiner Zeit.* Sudhoffs Archiv, Beihefte 9. Wiesbaden: Steiner, 1968. 42 pp.

Contains an analysis of the *Arithmetica integra* (1544). Stifel is considered to be the greatest German algebraist of the sixteenth century. It was he who prepared the revised and enlarged ed. (1554) of Chr. Rudolff's *Coss* (1525), the first comprehensive work on algebra in German.

769. Klein, Jacob. *Greek Mathematical Thought and the Origin of Algebra.* With an Appendix containing Vièta's *Introduction to the Analyti-*

cal Art. Cambridge, Mass.: MIT Press, 1968. 360 pp. [Translated from the German original of 1934–36]

Pp. 150–85: The formalism of Viète and the transformation of the *arithmos* concept. Pp. 186–224: The concept of "number" in Stevin, Descartes, and Wallis. Reviewed in *Isis* 61 (1970), 132–33.

770. Baron, Margaret. *Origins of the Infinitesimal Calculus.* Oxford: Pergamon, 1969. 304 pp.

Study of the "transition from the geometrical method of exhaustion to the algorithms of the calculus" during the seventeenth century. Pp. 108–252: Pre-Newtonian techniques of infinitesimal analysis.

771. Dijksterhuis, E. J. *Simon Stevin: Science in the Netherlands Around 1600.* The Hague: Martinus Nijhoff, 1970. 145 pp. [Condensed English ed. Dutch original published 1943]

Surveys Stevin's contribution to: mathematics, mechanics, hydrostatics, astronomy, navigation, technology, military science, architecture, double-entry bookkeeping, Dutch linguistics. List of Stevin's works (including posthumous eds.). Reviewed in *Isis* 62 (1971), 544–45.

772. Rose, Paul Lawrence. "Renaissance Italian Methods of Drawing the Ellipse and Related Curves." *Physis* 12 (1970), 370–404.

On the interest in conic sections shown by mathematicians, engineers and artists. Describes some of the mechanical methods of construction.

773. Schmitt, Charles B. "A Fresh Look at Mechanics in 16th-Century Italy." [Essay Review of:] Stillman Drake and I. E. Drabkin, *Mechanics in Sixteenth-Century Italy* (Madison: Univ. of Wisconsin Press, 1969. 428 pp.)." *Studies in History and Philosophy of Science,* 1 (1970), 161–75.

774. Clulee, N. H. "John Dee's Mathematics and the Grading of Compound Qualities." *Ambix* 18 (1971), 178–211.

On Dee's attempt at grading based on the intension and remission of forms in his "Mathematical praeface" to H. Billingsley's English translation of Euclid's *Elements* (1570).

775. Hofmann, J. E. "Dürer als Mathematiker." *Humanismus und Technik* 15 (1971), (2), 1–16.

Brief survey of Dürer's study of mathematics and an analysis of his *Unterweisung* (1527).

776. Hofmann, J. E. "Über Auftauchen und Behandlung von Differentialgleichungen in 17. Jahrhundert." *Humanismus und Technik* 15 (1972), (3), 1–40.

Traces the history of Debeaune's differential equation and its solution from Descartes to Jakob Bernoulli.

777. Hofmann, Joseph E. "Bombellis Algebra—eine genialische Einzelleistung und ihre Einwirkung auf Leibniz." *Studia Leibnitiana* 4 (1972), 196–252.

An analysis of Bombelli's *Algebra* (1572), and a study of its influence on Leibniz.

778. Kline, Morris. *Mathematical Thought from Ancient to Modern Times.* New York: Oxford Univ. Press, 1972. 1,238 pp.

Pp. 216–399: The Renaissance; Mathematical contributions in the Renaissance; Arithmetic and algebra in the sixteenth and seventeenth centuries; The beginnings of projective geometry; Co-ordinate geometry; The mathematization of science; The creation of the calculus; Mathematics as of 1700. Bibliographies. Reviewed in *British Journal for the Philosophy of Science* 29 (1978), 68–87.

779. Jayawardene, S. A. "The Influence of Practical Arithmetics on the Algebra of Rafael Bombelli." *Isis* 64 (1973), 510–23.

Study of the problems of practical arithmetic in an unpublished version of the *Algebra* (written between 1557 and 1560), later replaced in the printed version (1572) by problems from the *Arithmetica* of Diophantus. Bombelli was employed as an architect by the Papal States. See the same author's article, "Rafael Bombelli, Engineer Architect," *Isis* 56 (1965), 298–306.

780. Mahoney, Michael Sean. *The Mathematical Career of Pierre de Fermat (1601–1665).* Princeton, N.J.: Princeton Univ. Press, 1973. 419 pp.

Well-documented account of Fermat's life and work. Pp. 26–71: Viète's analytic program and its influence on Fermat. Reviews: *Isis* 65 (1974), 398–400; *Revue d'histoire des sciences* 27 (1974), 335–46; *British Journal for the History of Science* 8 (1975), 81–84. See also critical essay review by A. Weil in *Bulletin, American Mathematical Society* 79 (1973), 1138–49.

781. Bos, Hendrik J. M. "Differentials, Higher-Order Differentials and the Derivative in the Leibnizian Calculus." *Archive for History of Exact Sciences* 14 (1974–75), 1–90.

Letter of 22 December 1561 from the Cardinal Chamberlain to the Prefect of Castel Sant'Angelo (Vatican Archives, Divers. Camer. 206, fol. 161) asking for two ropes to be loaned to Rafael Bombelli, architect in charge of the reconstruction of the *Ponte Santa Maria* (bridge on the Tiber in Rome, on the site of the present *Ponte Rotto*). Bombelli is better known today for his *L'Algebra* (Bologna, 1572). *Entries 777, 779.*

Study of the *differential*, the fundamental concept of the infinitesimal calculus, as it was understood and used by Leibniz and those mathematicians who developed the differential and integral calculus along the lines set out by him.

782. Folkerts, Menso. "Die Entwicklung und Bedeutung der Visierkunst als Beispiel der praktischen Mathematik der frühen Neuzeit." *Humanismus und Technik* 18 (1974), 1–41.

On gauging (barrel measurement). Sixty-six texts cited.

783. Hofmann, Joseph E. *Leibniz in Paris, 1672–1676. His Growth to Mathematical Maturity*. Cambridge: Cambridge Univ. Press, 1974. 372 pp. [Revised translation of the German original published in 1949]

Introduction to a study of the origins and development of Leibniz's mathematical ideas. Considers the disputes between Leibniz and Newton on priority and plagiarism in regard to the "invention of the calculus." Reviewed in *Isis* 67 (1976), 313–15.

784. Sheynin, O. B. "On the Prehistory of the Theory of Probability." *Archive for History of Exact Sciences* 12 (1974), 97–141.

Sketches the history of randomness and probability from antiquity to the seventeenth century (antique philosophy; jurisprudence; fine arts; games of chance; biology and medicine; astrology; astronomy; modern philosophy).

785. Hacking, Ian. *The Emergence of Probability*. Cambridge: Cambridge Univ. Press, 1975. 209 pp.

Analyses the early ideas about probability, induction, and statistical inference—from the fifteenth century up to Bernoulli's *Ars conjectandi* (1713). Reviewed in *Historia Mathematica* 7 (1980), 212–16, by E. Knobloch with a survey of recent studies.

786. Rose, Paul Lawrence. *The Italian Renaissance of Mathematics. Studies on Humanists and Mathematicians from Petrarch to Galileo.* Geneva: Droz, 1975. 316 pp.

Richly documented study illustrating the "interaction of humanists and mathematicians in Italy during the formative period before the Renaissance had given way to the Scientific Revolution." Contends that the leading mathematicians of the time were obsessed with the need to revitalize the mathematical sciences. Focuses attention on some forty mathematicians, mostly of the humanistic tradition and unfortunately ignores the *maestri d'abaco*. Exaggerates the role of the Italian algebraists in restoring Greek mathematics. Pp. 295–98: Index of MSS. Pp. 299–316: Index of names and topics. Reviews: *TLS* 17 Dec. 1976, p. 1578 (author's reply in *TLS* 29 April 1977, p. 523); *Annals of Science* 34 (1977), 628–29; *Historia Mathematica* 4 (1977), 106–7; *Isis* 69 (1978), 298–300.

787. Crombie, A. C. "Mathematics and Platonism in the Sixteenth-Century Italian Universities and in Jesuit Educational Policy." *Prismata: naturwissenschaftsgeschichtliche Studien. Festschrift für Willy Hartner.* Y. Maeyama and W. G. Saltzer, eds. Wiesbaden: Frank Steiner, 1977, pp. 63–94.

With notes and bibliography.

788. Folkerts, Menso. "Regiomontanus als Mathematiker." *Centaurus* 21 (1977), 214–45.

Brief sketch of the mathematical achievements of Regiomontanus.

789. Goldstine, H.H. *A History of Numerical Analysis from the Sixteenth through the Nineteenth Century.* New York: Springer, 1977. 348 pp.

Pp. 1–118 deal with the sixteenth and seventeenth centuries (logarithms, Viète, Kepler, the Age of Newton). Reviewed in *Historia Mathematica* 5 (1978), 479–82 and in *Janus* 65 (1978), 303–12.

790. Hofmann, Joseph E. *Register zu Gottfried Wilhelm Leibniz mathematische Schriften und Die Briefwechsel mit Mathematikern (herausgegeben von C. I. Gerhardt).* Hildesheim: Georg Olms, 1977. 312 pp.

Index to Gerhardt's ed. of Leibniz's mathematical writings (7 vols.) and correspondence (1 vol.), prepared by Hofmann for his own use, later edited by Christoph J. Scriba and others. Four separate indexes containing: names and titles of works; periodicals cited.

791. Rose, Paul Lawrence. "Erasmians and Mathematicians at Cambridge in the Early Sixteenth Century." *Sixteenth Century Journal* 8 (1977), [Suppl.], (2), 47–59.

On the revival of mathematical teaching.

792. Rose, Paul Lawrence. "A Venetian Patron and Mathematician of the Sixteenth Century: Francesco Barozzi (1537–1604)." *Studi veneziani* (n.s.), 1 (1977), 119–78.

Corrected and updated bio-bibliography, with excerpts from Barozzi's correspondence. Includes a discussion of: Barozzi and the mathematical collection of Pappus; The translations and texts of Hero Byzantinus.

793. Sheynin, O. B. "Early History of the Theory of Probability." *Archive for History of Exact Sciences* 17 (1977), 201–59.

Covers the period 1654–1713 (Pascal, Fermat, Huygens).

794. Pearson, Karl. *The History of Statistics in the Seventeenth and Eighteenth Centuries against the Changing Background of Intellectual, Scientific and Religious Thought.* Edited by E. S. Pearson. London: Charles Griffin, 1978. 744 pp.

Lectures given at University College, London from 1921 to 1933. Pp. 1–176 cover the seventeenth century. Reviewed in *NTM* 22 (1985), 114.

795. Sinisgalli, Rocco. *Per la storia della prospettiva (1405–1605). Il contributo di Simon Stevin allo sviluppo scientifico della prospettiva artificiale ed i suoi precedenti storici.* Rome: "L'Erma" di Bretschneider, 1978. 354 pp.

Pp. 25–122: Analysis of the works of Stevin's predecessors on perspective (from Brunelleschi to Kepler). Pp. 123–46: Study of Stevin's work. Review: *Revue d'histoire des sciences* 35 (1982), 70–71.

796. Victor, Joseph M. *Charles de Bovelles, 1479–1553: An Intellectual Biography.* Travaux d'humanisme et renaissance, 161. Geneva: Droz, 1978. 191 pp.

Study of Bovelles's life and works (theology, philosophy, and mathematics).

797. Edwards, C. H., Jr. *The Historical Development of the Calculus.* New York: Springer, 1979. 351 pp.

Work intended primarily for those who study, teach, and use calculus. Pp. 98–267: Early indivisibles and infinitesimal techniques; Early tangent constructions; Napier's wonderful logarithms; The arithmetic of the infinite; Newton; Leibniz. The author has made use of the recently published mathematical papers and correspondence of Newton. Reviewed in *Science* 208 (1980), 1139–40.

798. Lohne, J. A. "Essays on Thomas Harriot. 1. Billiard Balls and Laws of Collision 2. Ballistic Parabolas. 3. A Survey of Harriot's Scientific Writings." *Archive for History of Exact Sciences* 20 (1979), 189–312.

Studies on the theme: "Harriot and his older contemporary Gilbert declared war on the peripatetic physicists. Instead of commenting on Aristotle they examined problems vital to the English nation, such as magnetic needles and ballistic trajectories."

799. Clagett, Marshall. *Archimedes in the Middle Ages.* Vol. 4: *A Supplement on the Medieval Latin Traditions of Conic Sections (1150–1566).* Part 1: *Texts and Analysis.* Part 2: *Bibliography, Diagrams, and Indexes.* Memoirs of the American Philosophical Society, 137. Philadelphia, 1980. 2 vols. 566 pp.

Vol. 1, pp. 235–461: Medieval traditions in the sixteenth century (Giorgio Valla, Johann Werner, Albrecht Dürer, Finaeus, and Barozzi). Reviewed in *Isis* 73 (1982), 271–74.

800. Goldstine, Herman H. *A History of the Calculus of Variations from the Seventeenth through the Nineteenth Century.* New York: Springer, 1980. 410 pp.

Pp. 1–66: Discussion of the work of Fermat, Newton, Leibniz, and the Bernoullis. Reviewed in *Isis* 73 (1982), 297.

801. Hall, A. Rupert. *Philosophers at War: The Quarrel between Newton and Leibniz.* Cambridge: Cambridge Univ. Press, 1980. 338 pp.

An analysis of the controversy about priority in the discovery of the infinitesimal calculus. Reviewed in *Isis* 72 (1981), 683–84.

802. Tanner, R. C. H. "The Ordered Regiment of the Minus Sign: Offbeat Mathematics in Harriot's Manuscripts." *Annals of Science* 37 (1980), 127–58.

Discussion of Harriot's rough notes—"experimentations in the field of the sign-rule of multiplication in algebra."

803. Tanner, R. C. H. "The Alien Realm of the Minus: Deviatory Mathematics in Cardano's Writings." *Annals of Science* 37 (1980), 159–78.

Examines the *De aliza regula* ("algebraic logistics") and the *Sermo de plus et minus* (a commentary on Bombelli's *Algebra*) in the light of the algebraical work of Thomas Harriot.

804. Zetterberg, J. Peter. "The Mistaking of 'the Mathematicks' for Magic in Tudor and Stuart England." *Sixteenth Century Journal* 11 (1980), 83–97.

Discusses the reasons for mistaking mathematics for magic: partly gossip, misunderstanding and fable; most English mathematicians were astrologers. States that the persistence of such confusion throughout the period 1550–1650 was partly due to the mathematicians' own rhetoric.

[224]

diElum eft nuper in recenfione TraEtatus D. Cheynæi, *Medici* Scoti Londini *degentis. Conferri etiam poteft TraEtatus D.* Craigii Scoti *de Quadraturis, & ejufdem Theorema ad Quadraturas pertinens, nuper in bis AEtis exhibitum ; quæ faciunt etiam ut ipfis Theorematis* Newtonianis *recenfendis fuperfedeamus, quia paucis exponi non poffunt: quemadmodum nec ejufdem Theoremata quædam reduEtionis ad Quadraturas faciliores.*

His permotus D. Joannes Keill, *in Epiftola in* Philofophicis Tranfactionibus *A. C.* 1708, *menfibus* Septemb. & Octob. *impreffa, fcripfit in contrarium, quod* Fluxionum Arithmeticam, fine omni dubio, primus invenit Dominus *Newtonus,* ut cuilibet ejus Epiftolas a *Wallifio* editas legenti facile conftabit. Eadem tamen Arithmetica poftea mutatis Nomine & Notationis modo, a Domino *Leibnitio* in *AEtis Eruditorum* edita eft.

N°LXXX *Epiftola D.* Leibnitii *ad D.* Hans Sloane *Regiæ Societatis Secretarium,* 4° Martii *S. N.* 1711 *data.*

GRatias ago quod noviffimum Volumen præclari Operis *Tranfactionum Philofophicarum* ad me mififti; quamvis nunc demum mihi *Berolinum* excurrenti redditum fit. Itaque excufabis quod pro munere fuperioris anni nunc demum gratiæ dudum debitæ redduntur.

Vellem infpectio Operis me non cogeret nunc fecunda vice ad vos querelam deferre: Olim *Nicolaus Fatius Duillierius* me pupugerat in publico fcripto, tanquam alienum Inventum mihi attribuiffem. Ego eum in *AEtis Eruditorum Lipfienfibus* meliora docui ; & vos ipfi, ut ex Literis a Secretario Societatis veftræ inclytæ (id eft, quantum
me-

Letter from Leibniz to Hans Sloane complaining that Fatio de Duillier had accused him of publishing the arithmetic of fluxions invented by Newton

[225]

memini, a Teipſo) ſcriptis didici, hoc improba-
ſtis. Improbavit *Newtonus* ipſe vir excellentiſſi-
mus, (quantum intellexi) præpoſterum quorundam
hac in re erga veſtram gentem & ſe ſtudium. Et
tandem D. *Keillius* in hoc ipſo volumine, menſe
Sept. Octob. 1708, pag. 185, renovare ineptiſſimam
accuſationem viſus eſt, cum ſcripſit, *Fluxionum A-*
rithmeticam a Newtono *inventam, mutato nomine &*
notationis modo a me editam fuiſſe. Quæ qui legit,
& credit, non poteſt non ſuſpicari alterius inventum a me larvatum ſubdititiis nominibus characte-
ribuſque fuiſſe protruſum. Id quidem quam fal-
ſum ſit nemo melius ipſo D. *Newtono* novit. Cer-
te ego nec nomen Calculi Fluxionum fando audi-
vi, nec Characteres quos adhibuit D. *Newtonus* his
oculis vidi, antequam in *Walliſianis* Operibus pro-
diere. Rem etiam me habuiſſe, multis ante annis
quam edidi, ipſæ literæ apud *Walliſium* editæ de-
monſtrant. Quomodo ergo aliena mutata edidi
quæ ignorabam.

Etſi autem D. *Keillium* (a quo magis præcipiti
judicio quam malo animo peccatum puto) pro ca-
lumniatore non habeam; non poſſum tamen non
ipſam accuſationem in me injuriam pro calumnia
habere. Et quia verendum eſt ne ſæpe vel ab im-
probis vel ab imprudentibus repetatur; cogor re-
medium ab Inclyta veſtra Societate Regia petere.
Nempe æquum eſſe vos ipſi credo judicabitis, ut
D. *Keillius* teſtetur publice, non fuiſſe ſibi animum
imputandi mihi quod verba inſinuáre videntur.
quaſi ab alio hòc quicquid eſt Inventi didicerim &
mihi attribuerim. Ita ille & mihi læſo ſatisfaciet,
& calumniandi animum a ſe alienum eſſe oſten-
det; & aliis aliás ſimilia aliquando jactaturis fræ-
num injicietur. Quod ſupereſt vale & fave.

Dabam *Berolini* 4 *Martii* 1711.

Q *Epi-*

under his name, after altering the name and the style of notation. From *Com-*
mercium Epistolicum, ed. John Collins, London, 1722, pp. 224–25 (University
of London Library). *Entry 801.*

805. Franci, R. and L. Toti Rigatelli. "La trattatistica matematica del Rinascimento senese." *Atti dell'Accademia delle Scienze di Siena detta de' Fisiocritici*, (14), 13 (1981), 1–71.

Study of the life and work of three sixteenth-century Sienese *maestri d'abaco*—Giovanni Sfortunati, Pietro Cataneo, and Dionigi Gori. Pp. 45–50 deal with the algebra ("pratica d'alcibra") in Gori's work.

806. Schneider, Ivo. "Why Do We Find the Origin of a Calculus of Probabilities in the Seventeenth Century?" *Proceedings, 1978 Pisa Conference on the History and Philosophy of Science.* Edited by J. Hintikka, D. Gruender, and E. Agazzi. Dordrecht: Reidel, 1981, Vol. 2, pp. 3–24.

States that although the concept of probability existed, it remained impossible to connect the concepts of chance and probability until the seventeenth century.

807. Marinoni, Augusto. *La matematica di Leonardo da Vinci: una nuova immagine dell'artista scienziato.* Milan: Arcadia, 1982. 174 pp.

NOT SEEN.

808. Grössing, Helmuth. *Humanistische Naturwissenschaft. Zur Geschichte der Wiener mathematischen Schulen des 15. und 16. Jahrhunderts.* Saecula Spiritalia, hrsg. von Dieter Wuttke, Bd. 8. Baden-Baden: Verlag Valentin Koerner, 1983. 335 pp.

The humanist face of science in the Renaissance shown by a study of mathematics and astronomy in Vienna from 1450 to 1550. Mathematicians include: Regiomontanus (d. 1476); Conrad Celtis (d. 1508); Georg Tanstetter Collimitius (d. 1535). Pp. 303–21: Bibliography; pp. 335–55: Index (name and subject). Reviewed in *Journal for the History of Astronomy* 16 (1985), 52–54.

809. Giusti, E. "A tre secoli dal calcolo: la questione delle origini." *Bollettino della Unione Matematica Italiana*, (6), 3A (1984), 1–55.

An analysis of Leibniz's memoir of 1684—*Nova methodus*—and its history.

810. Sanders, P. M. "Charles de Bovelles's Treatise on the Regular Polyhedra (Paris 1511)." *Annals of Science* 41 (1984), 513–66.

An analysis of Bovelles's treatise and a study of its role as "a mathematical guide by means of which the mystery of the Trinity can be contemplated." With an introduction.

811. Aiton, E. J. *Leibniz: A Biography*. Bristol: Adam Hilger, 1985. 370 pp.

A biographical sketch based on letters and other sources. Reviewed in *History and Philosophy of Logic* 8 (1987), 92–100.

812. Franci, R., and L. Toti Rigatelli. "Towards a History of Algebra from Leonardo of Pisa to Luca Pacioli." *Janus* 72 (1985), 17–82.

Outline history of the algebraic method (*l'arte de la cosa*) from Leonardo's "Liber abaci" (1202) to Pacioli's *Summa de arithmetica* (1494). Pp. 56–66 deal with three practical arithmetics (*libri d'abaco*) of the second half of the fifteenth century (Raffaello Canacci, Piero della Francesca, and Pacioli). Pacioli's *Summa* was the first printed book to contain a treatment of the algebraic method.

813. Sasaki, Chikara. "The Acceptance of the Theory of Proportion in the Sixteenth and Seventeenth Centuries. Barrow's Reaction to the Analytic Mathematics." *Historia Scientiarum* 29 (1985), 83–116.

814. Van der Waerden, B. L. *A History of Algebra from al-Khwārizmī to Emmy Noether*. Berlin: Springer, 1985. 271 pp.

A survey by one of the leading algebraists of this century. Pp. 52–75: history of algebraic equations from Scipione dal Ferro to Bombelli and from Viète to Descartes.

815. Heinekamp, Albert, ed. *300 Jahre "Nova Methodus" von G. W. Leibniz (1684–1984)*. Symposion der Leibniz-Gesellschaft in Congresscentrum "Leewenhorst" in Noordwijkerhout (Niederlande) 28 bis 30 August 1984. Studia Leibnitiana, Sonderheft 14. Stuttgart: Franz Steiner, 1986. 268 pp.

Nineteen papers (documented) on the *Nova methodus*, its prehistory, its development by Leibniz, and its reception, read at a symposium commemorating the discovery of the infinitesimal calculus.

816. Edwards, A. W. F. *Pascal's Arithmetical Triangle*. London: Charles Griffin, 1987. 174 pp.

Pp. 1–56: Historical survey from ancient times. Pp. 57–86: Pascal's treatise on the arithmetical triangle. Pp. 87–111: The arithmetical triangle in the development of analysis.

816A. Swetz, Frank J. *Capitalism and Arithmetic. The New Math of the 15th Century. Including the Full Text of the* Treviso Arithmetic *of*

1478. Translated by David Eugene Smith. La Salle, Illinois: Open Court, 1987. 345 pp.

A study of the *Treviso Arithmetic*—the first dated printed arithmetic using Hindu-Arabic numerals. Written in Italian for the use of merchant apprentices, it was published in Treviso in 1478. Pp. 40–175: Translation into English. Includes a bibliography, notes and a tribute to D. E. Smith. A detailed study of this text from a bibliographical and historical viewpoint was made by Baldassarre Boncompagni: "Intorno ad un trattato d'aritmetica stampato nel 1478," *Atti dell'Accademia Pontificia de' Nuovi Lincei* 16 (1862–63), 1–64, 104–228, 301–64, 389–452, 503–630, 683–842, 909–1044.

817. Gamba, Enrico, and Vico Montebelli. *Le scienze a Urbino nel tardo Rinascimento.* Parte I. *Matematici urbinati del cinque-seicento.* Parte II. *Il problema dell'equilibrio della bilancia nella polemica di Guidobaldo dal Monte con il "Goto."* Urbino: QuattroVenti, 1988. 274 pp.

On scientists and mathematicians (Federico Commandino, Guido Baldo dal Monte, Bernardino Baldi, Muzio Oddi) and mathematical instrument making in the Duchy of Urbino during the years 1500–1700. Pp. 263–71: Name index. Includes a study of the correspondence between Guidobaldo and Christoph Clavius on the equilibrium of the balance.

818. Hay, Cynthia, ed. *Mathematics from Manuscript to Print, 1300–1600.* Oxford: Clarendon Press, 1988. 273 pp.

Proceedings of a conference held at Oxford in 1984, celebrating the quincentenary of the completion of Nicolas Chuquet's *Triparty*. Subjects covered include: Chuquet and French mathematics (six papers, of which one is on his mathematical executor, Estienne de la Roche); mathematics in the sixteenth century (three papers: the arithmetic of Christianus van Varenbraken; Snellius; Maurolico); mathematics and its ramifications (seven papers, five relating to: Boncompagni's *Bullettino di bibliografia e di storia delle scienze matematiche e fisiche*; recreational mathematics; mathematical magic; perspective; games of chance).

819. Parshall, Karen Hunger. "The Art of Algebra from al-Khwārizmī to Viète: a Study in the Natural Selection of Ideas." *History of Science* 26 (1988), 129–64.

Interprets "the development of the Arabic, algebraic line from the time of al-Khwārizmī to the sixteenth century in light of an evolutionary framework" and examines the "way in which natural selection may be thought of as having acted on and modified this approach in the presence of the reintroduced, Diophantine concepts." Discussion includes the *maestri d'abaco,* Pacioli, Cardano, Bombelli, and Viète but not Stevin and the German algebraists.

820. Biagioli, Mario. "The Social Status of Italian Mathematicians." *History of Science* 27 (1989), 41–95.

A preliminary study of the socio-professional world of the mathematical practitioners who worked before or during the time of Galileo. With a bibliography (570 items, indexed) and a list of mathematics teachers at Italian universities. The author believes that the mathematization of nature during the Scientific Revolution resulted in raising the social status of mathematical practitioners.

820A. Picutti, Ettore. "Sui plagi matematici di frate Luca Pacioli. Il confronto delle opere del frate minorita con trattati matematici medievali porta a dar ragione ai suoi decisi accusatori, storici dell'arte, contro i suoi decisi difensori, ecclesiastici e storici della matematica." *Le Scienze* 22 (1989), (246), 72–79.

Throws new light on the accusations of plagiarism, first made by Giorgio Vasari, against Pacioli. States, *inter alia*, that ff. 1–59v of the *Tractatus geometrie* of the *Summa de arithmetica* are a transcription of a late medieval text, MS Palat. 577, Biblioteca Nazionale Centrale, Florence, ff. 1–241.

821. Rambaldi, Enrico I. "John Dee and Federico Commandino: An English and an Italian Interpretation of Euclid during the Renaissance." *Rivista di storia della filosofia,* anno 44, NS, 2 (1989), 211–48.

A comparison between Commandino's "Prolegomena" to his Latin translation of Euclid's *Elements* (Pisa, 1572) and Dee's Preface to H. Billingsley's English ed. of 1570.

822. Roero, Clara Silvia. "The Passage from Descartes' Algebraic Geometry to Leibniz's Infinitesimal Calculus in the Writings of Jacob Bernoulli [1654–1705]." *Studia Leibnitiana,* Sonderheft 17 (1989), 140–50.

Retraces the cultural itinerary that led Bernoulli to the understanding of the Leibnizian infinitesimal calculus.

823. Feingold, Mordechai, ed. *Before Newton: The Life and Times of Isaac Barrow.* Cambridge: Cambridge Univ. Press, 1990. 380 pp.

Seven studies covering different aspects of Barrow's life and work: Divine, scholar, mathematician (Feingold); The *Optical Lectures* and the foundations of the theory of optical imagery (Alan E. Shapiro); Mathematics: between ancients and moderns (Michael S. Mahoney); Academic milieu: Interregnum and Restoration Cambridge (John Gascoigne); Scholar (Anthony Grafton); Preacher (Irene Simon); Library (Feingold), with a modern

Súma de Arithmetica Geo/ metria Proportioni ⁊ Pro/ portionalita.

Continentia de tutta lopera.

De numeri e misure in tutti modi occurrenti.

Proportioni ⁊ proportiõalita anotitia del. 5º de Eucli de.e de tutti li altri soi libri.

Chiaui ouero euidentie numero.13.p le ꝗtita conti/ nue ꝓportiõali del.6ºe.7º de Euclide extratte.

Tutte le pti delalgorismo:cioe releuare. prir. multi/ plicar.sumare.e sotrare cõ tutte sue ꝓue i sani e rot/ ti.e radici e progressioni.

De la regola mercantesca ditta del.3.e soi fõdamen/ ti con casi exemplari per c:m: 8.G.guadagni:perdi te:transportationi:e inuestite.

Partir.multiplicar.summar.e sotrar de le proportio ni e de tutte sorti radici.

De le.3.regole del cata yn ditta positiõe e sua origie.

Euidentie generali ouer conclusioni nº66.absoluere ogni caso che per regole ordinarie nõ si podesse.

Tutte sorte binomii e recisi e altre linee irratiõali del decimo de Euclide.

Tutte regole de algebra ditte de la cosa e lor fabri/ che e fondamenti.

Compagnie i tutti modi.e lor partire.

Socide de bestiami. e lor partire

Fitti:pesciõl: cottimi: liuelli: logagioni:e godimenti.

Baratti i tutti modi semplici:composti:e col tempo.

Cambi reali.secchi.fittitii.e diminuti ouer comuni.

Meriti semplici e a capo danno e altri termini.

Resti.saldi.sconti.de tempo e denari ede recare a un di piu partite·

Or.argẽti.elloro affinare. e carattare

Molti casi e ragioni straordinarie varie e diuerse a tutte occurentie commo nella sequente tauola ap/ pare ordinatamente de tutte.·

Ordine a saper tener ogni cõto e scripture e del qua derno in vinegia.

Tariffa de tutte vsange e costumi mercanteschi in tut to el mondo.

Pratica e theorica de geometria e de li:5.corpi regu lari e altri dependenti.

E molte altre cose õ grandissimi piaceri e frutto cõ/ mo difusamente per la sequente tauola appare.

First printed page of Luca Pacioli's *Summa de arithmetica*, Venice, 1494 (Bodleian Library, Oxford, Auct.2Q inf.l.25). ***Entry 820A.***

rendition of the 1677 catalogue. Includes a list of Newton's books that may possibly have belonged to Barrow. No bibliography but well documented.

824. Hald, Anders. *A History of Probability and Statistics and Their Applications before 1750.* New York: John Wiley and Sons, 1990. 586 pp.

Pp. 13–190 cover the history from Cardano to 1708. Pp. 549–70: Bibliography. Pp. 571–86: Name and subject index. Aims to correct and complement Todhunter's *History.* See item 732.

825. Edgerton, Samuel Y., Jr. *The Heritage of Giotto's Geometry. Art and Science on the Eve of the Scientific Revolution.* Ithaca, N.Y.: Cornell Univ. Press, 1991. 319 pp.

Explores the relationship between the Scientific Revolution of the seventeenth century and the artistic revolution of the Renaissance begun by Giotto three centuries earlier. Pp. 291–319: Bibliography and index. Criticized in *Journal for the History of Astronomy* 23 (1992), 227–29.

826. Willmoth, Frances. *Sir Jonas Moore: Practical Mathematics and Restoration Science.* Woodbridge, Suffolk: The Boydell Press, 1993. 244 pp.

Biography of Jonas Moore (1617–79), practical mathematician, teacher, author, surveyor, cartographer, Ordnance officer, courtier, and patron of learning. He made a substantial fortune from mathematical practice. Pp. 216–33: Bibliography (primary and secondary sources). See also item 706.

Astronomy

BIBLIOGRAPHIES

827. Lalande, Jérome de. *Bibliographie astronomique; avec l'histoire de l'astronomie depuis 1781 jusqu'à 1802.* Amsterdam: J. C. Gieben, 1970. 966 pp. [Reprint of the 1803 Paris ed.]

Pp. 9–339: Books published from 1471 to 1700, arranged alphabetically by author, with occasional annotations. Pp. 881–966: Author and subject indexes.

828. Knobel, E. B. "The Chronology of Star Catalogues." *Memoirs, Royal Astronomical Society* 43 (1877), 1–76.

Table of 530 Star Catalogues of which forty-six were published before 1700.

829. Houzeau, J. C. *Vade-mecum de l'astronomie.* Appendice à la nouvelle série des *Annales astronomiques.* Brussels: F. Hayez, imprimeur de l'Académie Royale de Belgique, 1882. 1,144 pp. [First published in 1878]

Revised version of a manual of fundamental astronomical data, arranged in the form of a bibliography of 3,447 titles in 29 chapters (366 sections). Historical introductions to each section and occasional annotations. Author-title and subject indexes. Useful as a complement to item 833, which is incomplete in some sections.

830. Royal Observatory, Edinburgh. *Catalogue of the Crawford Library.* Edinburgh, 1890. 497 pp. *Supplement.* [A Computer-produced Index to the Whole Collection]. 1977. 112 pp.

Catalogue of a rich collection of some 2500 books and MSS on astronomy and mathematics, the nucleus of which was the library of Charles Babbage (1792–1871). The library is described by E. G. Forbes in *British Journal for the History of Science* 6 (1972–73), 459–61.

831. Zinner, Ernst. *Verzeichnis der astronomischen Handschriften des deutschen Kulturgebietes.* Munich: C. E. Beck, 1925. 544 pp. [Reproduced from handwriting]

Index of some 12,750 MSS (pre-1850) in Latin and Germanic languages in the libraries of Germany and neighbouring countries. With notes. Reviewed in *Isis* 15 (1931), 193–95.

832. Johnson, Francis R. *Astronomical Thought in Renaissance England. A Study of the English Scientific Writings from 1500 to 1645.* Baltimore: Johns Hopkins Press, 1937. 357 pp.

Includes a chronological list of books dealing with astronomy printed in England to 1640. Reviewed in *Isis* 28 (1938), 514–16.

833. Houzeau, J. C., and A. Lancaster. *Bibliographie générale de l'astronomie jusqu'en 1880. General Bibliography of Astronomy to the Year 1880.* New ed. with introduction by D. W. Dewhirst. London: Holland Press, 1964. 2 vols. in 3 pts. [First published 1882–89]

Monumental work, unfortunately incomplete (see below). Excellent coverage of the period 1450–1700. Classification is chronological, ethnographical, and systematic. Except for biographies, the arrangement within each section is chronological. Vol. 1 (Pts. 1, 2: Books and MSS). Pp. 1–325: Introductory essay on the history of astronomy. Pp. 327–680: Historic works (includes: histories of astronomy; bibliographies and catalogues of astronomical works; works of Renaissance astronomers). Pp. 681–858: Astrology. Pp. 859–986: Biography and correspondence. Pp. 987–1138: General and didactic works (including encyclopedias and dictionaries). Pp. 1139–1310: Spherical astronomy (includes gnomonics). Pp. 1311–1623: Theoretical astronomy (includes astronomical tables, chronology, calendar, ephemerides, and almanacs). Pp. 1631–33: Guide to the MSS. Pp. 1635–1722: Author index. Vol. 2. Memoirs and papers in Academy publications and periodicals (arranged in nine main sections, including history of astronomy and biography). Pp. 1–55: List of periodicals. Pp. xi–xxxix: List of subjects. Author index. (NOT PUBLISHED. Vol. 1, pt. 3: Celestial mechanics, physical astronomy, practical astronomy, descriptive astronomy, cosmology. Vol. 3: Astronomical observations, star catalogues and atlases; journal articles on observatories, meteors, etc.) See items 829 and 834.

834. Zinner, Ernst. *Geschichte und Bibliographie der astronomischen Literatur in Deutschland zur Zeit der Renaissance.* Zweite Aufl. Stuttgart: Hiersemann, 1964. 480 pp. [First published 1941]

Pp. 3–88: historical survey; pp. 89–480: bibliography. Includes corrections to Houzeau and Lancaster (833). Reviews: *Isis* 36 (1946), 261–66; *The Library* (5), 20 (1965), 331–32.

835. Burmeister, Karl Heinz. *Georg Joachim Rheticus, 1514–1574. Eine Bio-Bibliographie. I. Humanist und Wegbereiter der modernen Naturwissenschaften. II. Quellen und Bibliographie. III. Briefwechsel.* Wiesbaden: Guido Pressler, 1967–68. 3 vols.

Biography; bibliography of writings by and about Rheticus (includes: archival documents, MSS, and correspondence); Correspondence (51 letters). Reviewed in detail by Edward Rosen in: *Isis* 59 (1968), 231–33; 60 (1969), 117–19; and 61 (1970), 137–39.

836. Caspar, Max. *Bibliographia Kepleriana. Ein Führer durch das gedruckte Schrifttum von Johannes Kepler.* 2. Aufl. besorgt von Martha List. Munich: C. H. Beck'sche Verlag, 1968. 181 pp.

Annotated bibliography of Kepler's works, including a list of writings about him. Reviewed in *Isis* 60 (1969), 567–68. Continued in 838 and 839.

837. Rosen, Edward. *Three Copernican Treatises. The* Commentariolus *of Copernicus, the* Letter against Werner, *the* Narratio prima *of Rheticus.* Translated with introduction and notes by Edward Rosen. 3rd ed. revised with *A Biography of Copernicus and Copernicus Bibliographies, 1939–1958 and 1959–1970.* New York: Octagon Books, 1971. 425 pp.

It was by the *Commentariolus* and the *Narratio prima* that "the learned world was first apprised of the revolution in the conceptual structure of the universe." Pp. 199–312: Bibliography of primary and secondary literature (1,092 items). Pp. 313–408: Biography.

838. List, Martha. "Bibliographia Kepleriana, 1967–1975." *Vistas in Astronomy* 18 (1975), 955–1010.

Continuation of 836.

839. List, Martha. "Bibliographia Kepleriana: Supplements and Continuation, 1975–78." *Vistas in Astronomy* 22 (1978),1–18.

See items 836 and 838.

840. Warner, Deborah J. *The Sky Explored: Celestial Cartography 1500–1800.* Amsterdam: Theatrum Orbis Terrarum, B.v., 1979. 293 pp.

List of star maps of some 170 cartographers, of whom about one hundred lived in the sixteenth and seventeenth centuries. Annotated. Reviewed in *Journal for the History of Astronomy* 11 (1980), 73–74 and in *Isis* 72 (1981), 663.

841. DeVorkin, David H. *The History of Modern Astronomy and Astrophysics. A Selected, Annotated Bibliography.* Bibliographies of the History of Science and Technology, Vol. 1. Garland Reference Library of the Humanities, Vol. 304. New York: Garland Publishing, 1982. 434 pp.

Some 1,400 citations relating to the period beginning with the invention of the telescope and its applications to astronomy. Inevitably they include many works whose coverage extends to earlier periods (see under: Bibliographies; General histories; National and Institutional histories; Instrumentation; Descriptive, theoretical and positional astronomy; Biographical, autobiographical, and collected works). Detailed list of contents on pp. xxi–xxvii. Pp. 397–434: Name and subject index. Reviews: *Sky and Telescope* 65 (1983), 140–42; *Journal for the History of Astronomy* 14 (1983), 146–47; *Isis* 74 (1985), 418.

842. Grassi, Giovanna. *Union Catalogue of Printed Books of the Fifteenth, Sixteenth and Seventeenth Centuries in European Astronomical Observatories.* Manziana (Rome): Vecchiarelli, 1989. 1,040 pp.

Catalogue of some five thousand titles (in 51 observatories) arranged alphabetically by author, with added entries under COMETS and EPHEMERIDES. Pp. 753–948: Chronological index. Pp. 949–1040: Index of printers and publishers (fifteenth and sixteenth centuries). Subjects include: astronomy, astrology, chemistry, mathematics, physics, natural history. An ed. containing only pre-1600 books was published in 1977.

MONOGRAPHS AND ARTICLES

843. Delambre, Jean. *Histoire de l'astronomie moderne.* Paris: Courcier, 1821. 2 vols. 715 pp.; 804 pp.

Describes in detail the works of the principal astronomers from Copernicus to Cassini, with a chapter on calendar reform. Pp. lxvi–lxxxii: List of contents.

844. Hoefer, Ferdinand. *Histoire de l'astronomie depuis ses origines jusqu'à nos jours.* Paris: Hachette, 1873. 631 pp.

General history of astronomy. No bibliography, no index. One of several histories of science written by the author. According to Sarton, they contain a wealth of information but are not dependable. See *Bulletin of the History of Medicine* 8 (1940), 419–32.

845. Dreyer, J. L. E. *Tycho Brahe. A Picture of Scientific Life and Work in the Sixteenth Century.* Edinburgh: Adam and Charles Black, 1890. 405 pp. [Reprinted 1963 by Dover, New York]
Life and work of the reformer of observational astronomy.

846. Dreyer, J. L. E. *History of the Planetary Systems from Thales to Kepler.* Cambridge: University Press, 1906. 432 pp.
Pp. 305–424 deal with the period from Copernicus to Kepler. 2nd ed. (1953) reviewed in *Isis* 44 (1953), 396–97.

847. Orchard, Thomas N. *Milton's Astronomy. The Astronomy of "Paradise Lost."* London: Longmans, Green, and Co., 1913. 288 pp.
Study of the astronomical allusions in Milton's poems. With index (pp. 281–88). Based on an earlier work, *The Astronomy of Milton's 'Paradise Lost'* (1896).

848. Stimson, Dorothy. *The Gradual Acceptance of the Copernican Theory of the Universe.* Hanover, N.H., 1917. 147 pp.
Elementary account of the Copernican system and its reception, preceded by a brief review of astronomical literature before Copernicus. Ph.D. thesis, Columbia University, 1917.

849. Johnson, Francis R., and Sanford V. Larkey. "Thomas Digges, the Copernican System, and the Idea of the Infinity of the Universe in 1576." *Huntington Library Bulletin* 5 (1934), 69–117.
Digges, an ardent supporter of Copernicus, was the first among English astronomers to advance the idea of an infinite universe as a corollary to the Copernican system.

850. McColley, Grant. "The Astronomy of *Paradise Lost.*" *Studies in Philology* 34 (1937), 209–47.
In *Paradise Lost* Milton followed his age, comparing the Ptolemaic and Copernican hypotheses. He favored the Copernican theory, but for poetic or practical reasons, he employed the Ptolemaic system, postulating ten solid spheres to explain celestial motions, and utilized these spheres in his cosmology.

851. Zinner, Ernst. *Entstehung und Ausbreitung der coppernicanischen Lehre. Zum 200 jährigen Jubiläum der Friedrich-Alexander-Universität zu Erlangen.* Sitzungsberichte der physikalisch-medizinischen Societät zu Erlangen, Bd. 74. Erlangen: Mencke, 1943. 594 pp.

[Reprinted by Topos Verlag, Vaduz in 1978 and by Verlag C. H. Beck, Munich in 1988]

"Panoramic survey of the development of astronomy" from antiquity to the eighteenth century, with special emphasis on Copernican astronomy and its development. Appendices include: a catalogue of Copernicus's library; a list of his astronomical observations; and a list of the correspondence of German Jesuit astronomers (1600–1660). Critical review by Edward Rosen in *Isis* 36 (1946), 261–66.

The 1988 reprint edited by Herbert M. Nobis and Felix Schmeidler contains a new commentary and a revised bibliography. Reviewed in *Isis* 82 (1991), 129–30.

852. Hellman, C. Doris. *The Comet of 1577. Its Place in the History of Astronomy*. New York: Columbia Univ. Press, 1944. 488 pp. [Reprinted in 1971 with additional material]

Study of the observations of and literature about the comet of 1577, with a bibliography of tracts and treatises. Includes an outline of earlier developments of cometary theory. Reviewed in *Isis* 36 (1946), 266–70 and in *Journal for the History of Astronomy* 4 (1973), 61–62.

853. Singer, D. *Giordano Bruno, His Life and Thought*. With annotated translation of his work *On the Infinite Universe and Worlds*. New York: Henry Schuman, 1950. 389 pp.

Reviewed in *Isis* 42 (1951), 247–48 and in *Centaurus* 4 (1955), 86–89.

854. Bernard Maitre, Henri. *La science européenne au Tribunal astronomique de Pékin (XVIIe–XIXe siècles)*. Université de Paris, Conférence du Palais de la Découverte, série D no. 9. Paris, 1951. 40 pp.

Notes on Chinese astronomy from the arrival of the Jesuits to 1800. Includes a catalogue of the astronomical section (European works) of the Pei-Thang Library.

855. Johnson, F. R. "Astronomical Textbooks in the Sixteenth Century." *Science, Medicine and History; Essays in Honour of Charles Singer*. Edited by E. Ashworth Underwood. London: Oxford Univ. Press, 1953, Vol. 1, pp. 258–302.

Examines the reasons for the slow acceptance of heliocentrism by textbook writers and teachers. Surveys the multifarious types of astronomical textbooks published in the sixteenth century.

856. Macomber, Henry P. "A Census of the Owners of Copies of the 1687 First Edition of Newton's 'Principia.' A Census of the Owners of Copies of the 1726 Presentation Issue of Newton's 'Principia.'" *Papers, Bibliographical Society of America* 47 (1953), 269–300.

Locations of 189 copies of the 1st ed. (Europe 124, USA 60, Canada 3, South Africa 1, Australia 1) and of the 1726 presentation issue (34 out of 50 copies).

857. Taton, René. "Inventaire des exemplaires des premières éditions des 'Principia' de Newton." *Revue d'histoire des sciences* 6 (1953), 60–63.

Supplements Macomber (856) by giving locations of copies in France.

858. Caspar, Max. *Kepler.* Translated and edited by C. Doris Hellman. London: Abelard-Schuman, 1959. 401 pp.

Biography by the dean of Kepler scholars. Reviewed in *American Historical Review* 66 (1960–61), 150–51. The German original (1948) was reviewed in *Isis* 41 (1950), 216–19.

859. De Santillana, Giorgio, and Stillman Drake. "Arthur Koestler and His Sleepwalkers." *Isis* 50 (1959), 255–60.

Essay review of Koestler's *Sleepwalkers* (item 860). His reply to criticism is in *Isis* 51 (1960), 74–79.

860. Koestler, Arthur. *The Sleepwalkers. A History of Man's Changing Vision of the Universe.* London: Hutchinson, 1959. 624 pp.

Pp. 119–511 cover Copernicus, Kepler, Tycho Brahe, and Galileo. The many reviews of this work are listed in *Isis* 51 (1960), 388 and 54 (1963), 540.

861. Needham, Joseph. "The Time of the Jesuits." *Science and Civilisation in China.* Cambridge: University Press, 1959. Vol. 3, pp. 437–58.

Note on the arrival of the Jesuits in China in the seventeenth century and their contribution to Chinese astronomy.

862. Hellman, C. Doris. "The Gradual Abandonment of the Aristotelian Universe: A Preliminary Note on Some Sidelights." *Mélanges Alexandre Koyré*, Vol. 1, *L'Aventure de la science*. Histoire de la Pensée, 12. Paris: Hermann, 1964, pp. 283–93.

"The dramatic appearances of comets and novae" and their role in the transition from the Aristotelian to the Copernican universe.

863. Duhem, Pierre. *To Save the Phenomena. An Essay on the Idea of Physical Theory from Plato to Galileo.* Chicago: Univ. of Chicago Press, 1969. 120 pp. [Translated from the French (1908)]

Pp. 46–120: The Renaissance before Copernicus; Copernicus and Rheticus; from Osiander's Preface, to the Gregorian reform of the calendar and to the condemnation of Galileo. With an introductory essay by Stanley L. Jaki (pp. ix–xxvi).

864. Meadows, A. J. *The High Firmament: A Survey of Astronomy in English Literature.* Leicester: Leicester Univ. Press, 1969. 207 pp.

Pp. 1–148 cover the period from about 1400 to the Scientific Revolution. Documented but no bibliography. With index of names and subjects. Reviewed in *TLS* 12 June 1969, p. 634.

865. Wilson, C. A. "From Kepler's Laws, So-called, to Universal Gravitation: Empirical Factors." *Archive for History of Exact Sciences* 6 (1970), 89–170.

Contends that the Keplerian revolution rests, in the last resort, upon conjecture, and that the "so-called" Keplerian laws cannot be verified satisfactorily. Examines Newton's arguments for universal gravitation.

866. Cohen, I. Bernard. *Introduction to Newton's 'Principia.'* London: Cambridge Univ. Press, 1971. 380 pp.

Introduction to the variorum ed. of the *Principia* by Koyré and Cohen. Contains a detailed history of the work from the completion of the MS for the 1st ed. (1685–87) printing to the printing of the 3rd ed. (1726). Reviews: *Isis* 63 (1972), 439–40; *Journal of Modern History* 46 (1974), 116–19.

867. Rauffner, J. A. "The Curved and the Straight: Cometary Theory from Kepler to Hevelius." *Journal for the History of Astronomy* 2 (1971), 178–94.

On the change from circular to rectilinear trajectories in cometary theory.

868. Aiton, E. J. *The Vortex Theory of Planetary Motions.* London: Macdonald, 1972. 282 pp.

Pp. 30–151: History of the theory (replacement of the solid spheres by a system of fluid vortices) from Descartes to Leibniz.

869. Dobrzycki, Jerzy, ed. *The Reception of Copernicus' Heliocentric Theory.* Proceedings of a Symposium Organized by the Nicolas Copernicus Committee of the International Union of the History and Philosophy of Science, Torun, Poland, 1973 [*sic*]. Dordrecht: Reidel, 1972. 368 pp.

Eleven papers on the impact of Copernicus' theory on scientific developments in centers of learning abroad. Reviewed in *Archives internationales d'histoire des sciences* 26 (1976), 177–79.

870. Gingerich, O. "Johannes Kepler and the New Astronomy." *Quarterly Journal of the Royal Astronomical Society* 13 (1972), 346–73.

On Kepler's *Astronomia nova*. Includes a translation (with William Walderman) of the preface to the *Tabulae Rudolphinae* (1627). The George Darwin lecture, 1971.

871. "Immortali Memoriae Nicolai Copernici Canonici Warmiensis Astronomi Clarissimi. [Copernicus Gedächtnisband]." *Zeitschrift für die Geschichte und Altertumskunde Ermlands,* Band 36, Heft 97 (1972), 1–238

Entire issue devoted to the Copernicus quincentenary (six articles and fifteen notices of books). Of special interest are the articles by B. Poschmann (The contribution of Warmian historians to the research on Copernicus, pp. 11–32), B. M. Rosenberg (Attempt at reconstructing Copernicus's library, pp. 134–60), and W. Thimm (On Sikorski's Copernicus chronology, pp. 173–98; Chronological table, pp. 233–38). In German with English summaries.

872. Bieńkowska, Barbara, ed. *The Scientific World of Copernicus. On the Occasion of the Five-Hundredth Anniversary of His Birth, 1473–1973.* Dordrecht: Reidel, 1973. 143 pp.

Nine essays by Polish scholars on the life and work of Copernicus and the impact of heliocentrism on European science and culture. Reviewed in *Isis* 66 (1975), 576–77.

873. *Internationales Kepler-Symposium, Weil der Stadt 1971.* Referate und Diskussionen herausgegeben von Fritz Krafft, Karl Meyer, Bernhard Sticker. Arbor scientarum, Beiträge zur Wissenschaftsgeschichte, Reihe A: Abhandlungen, Bd. 1. Hildesheim: Gerstenberg, 1973. 490 pp.

Seventeen papers under five topics (with discussions) read at a symposium held in Kepler's native town, in which attention was focussed on the lesser-known aspects of his work. Concludes with a study of his character and personality by B. Sticker (pp. 455–74). Reviews: *Studies in History and Philosophy of Science* 4 (1974), 387–92; *Studia Leibnitiana* 6 (1974), 150–53.

874. Michel, Paul Henry. *The Cosmology of Giordano Bruno.* Paris: Hermann, 1973. 306 pp. [Translated from the 1962 French ed.]

"Internalist study of the cosmological ideas in Bruno's extant writings." Reviewed in *British Journal for the History of Science* 7 (1974), 293–94.

875. "Symposium on Copernicus." *Proceedings, American Philosophical Society* 117 (1973), 413–550.

Four papers including: "The Derivation and First Draft of Copernicus's Planetary Theory. A Translation of the *Commentariolus* with Commentary" by Noel M. Swerdlow, and "Michael Maestlin's Account of Copernican Planetary Theory" by Anthony Grafton.

876. Hall, A. Rupert. "Newton and His Editors." *Proceedings, Royal Society,* A, 338 (1974), 397–417.

On the publication of the *Principia* and on Newton's relationships with Halley, Cotes, and Pemberton.

877. Maeyama, Y. "The Historical Development of Solar Theories in the Late Sixteenth and Early Seventeenth Centuries." *Vistas in Astronomy* 16 (1974), 35–60.

Study of the order of accuracy of planetary theories from Tycho Brahe to Newton, Flamsteed, and Halley.

878. Beer, A., and K. Aa Strand, eds. "Copernicus Yesterday and To-day. Proceedings of the Commemorative Conference Held [in December 1972] in Washington in Honour of Nicolaus Copernicus." *Vistas in Astronomy* 17 (1975), 1–225.

Fifteen papers presented at five sessions: Scientific significance of Copernicus; Humanistic significance of Copernicus; Nature of scientific revolutions; Humanistic significance of the Copernican heritage; Impact of Copernicus and his work. Reviewed in *Annali, Istituto e Museo di Storia della Scienza, Firenze* 1 (1976), 102–7 and in *Physis* 17 (1975), 305–8.

879. Beer, Arthur, and P. Beer, eds. "Kepler: Four Hundred Years. Proceedings of Conferences Held in Honour of Johannes Kepler." *Vistas in Astronomy* 18 (1975), 1–1034.

Texts in English of symposia held (in Philadelphia, Leningrad, Graz, Linz, London, Paris, Berlin, and Evanston) to commemorate the fourth centenary of Kepler's birth. Arranged by topic. Subjects include: Kepler's life, his wanderings across Europe, his relations with his scientific contemporaries, his influence on the new astronomy; study of his thoughts and beliefs; Kepler's mathematics and physics, and the development of his astronomical theories. With a bibliography (see item 838). Reviewed in: *Annali, Istituto e Museo di Storia della Scienza, Firenze* 2 (1977), 119–22; *Physis* 18 (1976), 217–20; *Journal for the History of Astronomy* 7 (1976), 57–61.

880. Bennett, J.A. "Christopher Wren: Astronomy, Architecture, and the Mathematical Sciences." *Journal for the History of Astronomy* 6 (1975), 149–84.

Wren's intellectual background and training in mathematics and astronomy is important for an understanding of his mature work in architecture.

881. Bennett, J. A. "Hooke and Wren and the System of the World: Some Points towards an Historical Account." *British Journal for the History of Science* 8 (1975), 32–61.

Examines the early development of the ideas of Hooke and Wren on planetary motion.

882. Forbes, Eric G., ed. *The Gresham Lectures of John Flamsteed.* London: Mansell, 1975. 479 pp.

Edition of the hitherto unpublished MS with an introduction (pp. 1–78). The lectures provide a "deeper insight into, and clearer understanding of, the problems confronting practical astronomy in the very decade when Newton conceived his *Principia* . . . (1687)." Much of the information found here about "the early phase of the first Astronomer Royal's scientific career has never been incorporated into any account (including Baily's) of his life and work" (Preface).

883. Swerdlow, Noel M. "The Planetary Theory of François Viète. 1. The Fundamental Planetary Models." *Journal for the History of Astronomy* 6 (1975), 185–208.

Viète's unpublished studies of the planetary theory of Ptolemy and Copernicus.

884. Westman, Robert S., ed. *The Copernican Achievement.* UCLA Center for Medieval and Renaissance Studies, Contributions 7. Berkeley and Los Angeles: Univ. of California Press, 1975. 405 pp.

Nine papers (with introduction and commentaries) presented at one of the quincentenary symposia. Topics vary from "technical details of Copernicus's astronomy to the philosophical assessment of its revolutionary character." Name and subject index (pp. 393–405). Reviews: *Journal for the History of Astronomy* 8 (1977), 211–13; *Polish Review* 21 (1976), 225–35.

885. Aiton, E. J. "Johannes Kepler in the Light of Recent Research." *History of Science* 14 (1976), 77–100.

A critical survey of symposia, exhibitions, and lectures commemorating the quatercentenary of Kepler's birth.

886. Heninger, S. K., Jr. *The Cosmographical Glass: Renaissance Diagrams of the Universe.* San Marino, Calif.: Huntington Library, 1977. 209 pp.

A collection of diagrams from 117 pre-1700 texts depicting the world, in one way or another, as someone in the Renaissance saw it. Reviewed in *Isis* 70 (1979), 617.

887. Czartoryski, Pawel, ed. "Science and History: Studies in Honor of Edward Rosen." *Studia Copernicana* 16 (1978), 1–553.

Thirty papers mostly dealing with Copernicus and the astronomy of his time. Of these, four deal with science and society. Pp. 9–22: Bibliography of Rosen.

888. Jaki, Stanley T. *Planets and Planetarians: A History of Theories of the Origin of Planetary Systems.* New York: Wiley, 1978. 266 pp.

Renaissance to the seventeenth century covered on pp. 12–86. Reviewed in *Isis* 72 (1981), 118.

889. Righini, Guglielmo. *Contributo alla interpretazione scientifica dell' opera astronomica di Galileo.* Istituto e Museo di Storia della Scienza, Firenze, Monografia 2. Florence: Museo di Storia della Scienza, 1978. 115 pp.

Galileo's work viewed by a modern astronomer. Reviewed in *Isis* 70 (1979), 621–22.

890. Shumaker, Wayne, ed. *John Dee on Astronomy.* Propaedeumata Aphoristica (*1558 and 1568*), *Latin and English.* Edited and translated, with general notes . . . . With an introductory essay on Dee's mathematics and physics and his place in the Scientific Revolution by J. L. Heilbron. Berkeley and Los Angeles: Univ. of California Press, 1978. 264 pp.

Pp. 1–49: Dee's role in the Scientific Revolution. Pp. 50–99: *Propaedeumata Aphoristica* (astrology as applied mathematics. Physics in the *Propaedeumata.* Geocentric astronomy. Astrology). Pp. 105–99: Text (Latin and English). Pp. 203–64: Notes, bibliography, and index. Essay review in *British Journal for the History of Science* 13 (1980), 255–58.

891. Westman, R. S. "The Astronomer's Role in the Sixteenth Century: A Preliminary Study." *History of Science* 18 (1980), 105–47.

Topics: Disciplinary communities and roles in the early modern period; The astronomer's role according to Osiander and Copernicus; The astronomer in the university; The court astronomer; The academic reformers (Ramus, Savile, and Clavius).

892. Donahue, William H. *The Dissolution of the Celestial Spheres, 1595–1650.* The Development of Science, Sources for the History of Science. New York: Arno Press, 1981. 330 pp.

Shows how and suggests why "cosmology as taught in the universities changed, successfully abandoning the planetary sphere and the stellar sphere." Based on the author's Ph.D. thesis, University of Cambridge, 1972.

893. Gingerich, Owen. "The Censorship of Copernicus' *De revolutionibus.*" *Annali, Istituto e Museo di Storia della Scienza, Firenze* 6 (1981), (2), 45–61.

A survey of some five hundred copies of the book reveals that about eight percent of all sixteenth-century copies were censored according to the decree of the Holy Office in 1620. Contains a map showing current locations of the copies.

894. Schofield, Christine Jones. *Tychonic and Semi-Tychonic World Systems.* New York: Arno Press, 1981. 398 pp.

Study of the "many ramifications—often scientifically abortive—of the work of Copernicus." Arranged under: (1) The state of astronomy by the end of the sixteenth century. (2) Tycho Brahe and the geoheliocentric planetary system. (3) Other forms of geoheliocentric planetary system proposed by Tycho's contemporaries. (4) Seventeenth-century debate over the world system. Pp. 377–98: Select bibliography.

895. Cavazza, Marta. "La cometa del 1680–1681: astrologi e astronomi confronto." *Studi e memorie per la storia dell'Università di Bologna* 3 (1983), 409–66.

NOT SEEN

896. Coyne, G. V., M. A. Hoskin, and O. Pedersen, eds. *Gregorian Reform of the Calendar. Proceedings of the Vatican Conference to Commemorate Its Four-Hundredth Anniversary, 1582–1982.* Vatican City: Specola Vaticana, 1983. 321 pp.

Fourteen papers of which ten relate to the sixteenth and seventeenth centuries. Reviewed in *Isis* 75 (1984), 732–33 and in *Sky and Telescope* 67 (1984), 330–31.

897. Grant, Edward. *In Defense of the Earth's Centrality and Immobility: Scholastic Reaction to Copernicanism in the Seventeenth Century.* Transactions, American Philosophical Society, Vol. 74(4). Philadelphia, 1984. 69 pp.

Study of scholastic Aristotelians who opposed Copernicanism. Reviews: *Isis* 76 (1985), 378–79; *British Journal for the History of Science* 19 (1986), 126.

898. Swerdlow, Noel, and Otto Neugebauer. *Mathematical Astronomy in Copernicus's* De Revolutionibus. Studies in the History of Mathematics and Physical Sciences, 10. New York: Springer-Verlag, 1984. 2 vols. 711 pp.

Study of Copernicus's mathematical astronomy from a "technical, rather than a philosophical or sociological point of view." Vol. 1, pp. 3–85 contains a biographical stretch and an account of his physical astronomy. "It is our, perhaps naive, belief that when Copernicus wrote that mathematics is written for mathematicians, he meant it, and that his work may best be understood within the history of mathematical astronomy" (p. 94). Reviewed in: *Renaissance Quarterly* 40 (1987), 109–12; *Annals of Science* 43 (1986), 596–97.

899. Van Helden, A. *Measuring the Universe. Cosmic Dimensions from Aristarchus to Halley.* Chicago: Univ. of Chicago Press, 1985. 203 pp.

Pp. 41–163 deal with the sixteenth and seventeenth centuries. Reviewed in: *Annals of Science* 43 (1986), 557–61; *Journal for the History of Astronomy* 18 (1987), 132–33.

900. Gingerich, Owen. "Copernicus's *De Revolutionibus*: An Example of Renaissance Scientific Printing." *Print and Culture in the Renaissance. Essays on the Advent of Printing in Europe.* Gerald P. Tyson and Sylvia S. Wagonheim, eds. Newark: Univ. of Delaware Press, 1986, pp. 55–73.

Study based on the examination of some 540 copies of the two eds. This "erudite and potentially inaccessible work was widely distributed and appreciated by a surprisingly large number of scholars, scientists, and other educated owners."

901. Kunitzsch, Paul. *Peter Apian und Azorphi: Arabische Sternbilder in Ingolstadt in frühen 16. Jahrhundert.* Bayerische Akademie der

Wissenschaften, philos. hist. Kl., Sitzungsberichte, 1986, Heft 3. Munich, 1986. 71 pp.

Study of Apian's use of a treatise on the constellations by the tenth-century Persian astronomer al-Sufi. For a shorter version of this study (in English) see *Journal for the History of Astronomy* 18 (1987), 117–24.

902. Rosen, Edward. *Three Imperial Mathematicians: Kepler Trapped between Tycho Brahe and Ursus.* New York: Abaris, 1986. 384 pp.

"Fascinating story of the three Imperial mathematicians" explored as far as possible in the words of the protagonists. With biographical information on many lesser-known figures in the Brahe circle. Reviews: *Isis* 79 (1988), 297–99; 80 (1989), 486–87 (letters); *British Journal for the History of Science* 20 (1987), 235–36.

903. Ariew, Robert. "The Phases of Venus before 1610." *Studies in History and Philosophy of Science* 18 (1987), 81–92.

On recent papers about the discovery of the phases of Venus. "By themselves, the phases of Venus are not sufficient to argue for the new cosmology. Galileo's achievement and the real novelty in our episode is not the discovery of the phases of Venus, but the discovery that the telescope reveals the true light of the heavenly bodies" (p. 92).

904. Dekker, E. "Early Exploration of the Southern Celestial Sky." *Annals of Science* 44 (1987), 439–70.

Outlines the background and content of the explorations by Dutch navigators at the end of the sixteenth century.

905. Harrison, Edward. *Darkness at Night. A Riddle of the Universe.* Cambridge, Mass.: Harvard Univ. Press, 1987. 293 pp.

The author attempts to answer the question: Why is the sky dark at night? In a universe that perhaps stretches away witnout limit and contains an unlimited number of stars, why do we not see starlight at every point of the sky? Pp. 34–80: From Thomas Digges to Edmund Halley. Reviewed in *TLS* 22 July 1988, p. 800 and *Isis* 79 (1988), 703–4.

906. Picolet, Guy, ed. *Jean Picard et les débuts de l'astronomie de précision au XVIIe siècle.* Paris: Editions du CNRS, 1987. 382 pp.

Seventeen papers, on different aspects of Picard's work, presented at a symposium commemorating the third centenary of his death (biography; astronomy; geodesy; cartography; surveying). Reviewed in *Annals of Science* 46 (1989), 433–35.

907. Stephenson, Bruce. *Kepler's Physical Astronomy*. Studies in the history of mathematics and physical sciences, 13. Berlin: Springer Verlag, 1987. 216 pp.

Claims "to explore and explain the development of Kepler's planetary theory, and of the physical hypotheses integral to that theory, more faithfully than has yet been done." Reviewed in: *British Journal for the History of Science* 21 (1988), 372–74; *Journal for the History of Astronomy* 19 (1988), 280–82.

908. McMullin, Ernan. "Bruno and Copernicus." *Isis* 78 (1987), 55–74.

"Bruno constructed the outline of an original natural philosophy, drawing on the rich resources of the Neoplatonic tradition to offer a radical alternative to the natural philosophy of the 'mathematicians.'"

909. Field, J. V. *Kepler's Geometrical Cosmology*. London: Athlone Press, 1988. 243 pp.

Aims at placing Kepler's cosmological works in the context of his work as a whole—in particular, his astronomy. Pp. 220–35: Notes, glossary, and bibliography. Reviewed in: *Isis* 79 (1988), 539–40; *Renaissance Quarterly* 42 (1989), 112–14; *Review of Metaphysics* 42 (1989), 826–28. Also *Centaurus*, 32 (1989) B. Stephenson.

910. Gingerich, Owen, and Robert S. Westman. *The Wittich Connection: Conflict and Priority in Late Sixteenth-Century Cosmology*. Transactions, American Philosophical Society, Vol. 78 (7). Philadelphia, 1988. 148 pp.

Discusses the "catalytic role" played by Paul Wittich in the development of late sixteenth-century cosmologies. Provides a new look at the origins of Tycho's planetary system.

911. Taton, René, and Curtis Wilson, eds. *Planetary Astronomy from the Renaissance to the Rise of Astrophysics*. Part A: *Tycho Brahe to Newton*. The General History of Astronomy, General editor, Michael Hoskin, Vol. 2. Published under the auspices of the International Astronomical Union and the International Union for the History and Philosophy of Science. Cambridge: Cambridge Univ. Press, 1989. 274 pp.

In four sections: Tycho, Gilbert and Kepler; The impact of the telescope; Planetary, lunar, and cometary theories between Kepler and Newton; The Newtonian achievement in astronomy. With index of names and subjects. No notes, but each chapter has a list of works for further reading.

912. Dekker, Elly. "The Light and the Dark: A Reassessment of the Discovery of the Coalsack Nebula, the Magellanic Clouds and the Southern Cross." *Annals of Science* 47 (1990), 529–60.

An analysis of early observations of the southern celestial sky as reported in some sixteenth-century books and compilations of voyages of discovery.

913. Thoren, Victor E. *The Lord of Uraniborg. A Biography of Tycho Brahe.* With contributions by John R. Christianson. Cambridge: Cambridge Univ. Press, 1990. 523 pp.

A comprehensive study of the life and work of Tycho Brahe taking into account the results of recent scholarship. Reviews: *Observer* (London) 9 June 1991, p. 59; *Isis* 83 (1992), 658–60.

914. Zinner, Ernst. *Regiomontanus: His Life and Work.* Trans. by Ezra Brown. Studies in the History and Philosophy of Mathematics, Vol. 1. Amsterdam: North-Holland, 1990. 402 pp. [Trans. of the 1968 German 2nd ed. 1st ed. 1938]

Includes supplementary articles and bibliographies (pp. 289–402) by Wolfgang Kaunzner, Rudolf Mett, Hans Wussing, Felix Schmeidler, Armin Gerl, Karin Reich and Menso Folkerts. Pp. 195–260: Notes; pp. 261–69: Bibliography; pp. 271–87: Index.

915. Kelly, John T. *Practical Astronomy during the Seventeenth Century. Almanac-Makers in America and England.* New York: Garland Publishing, 1991. 319 pp.

Study in five chapters: 1. The prominence of almanacs in the history of astronomy in colonial America. 2. Computational astronomy in England 1500–1660: a study of the role of almanacs and their authors. 3. American almanac-making during the seventeenth century: a technical assessment. 4. The influence of universities on the development of computational astronomy in England and America: a study of practical astronomy at Cambridge, Oxford, and Harvard. 5. Comparison of the development of computational astronomy in Continental Europe, England, and America, 1500–1800. An overview. Pp. 285–319: Bibliography.

916. Whiteside, D. T. "The Prehistory of the Principia from 1664 to 1686." *Notes and Records, Royal Society of London* 45 (1991), 11–61.

Combined text of two talks given in 1987 to commemorate the tercentenary of the publication of the *Principia* (1687).

917. Yeomans, Donald K. *Comets. A Chronological History of Observation, Science, Myth, and Folklore.* New York: John Bailey and Sons, 1991. 485 pp.

Study of the development of cometary ideas from antiquity to 1986. Pp. 24–119 cover the years from Paolo Toscanelli (1397–1482) to Halley. Pp. 425–73: Bibliography; pp. 473–85: Index.

917A. Grant, Edward. *Planets, Stars and Orbs. The Medieval Cosmos, 1200–1687.* Cambridge: Cambridge University Press, 1994. 816 pp.

Describes the extraordinary range of ideas that constituted scholastic cosmology for some five hundred years and the ways in which scholastic natural philosophy of the sixteenth and seventeenth centuries responded to rival interpretations, especially Copernicanism. Pp. 681–741: Catalog of questions on medieval cosmology, 1200–1687; pp. 776–97: Bibliography; pp. 798–816: Index.

Physics

BIBLIOGRAPHIES

918. Mottelay, Paul Fleury. *Bibliographical History of Electricity and Magnetism, Chronologically Arranged.* London: Griffin, 1922. 673 pp.

Chronological narrative from 2637 B.C. to a.d. 1821. Pp. 64–148 cover the period a.d. 1490 to 1700.

919. Schaffer, Simon. "History of Physical Science." *Information Sources in the History of Science and Medicine.* Editors: Pietro Corsi and Paul Weindling. London: Butterworth Scientific, 1983, pp. 285–314.

Brief survey from the Scientific Revolution to the end of classical physics. Twelve-page bibliography.

920. Home, R. W. *The History of Classical Physics. A Selected Annotated Bibliography.* With the assistance of Mark J. Gittins. Bibliographies of the History of Science and Technology, 8. New York: Garland, 1984. 324 pp.

Although intended as a bibliography of the history of post-1700 physics, many titles cover the earlier years.

MONOGRAPHS AND ARTICLES

921. Mach, Ernst. *The Science of Mechanics. A Critical and Historical Exposition of Its Principles.* London: Watts and Co., 1893. 534 pp. [Translated from the second German ed. (1888)]

Pp. 24–335: The principles of statics and of dynamics and their extension, from Stevin to Jakob Bernoulli I and Brook Taylor.

922. Boutroux, Pierre. "L'enseignement de la mécanique en France au XVIIe siècle." *Isis* 4 (1921–22), 276–94.

Points out that seventeenth-century textbooks did not reflect the pattern of scientific change that took place during the century.

923. Charbonnier, Prosper Jules. *Essais sur l'histoire de la balistique.* Paris: Société d'Editions Géographiques, Maritimes et Coloniales, 1928. 338 pp.

Written for artillery officers. Treatment mathematical. Pp. 1–103 cover the period from Tartaglia (1537) to Johann Bernoulli (1719). Bibliography.

924. Zilsel, Edgar. "The Origins of William Gilbert's Scientific Method." *Journal of the History of Ideas* 2 (1941), 1–32.

Analysis of the *De magnete* and the *De mundo*. "When reading *De magnete* we must never forget that twelve years before its publication English ships and English iron guns annihilated the Spanish Armada, then the most powerful fleet in the world. England, the country of iron mines and advancing navigation, produced the first learned book on experimental physics."

925. Whittaker, Edmund T. *A History of the Theories of Aether and Electricity. The Classical Theories.* Revised ed. London: Thomas Nelson, 1951. 434 pp. [First published 1910]

Pp. 1–40 cover the history up to the death of Newton. Reviewed in *Isis* 2 (1914), 222–24.

926. Hall, A. Rupert. *Ballistics in the Seventeenth Century: A Study in the Relations of Science and War with Reference Principally to England.* Cambridge: Cambridge Univ. Press, 1952. 186 pp.

Study of ballistics in the light of the suggestion that practical needs helped its development. Pp. 172–81: Bibliography. Reviewed in *Isis* 44 (1953), 284–85.

927. Koyré, Alexandre. "A Documentary History of the Problem of Fall from Kepler to Newton." *Transactions, American Philosophical Society* 45 (1955), (4), 329–55.

Study of attempts to determine the trajectory of a falling body under the influence of a gravitational force on a rotating earth. Review: *Isis* 48 (1957), 91–93.

928. Dugas, René. *Mechanics in the Seventeenth Century. (From the Scholastic Antecedents to Classical Thought).* Neuchâtel: Griffon, 1958. 612 pp. [Translated from the 1954 French ed.]

"Attempts to show how not only the science of mechanics but also the mechanistic views of science and philosophy developed and influenced one another." Reviewed in *Isis* 47 (1956), 449–52.

929. Boyer, Carl B. *The Rainbow: From Myth to Mathematics.* New York: Thomas Yoseloff, 1959. 376 pp.

Study of man's recorded ideas on the subject. Pp. 143–268 cover the sixteenth century, Kepler and his contemporaries, Cartesian theory, and the Age of Newton. Reviewed in *Isis* 55 (1964), 218–20.

930. Dijksterhuis, E. J. "The Origins of Classical Mechanics from Aristotle to Newton." *Critical Problems in the History of Science.* Edited by Marshall Clagett. Madison: Univ. of Wisconsin Press, 1959, pp. 163–96.

Paper read at a symposium, with comments by Carl B. Boyer and A. Rupert Hall. See item 506.

931. Hesse, M. B. *Forces and Fields: The Concept of Action at a Distance in the History of Physics.* London: Thomas Nelson and Sons, 1961. 318 pp.

Pp. 74–163 deal with: Greek inheritance; corpuscular philosophy; theory of gravitation; action at a distance. Reviewed in *Philosophy of Science* 29 (1962), 434–35.

932. Maier, Annaliese. "Galilei und die scholastische Impetustheorie." *Ausgehendes Mittelalter.* Rome: Edizioni di Storia e Letteratura, 1967, Vol. 2, pp. 465–90.

Analyses "the version of the theory of impetus advanced by Galileo in his early unpublished treatise on motion *De motu* (ca. 1590) and shows how it differs from the standard scholastic treatment of impetus and conflicts with the modern concept of inertia." For a translation of this paper, see *On the Threshold of Exact Science: Selected Writings of Annaliese Maier on Late Medieval Philosophy.* Edited and translated with an introduction by Steven D. Sargent. Philadelphia: Univ. of Pennsylvania Press, 1982, pp. 103–23.

933. Sabra, A. I. *Theories of Light from Descartes to Newton.* London: Oldbourne, 1967. 365 pp.

Study of some important problems and controversies in the development of seventeenth-century theories about the nature of light. Pp. 343–53: Bibliography. Reviewed in *Isis* 63 (1972), 445–46.

934. Drake, Stillman, and I. E. Drabkin, eds. *Mechanics in Sixteenth-Century Italy. Selections from Tartaglia, Benedetti, Guido Ubaldo, and Galileo.* Madison: Univ. of Wisconsin Press, 1969. 428 pp.

Pp. 3–60: Introduction by Stillman Drake (Ancient and medieval traditions; the older traditions in the sixteenth century; The two Italian schools of mechanics; Niccolò Tartaglia; Girolamo Cardano; Giovanni Battista Benedetti; Federico Commandino; Guido Ubaldo del Monte; Bernardino Baldi and some minor mechanicians; Galileo). Pp. 391–420: Bibliography. Reviewed in *Isis* 60 (1969), 565–66. See also item 773.

935. Centore, F. F. *Robert Hooke's Contributions to Mechanics: A Study in Seventeenth Century Natural Philosophy.* The Hague: Martinus Nijhoff, 1970. 136 pp.

"Hooke represents the last great Baconian in the history of science."

936. Drake, Stillman. *Galileo Studies: Personality, Tradition, and Revolution.* Ann Arbor, Mich.: Univ. of Michigan Press, 1970. 289 pp.

Collection of essays in which the author "attempts to reconstruct Galileo's thought mainly on the basis of the internal evidence of his writings and psychological plausibility." Reviewed in *Isis* 62 (1971), 546–47.

937. Ronchi, Vasco. *The Nature of Light: An Historical Survey.* Cambridge, Mass.: Harvard Univ. Press, 1970. 288 pp. [Translation of the 2nd Italian ed. Originally published in Italian in 1939. French translation published in 1956]

Popular exposition of the history of optics (pp. 78–208: Sixteenth and seventeenth centuries). Criticized by David C. Lindberg in *Isis* 62 (1971), 522–24. Italian ed. reviewed by I. B. Cohen in *Isis* 33 (1941), 294–96. See item 944.

938. Rose, P. L., and Stillman Drake. "The Pseudo-Aristotelian *Questions of Mechanics* in Renaissance Culture." *Studies in the Renaissance* 18 (1971), 65–104.

On the place of the *Questions of Mechanics* in the development of modern science.

939. Westfall, R. S. *Force in Newton's Physics: The Science of Dynamics in the Seventeenth Century.* London: Macdonald, 1971. 579 pp.

Seventeenth-century dynamics as seen through the eyes of the men who created the science. Pp. 551–65: Bibliography of secondary sources. Reviewed in *Isis* 63 (1972), 242–44.

940. Ziggelaar, August, S.J. *Le physicien Ignace Gaston Pardies, S.J. (1636–1673).* Acta Historica Scientiarum Naturalium et Medicinalium, 26. Odense: Odense Univ. Press, 1971. 242 pp.

Detailed study of Pardies, one of the many "savants secondaires" of the seventeenth century, "best known for his opposition to Newton's early theory of light and colours." Reviewed in *Isis* 64 (1973), 268–69.

941. Ruestow, Edward G. *Physics at Seventeenth- and Eighteenth-Century Leiden: Philosophy and the New Science in the University.* Archives Internationales d'Histoire des Idées, series minor, 11. The Hague: Martinus Nijhoff, 1973. 174 pp.

Pp. 1–112 deal with the seventeenth century. Leiden developed as one of Europe's most effective centers for the propagation of the methods and attitudes of modern science. Reviewed in *Historia Mathematica* 3 (1976), 96–100.

942. Shapiro, A. E. "Kinematic Optics: A Study of the Wave Theory of Light in the Seventeenth Century." *Archive for History of Exact Sciences* 11 (1973), 134–266.

943. Clavelin, Maurice. *The Natural Philosophy of Galileo. Essay on the Origins and Formation of Classical Mechanics.* Cambridge, Mass.: MIT Press, 1974. 498 pp. [Translated from the 1968 French ed.]

Study of Galileo's creation of the modern science of motion. Reviewed in *Isis* 61 (1970), 275–77.

944. Ronchi, Vasco. "How the History of Science Should Be Approached." *Atti della Fondazione Giorgio Ronchi* 29 (1974), (1), 47–72.

Reply to Lindberg's criticism of Ronchi's *The Nature of Light.* See item 937.

945. Lindberg, David C. *Theories of Vision from Al-Kindi to Kepler.* Chicago: Univ. of Chicago Press, 1976. 324 pp.

Pp. 142–213: Artists and anatomists of the Renaissance; Johannes Kepler and the theory of the retinal image; The translation of optical works from Greek and Arabic into Latin.

946. Clagett, Marshall. *Archimedes in the Middle Ages.* Vol. 3: *The Fate of the Medieval Archimedes.* Part III; *The Medieval Archimedes in the Renaissance, 1450–1565.* Memoirs, American Philosophical Society, Vol. 125, part B. Philadelphia, 1978. Pp. 297–1246.

Pp. 321–475: Jacobus Cremonensis, Regiomontanus, Piero della Francesca, Luca Pacioli, Giorgio Valla. Pp. 477–523: Leonardo da Vinci. Pp. 524–635: The fate of William of Moerbeke's translation in the sixteenth century (Andreas Corner, Niccolò Tartaglia, Federico Commandino). Pp. 749–812: Francesco Maurolico. Pp. 1055–224: Archimedean problems in the first half of the sixteenth century. Pp. 1225–46: The medieval traditions of Archimedes in the Renaissance: a retrospective survey.

947. Drake, Stillman. *Galileo at Work: His Scientific Biography.* Chicago: Univ. of Chicago Press, 1978. 536 pp.

A chronicle of Galileo's scientific activity. Reviewed in *Isis* 70 (1979), 273–75.

948. Hunt, Frederick Vinton. *Origins in Acoustics. The Science of Sound from Antiquity to the Age of Newton.* New Haven, Conn.: Yale Univ. Press, 1978. 196 pp.

Pp. 73–155 cover the period 1450–1700. Reviewed in *Isis* 70 (1979), 287–88.

949. Galluzzi, Paolo. *Momento: studi galileani.* Rome: Ateneo e Bizzari, 1979. 435 pp.

Study of the notion of "momentum" in the works of Galileo. With a history of the term from antiquity to 1600.

950. Heilbron, J. L. *Electricity in the Seventeenth and Eighteenth Centuries: A Study of Early Modern Physics.* Berkeley: Univ. of California Press, 1979. 606 pp.

A survey of the work of some two hundred "electricians." Pp. 9–166: Early modern physics and its cultivators. Pp. 167–226: Electricity in the seventeenth century. Reviews: *Isis* 72 (1981), 480–89; *Annals of Science* 38 (1981), 477–82.

951. Lewis, Christopher. *The Merton Tradition and Kinematics in Late Sixteenth and Early Seventeenth Century Italy.* Padua: Antenore, 1980. 328 pp.

Deals with "the continuing influence of the scientific tradition stemming from Merton College, Oxford among a selection of Galileo's contemporaries as well as Galileo himself."

952. Cohen, I. Bernard. "Newton's Discovery of Gravity." *Scientific American* 244 (1981), (3), 123–33.

"How did he come to develop the concept that marked the beginning of modern science? In essence he did so by repetitively comparing the real world with a simplified mathematical representation of it."

953. Schmitz, E. H. *Handbuch zur Geschichte der Optik.* Band 1. *Von der Antike bis Newton.* Bonn: Verlag J. P. Wayenborgh, 1981. 450 pp.

Pp. 124–411: From Leonardo da Vinci to Newton. Pp. 413–41: Literature references.

954. Schneer, Cecil J. "Aspects of Form and Structure: The Renaissance Background to Crystallography." *The Analytic Spirit: Essays in the History of Science in Honor of Henry Guerlac.* Edited by Harry Woolf. Ithaca: Cornell Univ. Press, 1981, pp. 278–92.

Sketches some aspects of the fifteenth- and sixteenth-century background out of which Kepler was to build the views of the relationships between crystals and solid geometry expressed in his works.

955. Blay, Michel. *La conceptualisation newtonienne des phénomènes de la couleur.* Paris: Vrin, 1983. 307 pp.

Study of the colour phenomenon from the classical theories to Newton's mathematisation. Includes translations of material from Newton's notebooks of the 1660s. Pp. 201–71: Notes. Pp. 273–93: Bibliography. Reviewed in *Annals of Science* 41 (1984), 410–13.

956. Heilbron, J. L. *Physics at the Royal Society during Newton's Presidency.* Los Angeles: William Andrews Clark Memorial Library, Univ. of California, 1983. 123 pp.

Study based on the *Journal Book* and the correspondence of the active fellows.

957. McGuire, J. E., and Martin Tamny. *Certain Philosophical Questions: Newton's Trinity Notebook.* Cambridge: Cambridge Univ. Press, 1983. 519 pp.

A study and ed. of the notebook which Newton began during his first year at Cambridge. "It allows us to understand the beginnings of an intellectual journey that culminated with the publication of the great *Principia Mathematica* in 1686 [*sic*] and the *Opticks* in 1704" (authors). Pp. 3–325: Commentary. Pp. 328–465: Transcription and expansion of *Questiones quaedam philosophiae.* With appendix, glossary, and select bibliography. Reviewed in *History of Science* 22 (1984), 93–97.

958. Lindberg, David C. "Laying the Foundations of Geometrical Optics: Maurolico, Kepler, and the Medieval Tradition." *The Dis-*

course of Light from the Middle Ages to the Enlightenment. Papers Read at a Clark Library Seminar, 24 April 1982. By David C. Lindberg and Geoffrey Cantor. Los Angeles: William Andrews Clark Memorial Library, University of California, 1985, pp. 1–65.

Historical study of the principle of rectilinear propagation of light.

959. Shapin, Steven, and Simon Schaffer. *Leviathan and the Air-Pump. Hobbes, Boyle, and the Experimental Life*. Princeton, N.J.: Princeton Univ. Press, 1985. 440 pp.

Study of the controversies "over the status and value of experimental methods in natural philosophy." Essay review by P. B. Wood in *History of Science* 26 (1988), 102–8.

960. Bennett, J. A. "The Mechanics' Philosophy and the Mechanical Philosophy." *History of Science* 24 (1986), 1–28.

On the contemporary mechanical arts and the mathematical sciences and their influence on the emergence of a dominant mechanical tradition in the natural philosophy of the later seventeenth century.

961. Laird, W. R. "The Scope of Renaissance Mechanics." *Osiris* (2), 2 (1986), 43–68.

Describes "the changing scope of mechanics through the sixteenth and early seventeenth centuries in relation to the then-current categories of scientific knowledge." Based on the contemporary commentaries on the pseudo-Aristotelian *Mechanical Problems*.

962. Lindberg, David C. "The Genesis of Kepler's Theory of Light: Light Metaphysics from Plotinus to Kepler." *Osiris* (2), 2 (1986), 5–42.

A study of the tradition of Neoplatonic thought concerning light and of Kepler's place within it.

963. Wallace, William A., ed. *Reinterpreting Galileo*. Studies in Philosophy and the History of Philosophy, 15. Washington, D.C.: Catholic Univ. of America Press, 1986. 286 pp.

Nine essays (of which eight were presented as lectures in 1982) arranged in three parts: Historical influences on Galileo's work; Galileo's contributions to the science of his day; How Galileo's writings impinged on the Catholic Church's edict proscribing Copernicus's heliocentric system. No bibliography, but each essay is documented. "Though conceived and planned independently of Pope John Paul's call for a re-examination, with complete objectivity, of all aspects of the trial of the famous scientist in 1633, the volume may be seen as a frank and collaborative effort to set the record straight on this celebrated figure who has become, in the eyes of many, a symbol of conflict between science and religion" (Introduction). Reviewed in *Isis* 79 (1988), 543–44.

Music

964. Pirro, André. *Descartes et la musique.* Paris: Librairie Fischbacher, 1907. 127 pp.

On Descartes's *Musicae compendium* (1656) and later observations on music. Bibliography.

965. Locke, Arthur W. "Descartes and Seventeenth Century Music." *Musical Quarterly* 21 (1935), 423–31.

"Even though Descartes may not have had any direct or preponderant influence on the evolution of music during the seventeenth century, those elements of clarity and simplicity which are characteristic of that evolution unquestionably show a relation to—if they are not the result of—Descartes's rationalistic outlook."

966. Carpenter, Nan Cooke. *Music in the Medieval and Renaissance Universities.* Norman, Okla.: Univ. of Oklahoma Press, 1958. 394 pp.

An account of the history of music as a subject of higher learning from the foundation of the universities until the end of the Renaissance.

967. Palisca, Claude V. "Scientific Empiricism in Musical Thought." *Seventeenth Century Science and the Arts.* Edited by Hedley H. Rhys. Princeton: Princeton Univ. Press, 1961, pp. 91–137.

Analysis of the ways in which scientific thinking affected musical composition and music theory in the seventeenth century.

968. Ammann, Peter J. "The Musical Theory and Philosophy of Robert Fludd." *Journal of the Warburg and Courtauld Institutes* 30 (1967), 198–227.

On Fludd's musical works and the controversies with Kepler and Mersenne.

969. Blume, Friedrich. *Renaissance and Baroque Music. A Comprehensive Survey.* London: Faber and Faber, 1968. 180 pp. [Translated from the German (1949, 1963)]

An introduction to the understanding of music in the period 1450–1750 against the background of cultural history.

970. Drake, Stillman. "Renaissance Music and Experimental Science." *Journal of the History of Ideas* 31 (1970), 483–500.

On the influence of musical theory and practice upon early modern science.

971. Dostrovsky, Sigalia. "Early Vibration Theory: Physics and Music in the Seventeenth Century." *Archive for History of Exact Sciences* 14 (1974–75), 169–218.

972. Dickreiter, Michael. *Der Musiktheoretiker Johannes Kepler.* Neue Heidelberger Studien zur Musikwissenschaft, Bd. 5. Bern: Francke Verlag, 1974. 252 pp.

Technical study of Kepler's application of the theory of harmonic proportions to the explanation of the structure of the heavens. Reviews: *Isis* 67 (1976), 126–27; *Journal for the History of Astronomy* 7 (1976), 198–201.

973. Walker, D. P. *Studies in Musical Science in the Late Renaissance.* Studies of the Warburg Institute, Vol. 37. London, 1978. 174 pp.

Collection of previously published essays. Topics: Harmony of the spheres; Vincenzo Galilei and Zarlino; Galileo Galilei; Kepler's celestial music; The expressive value of intervals and the problem of the fourth; Mersenne's musical competition of 1640 and John Albert Ban; Seventeenth-century scientists' views on intonation. Reviewed in *Annals of Science* 37 (1980), 706–8.

974. Cohen, H. Floris. *Quantifying Music: The Science of Music at the First Stage of the Scientific Revolution, 1580–1650.* The University of Western Ontario Series in Philosophy of Science, 23. Dordrecht: Reidel, 1984. 308 pp.

On the "changing view of music from a discipline based on number to one founded on physical principles." Bibliography. Reviewed in *Isis* 76 (1985), 444–45.

975. Walker, D. P. *Music, Spirit and Language in the Renaissance.* Edited by Penelope Gouk. London: Variorum, 1985. 352 pp.

Sixteen studies published between 1941 and 1984 on musical theory in the sixteenth and seventeenth centuries and its place in the humanist revival of Orphic and Neoplatonist thought.

Chemistry

BIBLIOGRAPHIES

976. Fuchs, G. F. C. *Repertorium der chemischen Litteratur von 494 vor Christi Geburt bis 1806 in chronologischer Ordnung aufgestellt.* Mit einem Vorwort von Georg E. Dann. Hildesheim: Georg Olms Verlag, 1974. 2 vols. [Facsimile reprint of the 1806–1808 ed.]

Annotated bibliography, with name and subject indexes. Includes articles in journals. Vol. 1, pp. 39–214 relate to the sixteenth and seventeenth centuries.

977. Bolton, Henry Carrington. *A Select Bibliography of Chemistry. 1492–1892 (1902).* Smithsonian Miscellaneous Collections 850, 1170, 1253, 1440. Washington, D.C., 1893–1904. 4 Vols.

Bibliography in eight sections: Bibliographies; Dictionaries; History; Biography; Chemistry (pure and applied); Alchemy; Periodicals; Academic dissertations. Entries are arranged alphabetically by author except in the section for Biography where the arrangement is by biographee. The section for Dissertations (France, Germany, Russia, U.S.—mostly nineteenth century) contains a large number of dissertations from the *Ecoles de pharmacie* of Paris and Montpellier.

978. Ferguson, John. *Bibliotheca Chemica: A Catalogue of the Alchemical, Chemical and Pharmaceutical Books in the Collection of the Late James Young of Kelly and Durris, Esq.* Glasgow: James Maclehose and Sons, 1906. 2 vols. 514 pp.; 603 pp.

Catalogue of some fourteen hundred books bequeathed to Anderson College, Glasgow (now University of Strathclyde). Richly annotated. Reviewed in *Archives internationales d'histoire des sciences* 9 (1956), 155–56.

979. University of Glasgow Library. *Catalogue of the Ferguson Collection of Books, Mainly Relating to Alchemy, Chemistry, Witchcraft and Gipsies.* Glasgow: Maclehose, 1943–55. 2 vols. and suppl.

Catalogue of the working library of John Ferguson (see item 978). "Superior in scope, variety and rarity of constitutional elements to the Young collection." Not annotated.

980. Duveen, Denis Ian. *Bibliotheca Alchemica et Chemica. An Annotated Catalogue of Printed Books on Alchemy, Chemistry and Cognate Subjects in the Library of Denis I. Duveen.* London: E. Weil, 1949. 669 pp.

Some three thousand titles of which about 1,100 are pre-1700. Part of the collection has been acquired by the University of Wisconsin libraries. Arrangement alphabetical, by author. Reviewed in *Isis* 40 (1949), 387. See also *Journal of Chemical Education* 29 (1952), 244–47.

981. Neu, John, comp. *Chemical, Medical and Pharmaceutical Books Printed before 1800 in the Collections of the University of Wisconsin Libraries.* Madison, Wis.: Univ. of Wisconsin Press, 1965. 280 pp.

Alphabetical catalogue of 4,442 works, of which about 2,200 are pre-1700. Full entries but no annotations. No subject index.

982. Brock, W. H. "History of Chemistry." *Information Sources in the History of Science and Medicine.* Editors: Pietro Corsi and Paul Weindling. London: Butterworth Scientific, 1983, pp. 317–46.

Bibliographical essay.

983. Multhauf, Robert P. *The History of Chemical Technology: An Annotated Bibliography.* Bibliographies of the History of Science and Technology, 5; Garland Reference Library of the Humanities, 348. New York: Garland Publishing, 1984. 299 pp.

Includes many works of relevance to the history of technology before 1700. Pp. 253–99: Author and title indexes. Reviews: *Isis* 75 (1984), 723; *Technology and Culture* 26 (1985), 661–62; *Ambix* 31 (1984), 140–41.

984. Linden, Stanton J., ed. *William Cooper's* A Catalogue of Chymicall Books, 1673–88. *A Verified Edition.* Garland Reference Library of the Humanities, 670. New York: Garland Publishing, 1987. 159 pp.

William Cooper compiled the first catalogue of chemical, alchemical, medical and hermetic books in English. The present work is a single alphabetical listing of the catalogues of 1673, 1675, and 1688, with *STC* and Wing numbers, and British Library press-marks. Some five hundred titles. Pp. vii–xlvii: Introduction. Pp. 157–59. Subject index. Reviewed in *Bulletin of the History of Medicine* 63 (1989), 291–92.

MONOGRAPHS AND ARTICLES

985. Hoefer, Ferdinand. *Histoire de la chimie*. 2nd ed. Paris: Firmin Didot, 1866–69. 2 vols.

Vol. 2, pp. 1–331 cover the sixteenth and seventeenth centuries. See item 844.

986. Metzger, H. *Les doctrines chimiques en France du début du XVIIe à la fin du XVIIIe siècle*. Tome 1. Paris: Presses Universitaires de France, 1923. 496 pp.

Topics: Chemical theory as taught in the seventeenth century; Alchemy and seventeenth-century chemistry; Iatrochemistry; Mechanical philosophy and its influence on chemical theory; Lémery; Experimental chemistry; The triumph and spread of corpuscular theory and mechanical philosophy.

Metzger's *Chemistry*, tr. and annotated by Colette V. Michael (West Cornwall, CT: Locust Hill Press, 1991), first published in 1930, contains a panoramic view of the history of chemistry from 1700 to 1900. Pp. 3–54 relate to the 17th century.

987. Stillman, John Maxon. *The Story of Early Chemistry*. New York: Appleton, 1924. 566 pp.

Pp. 300–423 cover the sixteenth and seventeenth centuries.

988. Metzger, Hélène. *Newton, Stahl, Boerhaave et la doctrine chimique*. Paris: Librairie Félix Alcan, 1930. 332 pp.

Three studies (sequel to *Les doctrines chimiques en France*, item 986): "L'influence de la philosophie newtonienne sur le développement de la science chimique"; "La doctrine chimique de Stahl et de ses disciples"; "La doctrine chimique de Boerhaave." Pp. 309–22: Bibliography and index.

989. Contant, Jean-Paul. *L'enseignement de la chimie au Jardin royal des plantes de Paris*. Cahors: Impr. Coueslant, 1952. 133 pp.

Notes on seven professors who successively held two chairs during the sixty years since the formal opening of the *Jardin* in 1640. Bibliography.

990. Leicester, Henry M., and Herbert S. Klickstein. *A Source Book in Chemistry 1400–1900*. New York: McGraw-Hill Book Co., 1952. 554 pp.

Pp. 1–66 contain selections from the works of: Biringuccio, Agricola, Paracelsus, Libavius, Van Helmont, Valentine, Glauber, Boyle, Hooke, Lémery, Becher, Stahl, and Boerhaave. With introductory notes.

991. Urdgang, George. "How Chemicals Entered the Official Pharmacopoeias." *Archives internationales d'histoire des sciences* 7 (1954), 303–14.

Discusses the pharmacopoeias of the sixteenth and seventeenth centuries.

992. Read, John. *Through Alchemy to Chemistry. A Procession of Ideas and Personalities.* London: G. Bell and Sons, 1957. 206 pp.

Pp. 24–126 cover the period from Paracelsus to Boerhaave. Chapter headings: The beginnings; Emergence of alchemy; The philosopher's stone; Alchemical crypticism and symbolism; Strands in the alchemical web (religion; mythology; the Saturnine mysticism; number, harmony, and music); The diversity of alchemists; The parting of the ways; The swan song of alchemy. No bibliography.

"[Chemistry is the] most romantic of all the branches of science; and in its variegated history, stretching back through unnumbered generations of alchemists into an indefinite past, its present votaries have (if they but knew) a richly human and humanistic heritage" (preface).

993. Boas, Marie. *Robert Boyle and Seventeenth Century Chemistry.* Cambridge: Cambridge Univ. Press, 1958. 240 pp.

Study of chemistry immediately before Lavoisier. Pp. 5–47: The making of a scientist. Reviewed in *Isis* 51 (1960), 111–12.

994. Aitchison, Leslie. *A History of Metals.* London: Macdonald and Evans, 1960. 2 vols. 647 pp.

Pp. 305–444 cover the years 1100–1700. Reviewed in *Isis* 53 (1962), 242–44.

995. Partington, J. R. *A History of Chemistry.* Vols. 2 and 3. [1500–1800]. London: Macmillan, 1961–62.

Vol. 2 surveys the work of some two hundred "chemists" from Leonardo da Vinci to Boerhaave (French chemists excluded; pp. 115–51: Paracelsus; pp. 486–549: Boyle). Vol. 3, pp. 1–48: Chemistry in France, 1600–1700.

996. Multhauf, Robert P. *The Origins of Chemistry.* London: Oldbourne, 1966. 412 pp.

Study includes the work of a large number of not so-well-known scientists. Pp. 208–353 cover the years of the Scientific Revolution. Reviewed in *Isis* 59 (1968), 104–5.

997. Kangro, H. *Joachim Jungius' Experimente und Gedanken zur Begründung der Chemie als Wissenschaft. Ein Beitrag zur Geistesgeschichte des 17. Jahrhunderts.* Boethius Texte und Abhandlungen zur Geschichte der exakten Wissenschaften, 7. Wiesbaden: Franz Steiner, 1968. 479 pp.

Study of the researches and chemical concepts of Jungius (1587–1657). Comprehensive bibliography of primary and secondary sources. Reviewed in *Isis* 60 (1969), 570–72 and *Archives internationales d'histoire des sciences* 23 (1970), 299–302.

998. Hannaway, Owen. *The Chemists and the Word: The Didactic Origins of Chemistry.* Baltimore, Md.: Johns Hopkins Press, 1975. 165 pp.

On the emergence of chemistry as an independent discipline. Reviewed in *Isis* 68 (1977), 152–53.

999. Agassi, Joseph. "Who Discovered Boyle's Law?" *Studies in History and Philosophy of Science* 8 (1977), 189–250.

Essay on the historiography of Boyle's law. With bibliography.

1000. Debus, A. G. *The Chemical Philosophy: Paracelsian Science and Medicine in the Sixteenth and Seventeenth Centuries.* New York: Science History Publications, 1977. 2 vols.

Selective study of Paracelsianism with major emphasis on the Germanic lands and England. Pp. 555–95: Bibliography. Pp. 597–606: Index of names and subjects. Detailed reviews in *Annals of Science* 36 (1979), 663–666 and *Isis* 70 (1979), 588–92.

1001. Howard, Rio. "Guy de la Brosse: botanique et chimie au début de la révolution scientifique." *Revue d'histoire des sciences* 31 (1978), 301–26.

Describes the influence of La Brosse (ca. 1586–1641), founder of the Jardin des Plantes in Paris and one of the first French Paracelsians.

1002. Richards, J. F., ed. *Precious Metals in the Later Medieval and Early Modern Worlds.* Durham, N.C.: Carolina Academic Press, 1983. 502 pp.

NOT SEEN.

1003. Golinski, Jan V. "Robert Boyle: Scepticism and Authority in Seventeenth-Century Chemical Discourse." *The Figural and the Literal: Problems of Language in the History of Science and Philosophy, 1630–1800.* Edited by Andrew E. Benjamin, Geoffrey N. Cantor, and John R. R. Christie. Manchester: Manchester Univ. Press, 1986, pp. 58–82.

On the paradoxical position enjoyed by the *Sceptical Chemist* in the canon of chemical literature.

1004. Debus, Allen G. *Chemistry, Alchemy and the New Philosophy, 1550–1700. Studies in the History of Science and Medicine.* London: Variorum Reprints, 1987. 320 pp.

Collection of fourteen papers concerning the "role played by chemistry in the rise of modern science and medicine." Reviewed in *Isis* 79 (1988), 165–66.

1005. Golinski, Jan. "Hélène Metzger and the Interpretation of Seventeenth Century Chemistry." *History of Science* 25 (1987), 85–97.

Critical study of Metzger's *Les doctrines chimiques en France* (1923) and *La chimie* (1930). See item 986.

1006. Freudenthal, Gad, ed. *Etudes sur/Studies on Hélène Metzger* [1889–1944]. *Réunies et présentées. En appendice: Hélène Metzger, Extraits de lettres, 1921–1944.* Corpus: Revue de philosophies no. 8/9. Paris, 1988. 280 pp.

Fifteen articles re-examining the work of Metzger, well known for her non-positivist approach to the history of chemistry (1600–1800). With bibliography and a biographical note.

Earth Sciences

BIBLIOGRAPHIES

1007. Hellmann. G. "Die Meteorologie in den deutschen Flugschriften und Flugblättern des XVI Jahrhunderts. Ein Beitrag zur Geschichte der Meteorologie." *Abhandlungen der Preussische Akademie der Wissenschaften, phys.-math. Kl.*, Jahrgang 1921, Nr. 1. 96 pp.

Survey of German broadsides and pamphlets containing weather information (fifteenth and sixteenth centuries). Pp. 32–96: Bibliography of meteorological pamphlets and broadsides.

1008. Bibliotheca De Re Metallica. *The Herbert Clark Hoover Collection of Mining and Metallurgy.* Annotated by David Kuhner. Catalogued by Tania Rizzo. Introduction by Cyril Stanley Smith. Claremont, Calif.: Libraries of the Claremont Colleges, 1980. 219 pp.

Alphabetical catalogue of 912 books (mainly on geology). Includes some 330 pre-1700 titles of which 105 are annotated. Reviews: *Isis* 73 (1982), 283; *Technology and Culture* 23 (1982), 473–75.

1009. Halleux, Robert. "La littérature géologique française de 1500 à 1650 dans son contexte européen." *Revue d'histoire des sciences* 35 (1982), 111–30.

Although no systematic treatise on geology was published in France between 1500 and 1650, there were many books dealing with geological topics (reflecting the main European debates). Bibliographical essay.

1010. Porter, Roy. *The Earth Sciences: An Annotated Bibliography.* New York: Garland, 1983. 192 pp.

808 items in ten chapters. Aimed as a working guide "with a bias toward the great age of the growth of geology." Includes many works of relevance to the earlier period.

1011. Grewe, Klaus. *Bibliographie zur Geschichte des Vermessungs- wesens. Bibliography of the History of Surveying.* Stuttgart: Konrad Wittwer, 1984. 336 pp.

Pp. 1–73 lists handbooks of surveying from ancient times to 1870. The rest of the bibliography (apart from Festschriften and biographies) is arranged under fifteen subjects. 5,732 entries.

1012. Ward, Dederick C., and Albert V. Carozzi. *Geology Emerging: A Catalog Illustrating the History of Geology (1500–1850) from a Collection in the Library of the University of Illinois at Urbana-Champaign.* Urbana-Champaign: Univ. of Illinois Library and The Graduate School of Library and Information Science, 1984. 565 pp.

Alphabetical catalogue of 2,380 items (some three hundred pre-1700). Pp. 4–32: Introduction. Pp. 537–65: Index of subjects and places.

1013. Brush, Stephen G., and Helmut E. Landsberg. *The History of Geophysics and Meteorology: An Annotated Bibliography.* New York: Garland, 1985. 450 pp.

Classified bibliography in twenty-two chapters, with name and subject indexes. No chronological divisions. Includes many works relevant to the years before 1700.

1014. Molloy, Peter M. *The History of Metal Mining and Metallurgy: An Annotated Bibliography.* New York: Garland, 1986. 319 pp.

Classified bibliography of 2,225 items in eleven chapters. Chapters of interest: Bibliographies (items 6–115); General works (116–94); Histories of mining, classified by country (195–721); History of metallurgy (768–948); Assaying (1266–347); Metallurgy (1348–83); Descriptions of mining districts (1667–829); Economic geology (1830–2147).

1015. Bassett, Douglas A. "History of Geology." *Information Sources in the Earth Sciences.* Editors: David N. Wood, Joan E. Hardy, and Anthony P. Harvey. 2nd ed. London: Bowker-Saur, 1989, pp. 463–504.

Bibliographical essay.

MONOGRAPHS AND ARTICLES

1016. Butterfield, Arthur D. *A History of the Determination of the Figure of the Earth from Arc Measurements.* Worcester, Mass.: Davis Press, 1906. 168 pp.

Outline history from the Chaldeans to 1866. Pp. 7–20 cover the work of Fernel, Snell, Norwood, Riccioli, Grimaldi, Picard, and Cassini. Pp. 162–68: Bibliography.

1017. Hellmann, G. "Die Entwicklung der meteorologischen Beobachtungen in Deutschland von den ersten Anfängen bis zur Einrichtung staatlicher Beobachtungsnetze." *Abhandlungen der Preussische Akademie der Wissenschaften, phys.-math. Kl.,* Jahrgang 1926, Nr. 1. 25 pp.

Pp. 1–12 cover the years 1450 to 1700.

1018. Mitchell, A. Crichton. "Chapters in the History of Terrestrial Magnetism. II. The Discovery of the Magnetic Declination." *Terrestrial Magnetism and Atmospheric Electricity* 42 (1937), 241–80.

On the evolution of the concept of declination from 1187 to 1600.

1019. Adams, Frank Dawson. *The Birth and Development of the Geological Sciences.* London: Baillière, Tindall and Cox, 1938. 506 pp.

Traces the evolution of the geological sciences from classical antiquity to the early nineteenth century. Reviewed in *Isis* 32 (1947), 218–20.

1020. Chapman, Sydney, and Julius Bartels. *Geomagnetism.* Oxford: Clarendon Press, 1940. 2 vols.

Vol. 2, pp. 898–922: Historical notes (discovery of magnetic declination; discovery of magnetic inclination; discovery of the secular variation; earliest magnetic charts; William Gilbert).

1021. Brunet, Pierre. "Les premiers linéaments de la science géologique: Agricola, Palissy, George Owen." *Revue d'histoire des sciences* 3 (1950), 67–79.

On the sixteenth-century precursors of Steno.

1022. Lenoble, Robert. *La géologie au milieu du XVIIe siècle.* Conférence faite au Palais de la Découverte, Université de Paris, série D, no. 27. Paris, 1954. 36 pp.

On the seventeenth-century sources of Buffon's *Histoire naturelle* (1749), considered to be the first book on modern geology.

1023. Bargalló, Modesto. *La mineria y la metalurgia en la America española durante la época colonial.* Mexico: Fondo de Cultura Economica, 1955. 442 pp.

Study of the technical aspects of mining in Spanish America. Traces the history of mining from the pre-discovery knowledge of the Spaniards and Indians to the end of the colonial era. With bibliography and index. Reviewed in *Hispanic American Historical Review* 37 (1957), 96–97.

1024. Balmer, Heinz. *Beiträge zur Geschichte der Erkenntnis des Erdmagnetismus.* Aarau: Verlag H. R. Sauerländer, 1956. 892 pp.

On the history of geomagnetism from the earliest times to the present (pp. 77–182: sixteenth and seventeenth centuries). Contains translations into German of extracts from the classics of the subject. Discusses two popular magnetic myths—the existence of an enormous magnetic mountain and the possibility of telecommunication with the aid of magnets. Biographical notices.

1025. Heninger, S. K., Jr. *A Handbook of Renaissance Meteorology with Particular Reference to Elizabethan and Jacobean Literature.* Durham, N.C.: Duke Univ. Press, 1960. 269 pp.

In three parts: (1) Background for the study of meteors in Renaissance England. (2) Meteorological theory and its literary paraphrase. (3) Meteorological imagery in the major creative writers (Spenser, Marlowe, Ben Jonson, Chapman, Donne, Shakespeare). Appendices (Weather prognostication; Index of authors before 1558). Bibliography and indexes. Reviewed in *Renaissance News* 13 (1960), 248–49 and in *Revue d'histoire des sciences* 13 (1960), 360–61.

1026. Bialas, Volker. *Praxis geometrica. Zur Geschichte der Geodäsie am Beginn der Neuzeit.* Deutsche Geodätische Kommission bei der Bayerischen Akademie der Wissenschaften, Reihe E: Geschichte und Entwicklung der Geodäsie, Heft 11. Munich, 1970. 27 pp.

Historical outline, fifteenth to seventeenth centuries.

1027. Biswas, Asit K. *History of Hydrology.* Amsterdam: North-Holland Publishing Co., 1970. 336 pp.

From antiquity to 1900. Pp. 135–251 cover the sixteenth and seventeenth centuries. Bibliographies. Reviewed in *Isis* 64 (1973), 250–51.

1028. Scherz, Gustav, ed. *Dissertations on Steno as Geologist.* Acta Historica Scientiarum Naturalium et Medicinalium, edidit Bibliotheca Universitatis Hauniensis, Vol. 23. Odense: Odense Univ. Press, 1971. 319 pp.

Twelve papers, relating to Steno's research in paleontology, geology, and mineralogy, published to mark the tercentenary of the publication of his *Prodromus de solido.* Pp. 9–139: Introductory essay on Steno's travels (in German) by the ed. Reviewed in *Isis* 63 (1972), 446–48.

1029. Bialas, Volker. *Der Streit um die Figur der Erde: zur Begründung der Geodäsie in 17. und 18. Jahrhundert.* Deutsche Geodätische Commission bei der Bayerischen Akademie der Wissenschaften. Reihe E: Geschichte und Entwicklung der Geodäsie, Heft 14. Munich, 1972. 39 pp.

On the development of geodesy during the period 1670–1750, and the controversy about the figure of the earth.

1030. Goldstein, Thomas. "The Renaissance Concept of the Earth and Its Influence upon Copernicus." *Terrae Incognitae* 4 (1972), 19–51.

On the concept of the earth "as a solid, three-dimensional body with a diversified surface, made up of varied portions of land and sea." While it had a mathematical meaning for Copernicus, it owed its origin to explicit geographic evidence.

1031. Prieto, Carlos. *Mining in the New World.* New York: McGraw-Hill, 1973. 239 pp.

NOT SEEN. Reviewed in *Western Historical Quarterly* 5 (1974), 219–20.

1032. Roger, Jacques. "La théorie de la terre au XVIIe siècle." *Revue d'histoire des sciences* 26 (1973), 23–48.

Historical notes on the idea of a "theory of the earth."

1033. Oldroyd, David R. "Some Neo-Platonic and Stoic Influences on Mineralogy in the Sixteenth and Seventeenth Centuries." *Ambix* 21 (1974), 128–56.

"Early mineralogical writings in their proper philosophical context."

1034. Wallace, William J. *The Development of the Chlorinity/Salinity Concept in Oceanography.* Amsterdam: Elsevier Scientific Publishing Co., 1974. 227 pp.

Traces man's ideas about the sea's saltness from antiquity to the present. Pp. 1–38: Views of antiquity on salt and sea water; Robert Boyle; The beginnings of the systematic study of the sea (Halley and Marsigli). Pp. 209–20: Bibliography.

1035. Ricci, Virgilio. *L'andata di Leonardo da Vinci al Monboso oggi Monte Rosa e la teoria dell'azzurro del cielo.* Rome: Arti Grafiche Fratelli Palombi, 1977. 82 pp.

On Leonardo's Alpine voyages and his observations thereon.

1036. Bialas, Volker. *Erdgestalt, Kosmologie und Weltanschauung. Die Geschichte der Geodäsie als Teil der Kulturgeschichte der Menschheit.* Stuttgart: Konrad Wittwer, 1982. 365 pp.

Pp. 58–150 (Ch. 3 and 4): History of geodesy in the sixteenth and seventeenth centuries. Review: *Isis* 76 (1985), 249.

1037. Kravath, Fred F. *Christopher Columbus, Cosmographer. A History of Metrology, Geodesy, Geography, and Exploration from Antiquity to the Columbian Era.* Rancho Cordova, Calif.: Landmark Enterprises, 1987. 359 pp.

Pp. 1–40: Columbus, the man, the navigator, and the cosmographer. Pp. 165–88: Toscanelli, Behaim, and Columbus on the size and shape of the earth and the distance from Portugal to India. Pp. 189–240: Cosmography and cartography in the sixteenth century. Pp. 241–84 (Appendix A): European discoveries and exploration in the Americas through the year A.D. 1504.

1038. Laudan, Rachel. *From Mineralogy to Geology: The Foundations of a Science, 1650–1830.* Chicago: Univ. of Chicago Press, 1987. 278 pp.

Pp. 1–46: Conceptual foundations of geology; Mineralogy and cosmogony in the late seventeenth century.

1039. Willmoth, Frances. "John Flamsteed's Letter Concerning the Natural Causes of Earthquakes." *Annals of Science* 44 (1987), 23–70.

A comparative study of seventeenth-century earthquake theories.

1040. Ellenberger, François. *Histoire de la géologie.* Tome 1. *Des anciens à la première moitié du XVIIe siècle.* Paris: Technique et Documentation Lavoisier, 1988. 352 pp.

Pp. 111–315 deal with the Renaissance and the seventeenth century. Pp. 327–33: Bibliography. Reviewed in *Isis* 80 (1989), 510–11 and *Archives of Natural History* 16 (1989), 226–27.

1041. Ito, Yushi. "Hooke's Cyclic Theory of the Earth in the Context of Seventeenth Century England." *British Journal for the History of Science* 21 (1988), 295–314.

On conflicting traditions of Earth science in seventeenth-century England.

1042. Thompson, Susan J. *A Chronology of Geological Thinking from Antiquity to 1899*. Metuchen, N.J.: Scarecrow Press, 1988. 320 pp.

Collection of citations showing the development of geological thinking—from the *Vedas* to Charles Doolittle Walcott (1850–1927). With bibliography and author index. Pp. 17–58 cover the period 1475–1700 (three hundred citations from 161 books).

Marine Science. Geography.
Surveying. Cartography

BIBLIOGRAPHIES

1043. Winsor, Justin. *A Bibliography of Ptolemy's Geography.* Library of Harvard University, Bibliographical contributions, no. 18. Cambridge, Mass.: University Press, 1884. 42 pp.

"An annotated list of eds. of the original and augmented texts and translations, and of Wytfliet's continuation, with particular reference to the development of early American cartography; and with an enumeration of copies in American libraries." Seventy-six eds., from 1462 [?] to 1867, listed, of which 69 are pre-1700. Gives full title, bibliographical description, locations of copies in the U.S., and references to catalogues and relevant articles.

1044. Ruge, Sophus. *Die Literatur zur Geschichte der Erdkunde vom Mittelalter an.* Fortgesetzt durch Walter Ruge und Konrad Kretschmer. Hildesheim: Georg Olms, 1980. 300 pp. [First published in the *Geographisches Jahrbuch* 1895–1926, vols. 18, 20, 23, 26, 30, 41]

Bibliographical survey of the history of post-medieval geography.

1045. Royal Geographical Society, London. *Catalogue of the Library.* 3rd ed. Compiled by Hugh Robert Mill. London: John Murray, 1895. 833 pp.

The Society was founded in 1830. The subjects covered by its collections include: exploration and its history; cartography. Appendices: collections of voyages and travels (pp. 525–612); Government, anonymous and other miscellaneous publications (pp. 615–769); Transactions and periodicals (pp. 771–833).

1046. Nordenskiöld, A. E. *Periplus: An Essay on the Early History of Charts and Sailing Directions.* Stockholm, 1897. 208 pp. [Translated from the Swedish]

Pp. 56–69: List of hand-drawn portolanos from the fourteenth century to 1669. Pp. 71–80: List of printed sea-charts of the fifteenth and sixteenth centuries. Pp. 149–60: List of globes and maps of the south and east littorals of Asia (1492–1561). Pp. 177–83: List of the oldest maps of the new hemisphere.

1047. Stevens, Henry N. *Ptolemy's Geography: A Brief Account of All the Printed Editions down to 1730.* 2nd ed. London: Henry Stevens, Son and Stiles, 1908. 62 pp.

Essay on the origin and history of the various eds. With a chronological list of 62 eds. (59 pre-1700) in the Stevens-Ayer collection.

1048. Ortroy, Fernand van. *Bio-bibliographie de Gemma Frisius, fondateur de l'Ecole belge de géographie, de son fils Corneille et de ses neveux les Arsenius.* Mémoires, Académie Royale de Belgique, cl. lett., sc.mor. et polit., (2), 11 (1920). Brussels, 1920. 418 pp.

Pp. 1–142: Biography of Gemma (1508–1555), the Arsenius brothers, and of Cornelius. Pp. 143–402: Bibliography.

1049. Atkinson, Geoffroy. *La littérature géographique française de la Renaissance. Répertoire bibliographique.* Paris: Picard, 1927. 565 pp. *Supplément.* 1936. 87 pp.

Bibliography of 524 works published in France before 1610 dealing with non-European countries and their peoples. Chronologically arranged, with full bibliographical descriptions and locations. Pp. 529–64: Indexes.

1050. Fontoura da Costa, Abel. *Bibliografia náutica portuguesa até 1700.* Rev. ed. Lisbon: Agência geral das Colónias, 1940. 157 pp. [First published in *Anais do Club Militar Naval* (Lisbon), 64 (1934), 817–927]

Annotated bibliography of 123 MSS and 72 printed works (of which 25 published after 1700).

1051. Stahl, William Harris. "Ptolemy's *Geography*: a Select Bibliography." *Bulletin, New York Public Library* 55 (1951), 419–32, 484–95, 554–64, 604–14; 56 (1952), 18–41, 84–96.

Bibliography of studies of the *Geography* (1,464 items). (Arrangement:) Bibliography (general); Regional geography (Europe; Africa; Asia); Ptolemaic studies; Mathematical geography; Maps; Appendices (MSS; Textual criticism; Editions; Translations; Sources; Atlases; Gazeteers; Influence; Biography). Index of authors. Early printed eds. (pre-1730) not listed.

1052. Baltimore Museum of Art. *The World Encompassed: An Exhibition of the History of Maps Held at the . . . Museum . . ., October 7 to November 23, 1952.* Baltimore, Md.: Trustees of the Walters Art Gallery, 1952. [Not paginated]

Catalogue of 282 items (60 plates) including 220 maps of the period 1450–1700. Annotated, with bibliography.

1053. Sanz, Carlos. *La geographia de Ptolomeo, ampliada con los primeros mapas impresos de América (desde 1507). Estudio bibliografico y critico. Con el catalogo de las ediciones aparecidas desde 1475 a 1853. Comentado e ilustrado. Numerosos facsimiles.* Madrid: Victoriano Suarez, 1959. 281 pp.

Pp. 65–256: Bibliography of eds. of the *Geography*, 1475–1883 (56 titles, of which 49 before 1700), arranged in chronological order.

1054. Mariner's Museum, Newport News, Va. *Dictionary Catalog of the Library.* Boston, Mass.: G.K. Hall, 1964. 9 vols.

Catalogue of some 44,000 vols. Pertaining to shipbuilding, navigation, voyages and exploration, naval history, merchant shipping, and other maritime subjects. Entries under author, title, and subject. Vol. 9, pp. 651–766: Chronological list of books 1497–1825 (some five hundred pre-1700).

1055. Koeman, Cornelis, ed. *Atlantes Neerlandici. Bibliography of Terrestrial, Maritime and Celestial Atlases and Pilot Books, Published in the Netherlands up to 1880.* Amsterdam: NV Theatrum Orbis Terrarum, 1967–71. 5 vols.

Vol. 4 is a bibliography of celestial atlases and charts, rutters, sea atlases, and pilot books. Vol. 5: Indexes (Terrestrial atlases; Sea atlases, rutters and pilot books; Name index; Index of geographical names). Biographical notes included.

1056. National Maritime Museum, Greenwich. *Catalogue of the Library.* London: HMSO, 1968–76. 5 vols.

Catalogue (to be completed) of some 50,000 vols. covering every aspect of maritime affairs (1: Voyages and travel. 2: Biography. 3: Atlases and cartography. 4: Piracy and pirateering. 5: Naval history: Middle Ages to 1815.) With annotations, indexes (general; of ships; geographical), chronological tables. Vol. 2 has a name index to biographies (some 12,000 names of maritime interest) in twenty-one biographical sources. Vol. 3 includes some one hundred works on cartography and its history. Vol. 5, pp. 13–75: Naval history of the sixteenth and seventeenth centuries.

1057. Albion, Robert Greenhalgh. *Naval and Maritime History: An Annotated Bibliography.* 4th ed. Mystic, Conn.: Marine Historical Association, 1972. 370 pp.

Coverage includes maritime science and exploration. Subject index.

1058. Howse, D., and P. Billings. *Handlist of Manuscript Sea Charts and Pilot Books Executed before 1700.* Greenwich: National Maritime Museum, 1973. 55 pp.

Reprint of Section 6 of the Museum's *Inventory* (item 1225).

1059. Howse, Derek, and Michael Sanderson. *The Sea Chart. An Historical Survey Based on the Collections in the National Maritime Museum.* Newton Abbot: David and Charles, 1973. 144 pp.

Describes sixty charts, of which thirty-two are pre-1700. Bibliographies.

1060. Ristow, Walter W. *Guide to the History of Cartography: an Annotated List of References on the History of Maps and Map-making.* Washington, D.C.: Library of Congress, Geography and Map Division, 1973. 96 pp.

398 titles, mostly books. Includes: general works, carto-bibliographies, books on individual countries and on specialized aspects of cartography. Index covers subjects, geographical areas, and authors.

1061. Shirley, Rodney W. *The Mapping of the World: Early Printed World Maps, 1472–1700.* London: Holland Press, 1983. 669 pp.

Carto-bibliographical study showing the development of the printed world map. Pp. 1–628: carto-bibliography (639 entries, 440 plates). Appendices include: Chronological list of maps; Chronology of discovery (p. 652); List of plates; Name index.

1062. *Bibliographie zur Geschichte der deutschen Kartographie.* Bearbeitet von Lothär Zögner unter Mitarbeit von Evelyn Schulte. Bibliotheca Cartographica, Sonderheft 2. Munich: K.G. Saur, 1984. 267 pp.

3,319 entries. Coverage: works by and on German cartographers; works on the regional cartography of Germany and German-speaking countries. In three sections: general, regional, and biographical. Not annotated.

1063. Adams, Thomas R. *The Non-Cartographical Maritime Works Published by Mount and Page. A Preliminary Handlist.* Occasional publications of the Bibliographical Society, no. 1. London, 1985. 54 pp.

An effort to identify the titles and eds. of the non-cartographical books (navigation and practical mathematics). Includes works of 37 seventeenth-century authors. The firm, founded by William Fisher in 1656, dominated maritime publishing in England during the first half of the eighteenth century. Reviewed in *The Library* (6), 9 (1987), 303–12.

1064. Hodgkiss, A. G., and A. F. Tatham. *Keyguide to Information Sources in Cartography.* London: Mansell, 1986. 253 pp.

I. Cartography and maps: the subject and its literature. II/III. Annotated bibliography of reference sources: history of cartography (pp. 79–127); contemporary cartography. IV. Directory of organizations.

1065. Campbell, Tony. *The Earliest Printed Maps 1472–1500.* The British Library, 1987. 244 pp,.

Annotated catalogue (with bibliographies) of 222 maps—broadsheet maps, atlases, and maps in books. With introduction (pp. 1–20) and general bibliography.

1065A. De Silva, Daya. *The Portuguese in Asia. An Annotated Bibliography of Studies on Portuguese Colonial History in Asia, 1498–c. 1800.* Bibliotheca Asiatica 22. Zug: IDC, 1987. 313 pp.

Classified, annotated bibliography (2,773 items) in nine sections: Reference works; Historiography; Cartography; Navigation; Travel and description; Conquest, expansion and decline; Religion; Economic foundation; Impact on Asian society. Section 3, items 163–213: Cartography; Section 4, items 214–302A: Navigation (ships, nautical instruments, astronomy, rutters, etc.).

1066. Vereeniging Nederlandsch Historisch Scheepvaart Museum, Amsterdam. *The Crone Library. Books on the Art of Navigation Left by Dr. Ernst Crone to the Scheepvaart Museum in 1975 and Books on the Same Subject Acquired by the Museum Previously. Including a Biography, a Short History of the Art of Navigation in the Netherlands and a List of the Crone Collection of Nautical Instruments. A Descriptive Special Catalogue.* With annotations, indexes, and an introduction to the catalogue by Hubert J. M. W. Peters. Nieuwkoop: De Graaf, 1989. 805 pp.

Pp. xi–xxxii: Bio-bibliography of Dr. Ernst Crone (H. G. T. Crone and H. J. M. W. Peters). Pp. xxxiii–xlix: Survey of the history of the art of navigation in the Netherlands (C. Koeman). Pp. 1–610: Catalogue of books (1,223 items of which 357 are pre-1700).

1067. Perkins, C. R., and R. B. Parry. *Information Sources in Cartography.* London: Bowker-Saur, 1990. 540 pp.

Pp. 75–125: The history of cartography (general sources; area studies; map types, design and production).

1067A. Landwehr, John. *VOC. A Bibliography of Publications Relating to the Dutch East India Company 1602–1800.* Edited by Peter van der Krogt. Utrecht: HES Publishers, 1991. 840 pp.

Describes the publications relating to the Company from its foundation to 1800. 1,674 items arranged in 18 sections. Includes: (3) Voyages; (4) Geography, history, ethnography; (5) Natural history; (9) Sanitation and tropical medicine (includes pharmacopoeias); (12) Ships, shipping and navigation; (15) History of the VOC. Pp. 785–834: General index. Review: *The Library* (6), 16 (1994), 58–61.

1067B. Karrow, Robert W., Jr. *Mapmakers of the Sixteenth Century and Their Maps. Bio-bibliographies of the Cartographers of Abraham Ortelius, 1570. Based on Leo Bagrow's A.* Ortelii Catalogus Cartographorum. Chicago: Speculum Orbis Press, for the Newberry Library, 1993. 846 pp.

A re-working of Bagrow's *Catalogus,* describing the lives and works of eighty-eight cartographers. With a new bibliography of secondary literature. Includes descriptions of more than 2,000 maps with their locations. Review: *Annals of Science* 52 (1995), 101–2.

MONOGRAPHS AND ARTICLES

1068. Günther, Siegmund. *Studien zur Geschichte der mathematischen und physikalischen Geographie.* Halle: Louis Nebert, 1877–79. 6 pts. 407 pp.

Pp. 277–332: Johann Werner of Nuremberg (1468–1522) and his contribution. Pp. 333–407: History of the loxodrome. Indexes of names.

1069. Gallois, Lucien. *Les géographes allemands de la Renaissance.* Bibliothèque de la Faculté des Lettres de Lyon, tome 13. Paris: Ernest Leroux, 1890. 266 pp.

History of the German School of geography from Peuerbach and Regiomontanus to Sebastian Münster.

1070. Wauwermans, H. E. *Histoire de l'école cartographique belge et anversoise du XVIe siècle.* Brussels: Institut National de Géographie, 1895. 2 vols.

Vol. 1, ch. 6–14: Renaissance Ptoléméenne; Les questions géographiques du XVIe siècle; Anvers, ses institutions, ses moeurs, son commerce au XVIe siècle. Vol. 2: La cartographie belge.

1071. De Piero, Antonio. "Della vita e degli studi di Gio. Battista Ramusio." *Nuovo archivio veneto,* NS, 4 (1902), 5–112.

Includes a study of Ramusio's *Delle navigationi et viaggi* (1550–59) and of his relations with Cardinal Bembo, Andrea Navagero, Girolamo Fracastoro, and others.

1072. Mees, Jules. "Henri le Navigateur et l'Académie Portugaise de Sagres." *Boletim da Sociedade de Geographia de Lisbõa* 21 (1903), 33–51.

Argues that the legendary story that Henry founded a school of navigation at Sagres has no foundation.

1073. Almagià, R. "Sullo sviluppo delle conoscenze delle profondità marine." *Bollettino della Società geografica italiana* (4), 6 (1905), 427–41, 502–22.

Pp. 427–41 deal with the early history of oceanography up to Dampier (1702).

1074. Grande, S. "Le relazioni geografiche fra P. Bembo, G. Fracastoro, G. B. Ramusio, G. Gastaldi." *Memorie, Società geografica italiana, Roma* 12 (1905), 93–197.

On Ramusio, his *Delle navigationi et viaggi,* the cartographer Giacomo Gastaldi, and their relations with Girolamo Fracastoro and Cardinal Bembo.

1075. Heawood, Edward. *A History of Geographical Discovery in the Seventeenth and Eighteenth Centuries.* New York: Octagon Books, 1969. 475 pp. [Reprint of the 1912 ed.]

Pp. 1–211 (Ch. 1–8): The Arctic regions; the East Indies; Australia and the Pacific; North America; Northern and Central Asia; Africa; South America; the South Seas.

1076. Thompson, Silvanus P. "The Rose of the Winds. The Origin and Development of the Compass-Card." *Proceedings, British Academy* 6 (1913–14), 179–209.

On the origins of: the names of the winds; the Rose of Thirty-two points; the distinctive marks used on compass-cards. Covers many sailing charts of the sixteenth century.

1077. De Morais e Sousa, L. *A sciência náutica dos pilotos portugueses nos séculos XV e XVI.* Lisbon: Imprensa National, 1924. 2 pts. 201 + 215 pp.

Includes a note (Pt. l, pp. 182–200) on the School of Sagres and a bibliography of works on navigation in the sixteenth century (Pt. 2, pp. 15–44).

1078. Taylor, E. G. R. *Tudor Geography, 1485–1583.* London: Methuen, 1930. 290 pp.

Study of English geography. Topics include: Roger Barlow and Cabot; French influences; John Dee; practical surveying and navigation. Pp. 163–90: Catalogue of English geographical or kindred works (books and MSS) to 1583. Review: *Isis* 23 (1935), 289–94.

1079. Brebener, John Bartlet. *The Explorers of North America, 1492–1806.* London: Adam and Charles Black, 1933. 431 pp.

Introduction to the explorers, with bibliographical notes on their narratives. Pp. 1–289 cover the years 1492–1700.

1080. Boxer, C. R. "Portuguese Roteiros, 1500–1700." *Mariner's Mirror* 20 (1934), 171–86.

On an exhibition of twenty rutters held at the Naval College in Lisbon in January 1934.

1081. Taylor, E. G. R. *Late Tudor and Early Stuart Geography, 1583–1650: A Sequel to Tudor Geography, 1485–1583.* London: Methuen, 1934. 322 pp.

Continuation of the history of English geography, both practical and academic. Pp. 177–298: Bibliography of English geographical literature. Reviewed in *Isis* 23 (1935), 289–94.

1082. Cortesão, Armando. *Cartografia e cartógraficos portugueses dos séculos XV e XVI (contribuïção para um estudo completo).* Lisbon: Edição da "Seara Nova," 1935. 2 vols.

Study of Portuguese cartography before 1700. Reviews: *Archeion* 17 (1935), 476–89 (Fontoura da Costa); 489–98 (R. Almagià). The first review gives a brief description of the book, the chapter headings, and a list of Portuguese maps and cartographers. The second is a critical analysis of the work, noting the author's tendency to exaggerate the contribution of the Portuguese to cartography and the science of navigation.

1083. La Roncière, Charles de. *Histoire de la découverte de la terre, explorateurs et conquérants.* Paris: Larousse, 1938. 304 pp.

Survey of the history of exploration, covering all regions of the world. Summary table of contents. No index. Pp. 87–224 cover the years 1450 to 1700.

1084. Fontoura da Costa, A. *La science nautique des Portugais à l'époque des découvertes.* Lisbon: Agência Geral das Colónias, 1941. 48 pp.

On the creation, by the Portuguese, of the art of navigation.

1085. Kenney, Cyril Ernest. *The Quadrant and the Quill. A Book Written in Honour of Captain Sturmy, "A Tryed and Trusty Sea-man," and Author of The Mariner's Magazine, 1669.* London, 1947. 166 pp. [Limited ed. of 600 copies]

Contains a description of *The Mariner's Magazine or Sturmy's Mathematical and Practical Arts*, with notes on his life and his library. Richly illustrated (93 pls.). Reviewed in *Geographical Journal* 112 (1949), 239–40.

1086. Wroth, Lawrence C. *Some American Contributions to the Art of Navigation, 1519–1802.* Providence, R.I.: The Associates of the John Carter Brown Library, 1947. 41 pp.

Paper read before the Massachusetts Historical Society. Pp.1–16 relate to the years before 1700.

1087. Barbosa, António. *Novos subsídos para a história da ciencia náutica portuguesa da época dos descobrimentos.* Porto: Instituto para a alta cultura, 1948. 332 pp.

Essays on the history of navigational methods during the age of discovery, based on Portuguese and Spanish sources.

1088. Brown, Lloyd A. *The Story of Maps.* Boston, Mass.: Little, Brown and Co., 1950. 397 pp.

Deals mainly with the "story of earth-knowledge, surveying, and geodesy." Reviewed in *Isis* 41 (1950), 243.

1089. Almagià, R., A. Métraux, A. Cortesão, and L. Guyot. *Les conséquences de la découverte de l'Amérique par Christophe Colomb. Les Conférences du Palais de la Découverte, Histoire des sciences.* Paris, 1951. 67 pp.

Four lectures: Columbus and modern science; The origin of the American Indians; The discovery of America and the science of navigation; Natural sciences and the discovery of America.

1090. Coutinho, Gago. *A náutica dos descobrimentos. Os descobrimentos marítimos vistos por um navegador. Colectânea de artigos, conferências e trabalhos inéditos.* Lisbon: Agência Geral do Ultramar, 1951. 2 vols.

Collection of papers by the "Grand Old Man" of the Portuguese Navy. Topics include: astrology in the peninsula; beginnings of navigational astronomy; astrolabes and quadrants; rectangular maps; caravelles. Reviewed in *Isis* 44 (1953), 301–2.

1091. Penrose, Boies, the Younger. *Travel and Discovery in the Renaissance, 1420–1620.* Cambridge, Mass.: Harvard Univ. Press, 1952. 369 pp.

On the "exploration and exploitation of non-European areas by Europeans during the fifteenth and sixteenth centuries." Pp. 241–73: The cartography and navigation of the Renaissance. Pp. 274–326: The geographical literature of the Renaissance. Pp. 335–56: Bibliography.

1092. Cortesão, Armando. "Nautical Science and the Renaissance." *Science, Medicine and History. Essays . . . in Honour of Charles Singer.* Edited by E. A. Underwood. London: Oxford Univ. Press, 1953, Vol. 1, pp. 303–16.

On the leading role of the Portuguese.

1093. Taylor, E. G. R. *The Mathematical Practitioners of Tudor and Stuart England.* New York: Cambridge Univ. Press for the Institute of Navigation, 1954. 443 pp.

An account of the development of ideas, methods, and instruments. Includes biographical notes on some six hundred teachers, textbook writers, technicians, and craftsmen. Reviewed in *Isis* 48 (1957), 377–78.

1094. Parks, George B., comp. "The Contents and Sources of Ramusio's *Navigationi.*" *Bulletin, New York Public Library* 59 (1955), 279–313.

A short-title list of eds. and an analysis of the contents of *Delle navigationi et viaggi* (3 vols.) of Gio. Battista Ramusio (1485–1557).

1095. Taylor, E. G. R. *The Haven-finding Art. A History of Navigation from Odysseus to Captain Cook.* London: Hollis and Carter (for the Institute of Navigation), 1971. 310 pp. [First published 1956]

Pp. 151–212: Instruments and tables (The Portuguese pioneers. The errors of compass and plain chart. The English awakening). Pp. 215–63: Towards mathematical navigation (The true chart. The longitude solved).

1096. Mollat, Michel, ed. *Le navire et l'économie maritime du XVe au XVIIIe siècles.* Travaux du Colloque d'histoire maritime tenu, le 17 mai 1956, à l'Académie de Marine. Paris: SEVPEN, 1957. 139 pp.

Seven papers dealing with shipping, shipbuilding, and navigation.

1097. Skelton, R. A. *Explorers' Maps. Chapters in the Cartographic Record of Geographical Discovery.* London: Routledge and Kegan Paul, 1958. 337 pp.

Collection of fourteen articles written for *The Geographical Magazine* (London). Includes some two hundred maps, with annotations. Reviewed in *Isis* 51 (1960), 96–97.

1098. Waters, David W. *The Art of Navigation in England in Elizabethan and Early Stuart Times.* New Haven, Conn.: Yale Univ. Press, 1958. 696 pp.

Pp. 597–618: Bibliography. Pp. 629–96: General index. Criticized by S. E. Morison in *Isis* 51 (1960), 109–11 (excellent study by seafarer-historian but makes sparing use of works not available in English translation and ignores almost everything written on the subject in America.)

1099. British Museum, London. *Prince Henry the Navigator and Portuguese Maritime Enterprise. Catalogue of an Exhibition, September–October 1960.* London, 1960. 166 pp.

Catalogue of 326 exhibits (includes: instruments, MSS, books). Pp. 1–54: Discovery and conquest; nautical science and cartography.

1100. Boxer, C. R. *Four Centuries of Portuguese Expansion, 1415–1825: A Succinct Survey.* Johannesburg: Witwatersrand Univ. Press, 1961. 102 pp.

Text of a series of four public lectures delivered at the Ernest Oppenheimer Institute of Portuguese Studies, University of Witwatersrand, Johannesburg. (1. "From the Maghreb to the Moluccas, 1415–1521." 2. "The Clash of Colour, Caste, and Creed in the Sixteenth Century." 3. "The Struggle for

Spices, Sugar, Slaves and Souls in the Seventeenth Century." 4. "The Golden Age of Brazil in the Eighteenth Century." With a bibliographical note.

1101. D'Elia, Pasquale M. "Recent Discoveries and New Studies (1938–1960) on the World Map in Chinese of Fr. Matteo Ricci, S.J." *Monumenta Serica* [Peking, *later* Los Angeles] 20 (1961), 82–164.

Deals with recent discoveries and studies on the four eds. of the Chinese World Map (1554, 1600, 1602, and 1603) of Matteo Ricci.

1102. Stadtarchiv Duisburg. *Gerhard Mercator, 1512–1594: zum 450. Geburtstag.* Duisburger Forschungen, Bd. 6. Duisburg: Renckhoff, 1962. 297 pp.

Nine articles by different authors; Development of cartography in the sixteenth century; Mercator in Louvain; Mercator in Duisburg; The Mercator projection; Ships illustrated on maps; Mercator and English geography in the sixteenth century; Mercator and the House of Plantin; Mercator's comments on the Epistle to the Romans; Walter Ghim's biography of Mercator.

1103. Burmeister, Karl Heinz. *Sebastian Münster. Versuch eines biographischen Gesamtbildes.* Basler Beiträge zur Geschichtswissenschaft, Bd. 91. Basel: Helbing und Lichtenhahn, 1963. 211 pp.

Life and work of Münster, Hebrew scholar and geographer. Documented.

1104. Bagrow, Leo. *History of Cartography.* Revised and enlarged by R. A. Skelton. London: C. A. Watts and Co., 1964. 312 pp. [Original German eds. published in 1944 and 1951]

History of early maps produced in Europe and elsewhere. Richly illustrated. Reviewed in *Isis* 56 (1965), 224.

1105. Huard, Pierre, and Ming Wong. "Les enquêtes scientifiques françaises et l'exploration du monde exotique aux XVIIe et XVIIIe siècles," *Bulletin, l'Ecole française d'Extrème Orient* 52 (1964), (1), 143–55.

Pp. 143–44: Notes on French explorers. Pp. 150–51: List of scientific voyages and expeditions undertaken by Frenchmen during the years 1581–1700.

1106. Deacon, Margaret. "Founders of Marine Science in Britain: The Work of the Early Fellows of the Royal Society." *Notes and Records, Royal Society of London* 20 (1965), 28–50.

Notes on Francis Bacon, John Greaves, John Wallis, Boyle, Newton, Cooke, and Sir Robert Moray.

1107. Parry, J. H. *The Age of Reconnaissance.* 2nd ed. London: Weidenfeld and Nicolson, 1966. 366 pp. [First published in 1963]

On European geographical exploration, trade and settlement outside the bounds of Europe in the fifteenth, sixteenth, and seventeenth centuries.

1108. Cotter, Charles H. *A History of Nautical Astronomy.* London: Hollis and Carter, 1968. 387 pp.

Outlines the history of the diverse problems of nautical astronomy and the ways in which they were solved from the Babylonians to the present day. Reviewed in *Isis* 64 (1973), 560–63.

1109. Crone, G. R. *Maps and Their Makers. An Introduction to the History of Cartography.* 5th ed. Folkestone, Kent: Wm. Dawson and Son, 1978. 152 pp. [First published in 1953]

Outlines the main stages of the development of cartography in the West. Few illustrations. Reviewed in *Annals of Science* 39 (1982), 424 and in *Imago Mundi* 11 (1954), 179.

1110. Parias, L. H., ed. *Histoire universelle des explorations.* 2. *La Renaissance (1415–1600).* By Jean Amsler. 3. *Le temps des grands voiliers.* By Pierre Jacques Charliat. Paris: Nouvelle Librairie de France, 1968. 414 pp., 366 pp.

Vol. 3, pp. 7–105 deal with the seventeenth century.

1111. Morison, Samuel Eliot. *The European Discovery of America.* (1: *The Northern Voyages, A.D. 500–1600.* 2: *The Southern Voyages, A.D. 1492–1616*). New York: Oxford University Press, 1971–74. 2 vols. 712 pp.; 758 pp.

Bibliography and notes at the end of each chapter. The reviewer of Vol. 1 in *Terrae Incognitae* 4 (1972), 129–32, describes the work as "essentially a who got where book, saved from lesser stature by the eminence of the writer and the skill of his prose." Vol. 2 reviewed in *Terrae Incognitae* 7 (1975), 69–71. A new ed. of material selected from these volumes (dealing mainly with the voyages of Columbus, Magellan, and Drake) was published in one volume under the title: *The Great Explorers: the European Discovery of America* (1978, 752 pp.).

1112. Deacon, Margaret. *Scientists and the Sea, 1650–1900: A Study of Marine Science.* London: Academic Press, 1971. 445 pp.

Pp. 1–65: The background to the seventeenth-century movement to a science of the sea. Pp. 68–174: Marine science in the seventeenth century. Review: *Isis* 65 (1974), 92–95.

1113. Thrower, Norman J. W. *Maps and Man. An Examination of Cartography in Relation to Culture and Civilization.* Englewood Cliffs, N.J.: Prentice-Hall, 1972. 184 pp.

Pp. 43–83: the re-discovery of Ptolemy and cartography in Renaissance Europe; Cartography in the Scientific Revolution and the Enlightenment.

1114. Diffie, Bailey W., and George D. Winius. *Foundations of the Portuguese Empire, 1415–1580.* Europe and the World in the Age of Expansion, ed. Boyd C. Shafer, Vol. 1. Minneapolis: Univ. of Minnesota Press, 1977. 533 pp.

Pp. 113–22: Henry "the Navigator" who followed his stars. (Debunks some of the myths surrounding Henry.) Pp. 123–43: Portuguese haven-finding in the ocean sea. Pp. 465–74: Chronology of Portuguese expansion. Pp. 480–516: Bibliography.

1115. Tooley, R. V. *Maps and Map-Makers.* 6th ed. London: B. T. Batsford, 1978. 140 pp.

Preliminary guide for students and collectors. Fourteen chapters, arranged by schools of geography and by region. Chronological treatment in each chapter with bibliography.

1116. Tyacke, Sarah. *London Map-Sellers, 1660–1720. A Collection of Advertisements for Maps Placed in the* London Gazette 1668–1719 *with Biographical Notes on the Map-Sellers.* Tring, Herts.: Map Collector Publications, 1978. 160 pp.

Pp. xi–xxvi: Introduction. Reviewed in *The Library* (6), 1 (1979), 389–92.

1117. Veltman, Kim. "Military Surveying and Topography: the Practical Dimension of Renaissance Linear Perspective." *Revista da Universidade de Coimbra* 27 (1979), 329–68.

"Perspective theory had only a minimal impact on military practitioners who preferred to rely on handy perspective aids." Pp. 338–53: Perspective aids. Thirty-five texts cited.

1118. Broc, Numa. *La géographie de la Renaissance (1420–1620).* Paris: Bibliothèque Nationale, 1980. 261 pp.

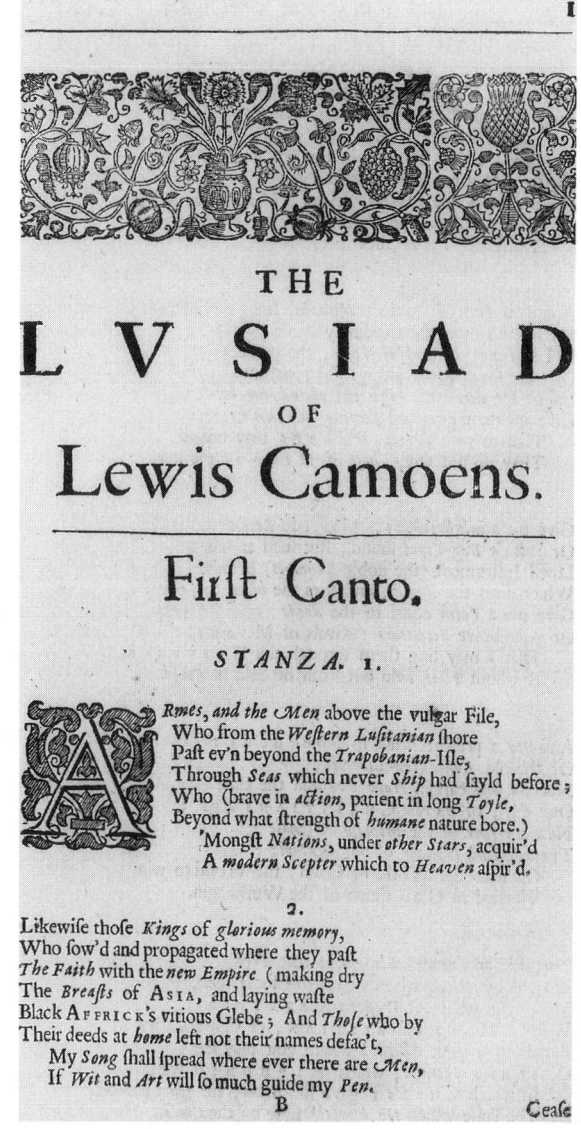

I

THE
L V S I A D
O F
Lewis Camoens.

Firſt Canto.

STANZA. 1.

Rmes, and the *Men* above the vulgar File,
Who from the *Weſtern Luſitanian* ſhore
Paſt ev'n beyond the *Trapobanian*-Iſle,
Through *Seas* which never *Ship* had ſayld before ;
Who (brave in *action*, patient in long *Toyle*,
Beyond what ſtrength of *humane* nature bore.)
'Mongſt *Nations*, under *other Stars*, acquir'd
A *modern Scepter* which to *Heaven* aſpir'd.

2.

Likewiſe thoſe *Kings* of *glorious memory*,
Who ſow'd and propagated where they paſt
The Faith with the *new Empire* (making dry
The *Breaſts* of ASIA, and laying waſte
Black AFFRICK's vitious Glebe ; And *Thoſe* who by
Their deeds at *home* left not their names defac't,
My *Song* ſhall ſpread where ever there are *Men*,
If *Wit* and *Art* will ſo much guide my *Pen*.

B

Ceaſe

First canto of *The Lusiad* by Luis de Camoens. From the English translation by Richard Fanshaw (London, 1655) of *Os Lusiadas* (Bodleian Library, Oxford, Fol.B.S.131(2)). This poem, the national epic of Portugal, celebrating the exploits of Vasco da Gama and other navigators, went through 45 editions between 1572 and 1702. **Entry 1114.**

Subjects covered include: Ptolemy's *Geography* and its influence; Voyages of discovery; Mercator; mapmakers; Geographical education of Europe. Reviewed in *Imago Mundi* 33 (1981), 117–18 and in *Isis* 72 (1981), 319.

1119. Tyacke, Sarah, and John Huddy. *Christopher Saxton and Tudor Map-making*. London: British Library, 1980. 64 pp.

The British Library's contribution to the four-hundredth anniversaries of the mapping of England and Wales (Atlas, 1579, Wall-Map, 1583). Tudor map-making (pp. 6–23); Christopher Saxton and the national survey (pp. 24–45); Christopher Saxton and local surveying (pp. 46–60), with a list of eighty-nine exhibits (fifty illustrations).

1120. Tyacke, Sarah, ed. *English Map-making 1500–1650: Historical Essays*. London: British Library, 1983. 125 pp.

Revised version of papers presented at a British Library Seminar (13 March 1981), representing some aspects of current research into the history of cartography in Britain of the period 1500–1650. Includes (pp. 93–106): Mathematical-instrument making in London in the sixteenth century (G. L'E. Turner).

1121. Milanesi, Marica. *Tolomeo sostituito: studi di storia delle conoscenze geografiche nel XVI secolo*. Studi e ricerche sul territorio, 14. Milan: Unicopli, 1984. 251 pp.

On "revising" Ptolemy's geographical lore. (1. The *Delle navigazioni et viaggi* (1550–59) of Ramusio and the geographic culture of the sixteenth century. 2. Sixteenth-century Europe's view of other regions of the world.)

1122. Russell, P. E. *Prince Henry the Navigator. The Rise and Fall of a Culture Hero*. Taylorian Special Lecture, 10 November 1983. Oxford: Clarendon Press, 1984. 30 pp.

Historiographical study of the culture-hero "who has undergone a succession of Protean changes as he has been made to function through the centuries as the symbol of the values and ideals of whatever group has selected him as its representative hero."

1123. "Actas da IV Reunião international de História Náutica e da Hidrografia." *Revista da Universidade de Coimbra* 32 (1986), 1–610.

Thirty-three papers mostly dealing with navigation and cartography in the sixteenth and seventeenth centuries. See item 1124.

1124. Albuquerque, Luís de. "Portuguese Books on Nautical Science from Pedro Nunes to 1650." *Revista da Universidade de Coimbra* 32 (1986), 259–78.

Notes on ten books written in Portuguese for the use of seamen. Authors: Pedro Nunes (1537), Teixeira da Mota (1592), Baptista Lavanha (1595), Manuel de Figueiredo (1608), Valentin de Sá, Antonio de Mariz Carneiro (1642), Simon de Oliveira (1606), Antonio de Naiera (1628). Includes two MSS (authors not known).

1125. Granzow, Uwe. *Quadrant, Kompass und Chronometer. Technische Implikationen des euro-asiatischen Seehandels von 1500 bis 1800.* Beiträge zur Kolonial- und Überseegeschichte, Bd. 36. Stuttgart: Steiner, 1986. 381 pp.

Discusses the five trading routes to the East and the technical problems of navigation. Four of these routes (pp. 1–289) relate to the years before 1700. Reviewed in *American Historical Review* 93 (1988), 1021.

1126. López Piñero, José María. *El arte de navegar en la España del renacimento.* 2nd ed. Barcelona: Editorial Labor, 1986. 285 pp.

Survey of the art of navigation in sixteenth-century Spain. Text consists almost entirely of a few sections, reproduced verbatim, from the author's *Ciencia y técnica en la sociedad española de los siglos XVI y XVII* (Barcelona, 1979)—review in *British Journal for the History of Science* 22 (1989), 99–100.

1127. Schilder, Günter, Peter van der Krogt, and Steven de Clercq, eds. *Marcel Destombes (1905–1983). Contributions sélectionnées à l'histoire de la cartographie et des instruments scientifiques. Selected contributions to the history of cartography and scientific instruments.* Utrecht: HES Publishers, 1987. 568 pp.

Destombes made significant contributions to the history of cartography and of navigational instruments. Thirty-four of his most important papers (the majority of which relate to the sixteenth and seventeenth centuries) are here. Includes a biographical note and a bibliography.

1128. Wallis, Helen, and Arthur H. Robinson, eds. *Cartographical Innovations. An International Handbook of Mapping Terms to 1900.* London: Map Collector Publications, 1987. 353 pp.

"Documents a wealth of innovative ideas and concepts which led to the advance of the art, craft, and science of mapmaking in the past." Contains some two-hundred terms arranged in eight groups: types of maps; maps of human occupation and activities; maps of natural phenomena; reference systems and geodetic concepts; symbolism; techniques and media; methods of duplication; atlases. Gives for each term: a definition; details of innovation and diffusion; and a bibliography.

1129. Livingstone, David N. "Science, Magic and Religion. A Contextual Reassessment of Geography in the Sixteenth and Seventeenth Centuries." *History of Science* 26 (1988), 269–94.

Sketches the history of geographical thought and practice against the background of the intellectual, ecclesiastical and political revolutions of the period.

Technology

BIBLIOGRAPHIES

1130. Cockle, Maurice J. D. *A Bibliography of English Military Books up to 1642 and of Contemporary Foreign Works.* Edited by H. D. Cockle. London: Simpkin, Marshall, Hamilton, Kent and Co., 1900. 268 pp.

166 English titles and 461 foreign. Full bibliographical descriptions with notes. Preface (pp. ix–xiii) contains notes on early military bibliographies.

1131. Spaulding, Thomas M., and Karpinski, Louis C. *Early Military Books in the University of Michigan Libraries.* Ann Arbor: University of Michigan Press, 1941. 371 pp.; 37 pls.

Books dealing with military science and art (military history, military medicine omitted) from 1493 to 1800. Out of 372 titles, 220 are pre-1700. Occasional annotations. P. 45: Index of military books by mathematicians and of mathematical works with sections on military science.

1132. Engineering Societies Library, New York. *Classed Subject Catalog.* Boston, Mass.: G. K. Hall, 1963. 13 vols. *Supplements 1–10.* 1964–73. 10 vols.

Some three hundred thousand entries (250,000 vols.) arranged by Universal decimal classification. Vol. l, pp. 19–667: Bibliographies. Vol. 12, pp. 287–524: Histories (unfortunately no chronological classification). Vol. 13: Index to the Classed Subject Catalog.

1133. Lacroix, Horst de. "The Literature on Fortification in Renaissance Italy." *Technology and Culture* 4 (1963), 30–50.

Surveys the literature from the work of Fra Egidio Colonna (1285) to Delbrueck's *Geschichte der Kriegswissenschaften* (1920). Bibliography of some two hundred works.

1134. "Current Bibliography in the History of Technology, 1962– ."
Technology and Culture 5 (1964)– [Annual]

Bibliography of books and articles published since 1962, classified under fifteen (now seventeen) subject groups. In 1967 chronological classifications were introduced (IV. From the Renaissance through the seventeenth century), subdivided into subject groups. Author and subject indexes.

1135. Ferguson, Eugene S. *Bibliography of the History of Technology*. Cambridge, Mass.: The Society for the History of Technology, 1968. 347 pp.

Extensively annotated bibliography in fourteen chapters: General works; General bibliographies and library lists; Directories; Early source books and MSS; Encyclopedias; Biography; Government publications; MSS; Illustrations; Travel and description; Periodicals; Societies and exhibitions; Technology and culture; Subject fields. Pp. 307–47: Name and subject index. Some seventeen hundred items under sixty headings. Each chapter has a brief introduction. Reviewed in *Isis* 60 (1969), 558–59.

1136. Russo, François. *Eléments de bibliographie de l'histoire des sciences et des techniques*. 2nd ed. Paris: Hermann, 1969. 214 pp.

Pp. 50–102: History of science and technology by periods (1500–1800). Pp. 160–68: History of technology. See item 19.

1137. Craig, Hardin, Jr. *A Bibliography of Encyclopedias and Dictionaries Dealing with Military, Naval and Maritime Affairs, 1577–1971*. 4th ed., revised. Houston, Tex.: Rice University, Department of History, 1971. 134 + 13 pp.

Includes some thirty pre-1700 titles. General encyclopedias and technical dictionaries have been excluded although they contain relevant material.

1138. Higham, Robin, ed. *A Guide to the Sources of British Military History*. London: Routledge and Kegan Paul, 1972. 630 pp.

Series of bibliographical surveys in twenty-five chapters, each with a list of references (not annotated). Ch. 1 (pp. 1–42): General bibliographical guidance; basic works of reference. Ch. 3 (pp. 65–83): Military developments of the Renaissance. Ch. 4 (pp. 84–119): The Navy to 1714. Ch. 7 (pp. 167–207): The scientific, technological and economic background to 1815. Ch. 23 (pp. 591–605): The evolution of military medicine. Ch. 24 (pp. 606–12): The evolution of naval medicine. No index. See also 1143.

1139. Mitcham, Carl, and Robert Mackey. *Bibliography of the Philosophy of Technology*. Chicago: Univ. of Chicago Press, 1973. 205 pp. [First published in *Technology and Culture*, Vol. 14]

Partly annotated bibliography in five sections (Comprehensive philosophical works; Ethical and political critiques; Religious critiques; Metaphysical and epistemological studies; Appendix). Concentrates on the literature of the years 1925–72. Reviewed in *Isis* 66 (1975), 271–73.

1140. "Current Bibliography in the Philosophy of Technology, 1973/74," *Research in Philosophy and Technology* 1 (1978), 313–90.

Continuation of 1139. Published at irregular intervals. Vol. 6 (1983), pp. 231–89 contains Bibliography 1977/78.

1141. Pacey, Arnold. "The History of Technology." *Information Sources in the History of Science and Medicine.* Editors: Pietro Corsi and Paul Weindling. London: Butterworth Scientific, 1983, pp. 44–60.

Historiographical essay with a select bibliography.

1142. Stapleton, Darwin H. *The History of Civil Engineering since 1600: An Annotated Bibliography.* Bibliographies of the History of Science and Technology, 14. New York: Garland, 1986. 232 pp.

Includes general works covering the years before 1600. Pp. 82–91: Renaissance and foundations of the Industrial Revolution, 1600–1750.

1143. Jordan, Gerald, ed. *British Military History: A Supplement to Robin Higham's Guide to the Sources.* Military history books, Vol. 10. New York: Garland Publishing, 1988. 586 pp.

Supplement to 1138. In 23 chapters, with bibliographies arranged alphabetically (no annotations). Ch. 1 (pp. 1–29): Introduction. Ch. 2 (pp. 31–100): The economic, scientific, and technological background, from the beginnings to 1914 (By W. H. G. Armitage). Ch. 3 (pp. 101–16): Military developments to 1485. Ch. 4 (pp. 117–34): The Army, 1485–1689. Ch. 5 (pp. 135–59): The Navy to 1714.

1144. Channell, David F. *The History of Engineering Science: An Annotated Bibliography.* Bibliographies of the History of Science and Technology, 16. New York: Garland, 1989. 311 pp.

1,494 entries, arranged in eight chapters (subdivided): General studies and reference works; Interactions between science and technology; Biographies and collected works; Institutions; The pre-history of engineering science; Applied mechanics; Thermodynamics and heat transfer; Fluid mechanics. No chronological divisions. Reviewed in *Technology and Culture* 32 (1991), 150–51.

GENERAL WORKS

1145. Feldhaus, Franz M. *Lexikon der Erfindungen und Entdeckungen auf den Gebieten der Naturwissenschaften und Technik in chronologischer Übersicht mit Personen und Sachregister.* Heidelberg: Carl Winter's Universitätsbuchhandlung, 1904. 144 pp.

Chronology of inventions and discoveries. Pp. 17–43 cover the years 1450 to 1700.

1146. Singer, Charles, E. J. Holmyard, A. R. Hall, and T. I. Williams, eds. *A History of Technology.* London: Oxford Univ. Press, 1954–1984. 8 vols.

Vol. 3. *From the Renaissance to the Industrial Revolution, c. 1500–c.1750.* 1957. 766 pp.

Vol. 8. *Consolidated Indexes.* Compiled by Richard Raper. 1984. 232 pp.

Vol. 3. Part I: Primary production (Food and drink; Metallurgy and assaying; coal mining and utilization; windmills). Part II: Manufacture (Tradesmen's tools; farm-tools, vehicles, and harness; spinning and weaving; figured fabrics; glass). Part III: Material civilization (Building construction; town planning; land drainage and reclamation; machines and mechanisms; military technology; printing). Part IV: Communications (Bridges; canals; shipbuilding; cartography, survey, and navigation). Part V: Approach to science (The calendar; precision instruments; scientific instruments; mechanical timekeepers; invention in chemical industries). Bibliographies. Vol. 8 contains: List of contents for vols. 1–7; Indexes of names, place names, and subjects. Vol. 3 reviewed in *Isis* 50 (1959), 163–65.

1147. Daumas, Maurice, ed. *A History of Technology and Invention; Progress through the Ages.* Vol. 2. *The First Stage of Mechanization.* New York: Crown Publishers, 1964. 694 pp. [Translated from the 1964 French ed.]

Covers the period 1450–1630 in twenty-four chapters. Bibliographies.

1148. Kranzberg, Melvin, and Carroll W. Pursell, Jr., eds. *Technology in Western Civilization.* Vol. 1. *The Emergence of Modern Industrial Society—Earliest Times to 1900.* New York: Oxford Univ. Press, 1967. 802 pp.

Comprehensive survey by a panel of specialists, resulting from a planning conference held at the University of Wisconsin in 1963. In five parts. Part 1: Emergence of technology (pp. 69–103: Early modern technology to 1600). Part 2: Background of the Industrial Revolution, 1600–1750 (pp. 107–214).

1149. Capocaccia, A. Agostino, ed. *Storia della tecnica.* Vol. 2. *Dalla rinascita dopo il Mille alla fine del Rinascimento.* Di Umberto Forti. Vol. 3. *Dal seicento al novecento.* Di Alberto Mondini. Turin: UTET, 1974–77. 446 pp.; 554 pp.

General history of technology with illustrations, notes, and index of names.

1150. Cipolla, Carlo M., ed. *The Sixteenth and Seventeenth Centuries.* The Fontana Economic History of Europe, Vol. 2. Hassocks, nr. Brighton: Harvester Press, 1977. 640 pp.

Essays by seven authors, on developments that took place in the technological and economic sphere: Population in Europe; Patterns and structure of demand; Technology; Rural Europe; European industries; European trade; Emergence of modern finance in Europe. With bibliographies.

1151. Jílek, František, Josef Kuba, and Jaroslavá Jílková. *The World Inventions in Dates: A Chronological Survey of Significant Events from the History of Creative Technological Work.* Prague: Národní Technické Muzeum, 1979. 237 pp.

Survey of inventions from the Stone Age to 1975. Pp. 11–173: Chronology; pp. 174–87: Tables; p. 189: References; pp. 191–222: Subject index; pp. 223–37: Index of inventions. Pp. 43–59 cover the years 1450–1700.

1152. Gille, Bertrand, ed. *The History of Techniques.* Vol. 1. *Techniques and Civilisations.* Vol. 2. *Techniques and Sciences.* New York: Gordon and Breach, 1986. [Originally published in French in 1978]

The period 1450–1700 is covered in: Vol. 1, pp. 502–88 (Classical systems); Vol. 2, pp. 1207–10 (Bibliography), 1272–80 (Table of events). Reviewed in *Isis* 80 (1989), 563–64.

1153. McNeil, Ian, ed. *An Encyclopaedia of the History of Technology.* London: Routledge, 1990, 1,062 pp.

Twenty-two chapters covering the development of technology (materials; power and engineering; transport; communication and calculation; technology and society) from Stone Age to Space Age. Pp. 1018–62: Indexes (names and topics).

MONOGRAPHS, COLLECTIONS OF ESSAYS, AND ARTICLES

1154. Charnock, John. *A History of Marine Architecture Including an Enlarged and Progressive View of the Nautical Regulations and Naval History, Both Civil and Military, of All Nations, Especially of Great Britain.* Vol. 3. London, 1801. 496 pp.

Covers the years 1450 to 1692.

1155. Fincham, John. *A History of Naval Architecture.* London: Scholar Press, 1979. 415 pp.

Reprint of the 1851 ed. with introduction by John Leather (pp. v–xi). The author was master shipwright at Portsmouth dockyard (1844–52). Pp. 31–68 cover the years 1500–1700, from the Tudors to the visit of Peter the Great to England.

1156. Hulme, E. W. "English Glass Making in the Sixteenth and Seventeenth Centuries." *Antiquary* 30 (1894), 210–14, 259–63; 31 (1895), 68–72, 102–106, 134–38.

Notes based on documents relating to licenses and patents.

1157. Campbell, Lily B. *Scenes and Machines on the English Stage during the Renaissance: A Classical Revival.* Cambridge: University Press, 1923. 302 pp.

The history of stage spectacle during the sixteenth and seventeenth centuries seen as a history of the Renaissance conceptions of the ancient classical stage, modified by modern conditions—a rapid development made possible by the growing interest in and knowledge of the mathematical sciences.

1158. Anderson, Romola, and R. C. Anderson. *The Sailing Ship: Six Thousand Years of History.* London: Harrap, 1926. 212 pp.

Pp. 120–62: Ships of the period, 1450–1700.

1159. Maks, C. S. *Salomon de Caus (1576–1626).* Amsterdam: Centen's Uitgevers-Maatschappij, 1935. 131 pp. [NOT SEEN]

De Caus, a native of Normandy, engineer, and architect, was mathematical tutor to Henry, Prince of Wales. See also *DNB*, Vol. 5 (1885–86), 715–16.

1160. Jenkins, Rhys. "The Heat Engine Idea in the Seventeenth Century: A Contribution to the History of the Steam Engine." *Transactions of the Newcomen Society* 17 (1936–37), 1–11.

Notes on Newcomen's predecessors (De Caus, David Ramsay, Marquis of Worcester, Samuel Borland, and Denis Papin).

1161. Parsons, William Barclay. *Engineers and Engineering in the Renaissance.* Baltimore, Md.: Williams and Wilkins Co., 1939. 661 pp. [Reprinted 1968]

From Leonardo da Vinci to Michelangelo. Reviews: *Isis* 32 (1949), 354–56; *Technology and Culture* 12 (1971), 337–40.

1162. Doorman, G. *Patents for Inventions in the Netherlands during the Sixteenth, Seventeenth and Eighteenth Centuries. With Notes on the Historical Development of Technics.* Abridged English version. The Hague: Martinus Nijhoff, 1942. 228 pp.

Pp. 11–38: History of patents in the Netherlands. Pp. 39–80. Notes on the history of selected subjects (Includes: Drainage mills; horology; navigation; paper-making; perpetual motion; printing; telescopes; windmills). Subject and name indexes.

1163. Frumkin, M. "The Early History of Patents for Invention." *Transactions, Chartered Institute of Patent Agents,* 1947–48, pp. 20–69.

Paper read to a meeting of patent agents, followed by a discussion. With a list of references (pp. 43–51).

1164. Forbes, R. J. *A Short History of the Art of Distillation. From the Beginnings up to the Death of Cellier Blumenthal.* Leiden: Brill, 1948. 405 pp.; illus.

Pp. 99–184: From Brunschwygk to Boyle. Pp. 363–96: Bibliography. Pp. 397–405: Name and subject indexes.

1165. Scoville, Warren C. "The Huguenots and the Diffusion of Technology." *Journal of Political Economy* 60 (1952), 294–311, 392–411.

Stresses the need to study the exodus of the Huguenots from France, preceding and following the Edict of Nantes (1598). The diffusion of technology resulting from its revocation (1685) was a contributory factor in the acceleration of economic development in Western Europe after 1720–30.

1166. Frumkin, M. "Les anciens brevets d'invention. Les pays du continent européen au XVIIe siècle." *Archives internationales d'histoire des sciences* 7 (1954), 315–23. [First published in 1950]

General survey.

1167. Gille, B. "Les problèmes techniques au XVIIe siècle." *Techniques et civilisations* 1954, pp. 177–208.

Outline for future research. Bibliography includes a list of seventeenth-century treatises (56 titles).

1168. Usher, Abbot Payson. *A History of Mechanical Inventions.* Boston, Mass.: Beacon Press, 1959. 450 pp. [Reprint of the 1954 ed. First published 1929]

Chapters VII–XII cover: Leonardo da Vinci; invention of printing; textile machinery; clocks and watches; power; machine tools. Reviews: *Isis* 46 (1955), 290–93; *Technology and Culture* 2 (1961), 34–37.

1169. Williams, Lionel. "Alien Immigrants in Relation to Industry and Society in Tudor England." *Proceedings of the Huguenot Society, London* 19 (1956), 146–69.

On the revolutionizing of the textile industry by alien immigrants. Thomas Wyatt, "Aliens in England before the Huguenots," *ibid.* 74–94 deals with alien craftsmen of French and Netherlands descent who arrived during the reign of Henry VIII.

1170. Coleman, D. C. *The British Paper Industry, 1495–1860. A Study in Industrial Growth.* Oxford: Clarendon Press, 1958. 367 pp.

Pp. 3–88 cover the sixteenth and seventeenth centuries.

1171. Klemm, Friedrich. *A History of Western Technology.* London: George Allen and Unwin, 1959. 401 pp. [Translated from the 1954 German ed.]

Based on a series of lectures. Pp. 109–227 deal with the Renaissance and Baroque period. Reviewed in *Isis* 51 (1960), 228–29.

1172. Gille, Bertrand. *Les ingénieurs de la Renaissance.* Paris: Hermann, 1964. 239 pp.; 233 illus.

A study of Renaissance engineers—Leonardo da Vinci, his predecessors and successors. With bibliography and a catalogue of MSS. Reviewed by Maurice Daumas in *Revue d'histoire des sciences* 17 (1964), 267–75.

1173. Nef, John Ulric, the Younger. *The Conquest of the Material World.* Chicago: Univ. of Chicago Press, 1964. 408 pp.

Collection of essays, originally published separately, of which six are relevant: Industrial Europe at the Reformation (pp. 67–117); Technology and large-scale industry in Great Britain 1540–1640 (pp. 121–43); Industrial growth in France and England (pp. 144–212); Protestant Reformation and industrial civilization (pp. 215–39); Prices and industrial capitalism in France and England (pp. 240–67); Genesis of industrialism and modern science (pp. 268–328). Reviewed in *Isis* 57 (1966), 136.

1174. Cipolla, Carlo M. *Guns and Sails in the Early Phase of European Expansion, 1400–1700.* London: Collins, 1965. 192 pp.

A study of what made the "Vasco da Gama era" possible. Pp. 165–82: Bibliography.

1175. Hale, J. R. "The Early Development of the Bastion: an Italian Chronology c. 1450–c. 1534." *Europe in the Late Middle Ages.* By John Hale, Roger Highfield, and Beryl Smalley. London: Faber and Faber, 1965, pp. 466–94.

"The most significant of all architectural forms evolved during the Renaissance was the angle bastion. By resisting the new artillery and providing platforms for heavy guns it revolutionized the defensive-offensive pattern of warfare."

1176. Klemm, Friedrich. "Die Rolle der Technik in der italienische Renaissance." *Technikgeschichte* 32 (1965), 221–43.

Notes on: Brunelleschi, Ghiberti, Alberti, Fioravante, Francesco di Giorgio Martini, Leonardo da Vinci, Biringuccio, Tartaglia, the Arsenal in Venice, Guidobaldo del Monte, Domenico Fontana.

1176A. Lane, Frederic C. *Navires at constructeurs à Venise pendant la Renaissance.* Paris: SEVPEN, 1965. 298 pp.

Text is essentially the same as that of *Venetian Ships and Shipbuilders of the Renaissance* (Baltimore: Johns Hopkins Press, 1934), but footnotes and appendix have been revised. In twelve chapters, four of which are devoted to the Arsenal. Review: *Journal of Modern History* 39 (1967), 308.

1177. Smith, Cyril Stanley, ed. *Sources for the History of the Science of Steel, 1532–1786.* Monographs in the history of technology and culture, no. 4. Cambridge, Mass.: MIT Press, 1968. 357 pp.

Pp. 1–62 contain extracts from English eds. of *Von Stahel und Eysen* (1532), Biringuccio's *De la pirotechnia* (1540), Della Porta's *Magia naturalis* (1589), and Jousse's *De l'art de serrurier* (1627). With introductory notes to the texts.

1178. Jensen, Martin. *Civil Engineering Around 1700. With Special Reference to Equipment and Methods.* Copenhagen: Danish Technical Press, 1969. 207 pp.

"An account of the building techniques which were used about the year 1700, intelligible to a structural engineer born later than 1650." Compiled from old texts and illustrations. Bibliography.

1179. Hart, Clive. *The Dream of Flight: Aeronautics from Classical Times to the Renaissance.* London: Faber and Faber, 1972. 200 pp.

An account of "the nature, use, and conceptual bases of aerodynamic devices in Europe from the earliest times to about 1600." Pp. 177–93: Bibliography.

1180. Guilmartin, John Francis. *Gunpowder and Galleys: Changing Technology and Mediterranean Warfare at Sea in the Sixteenth Century.* Cambridge: Cambridge Univ. Press, 1974. 321 pp.

Pp. 304–21: Sources, bibliography, and index. Reviewed in *Technology and Culture* 17 (1976), 373–75.

1181. Duffy, Christopher. *Siege Warfare; The Fortress in the Early Modern World, 1494–1660.* London: Routledge and Kegan Paul, 1979. 289 pp.

From the earliest artillery fortification to fortress warfare beyond the seas (Africas, India, Far East). Pp. 265–89: Bibliography, general index, and subject index.

1182. Qaisar, Ahsan Jan. *Indian Response to European Technology and Culture (A.D. 1498–1707).* Delhi: Oxford Univ. Press, 1982. 225 pp.

Study, based on Persian and English sources, of "the nature, degree and pattern of Indian contacts with, and response to, European technology and culture during the sixteenth and seventeenth centuries."

1183. Sanabria, Sergio Luis. "The Mechanization of Design in the Sixteenth Century: the Structural Formulae of Rodrigo Gil de Hontañón." *Journal of the Society of Architectural Historians* 41 (1982), 281–93.

Notes on the surviving fragments of an architectural booklet by Rodrigo Gil (ca. 1500/10–1577). This work provides (the work of Leonardo da Vinci aside) the principal extant evidence for the development of structural thinking among sixteenth-century master masons.

1184. Eamon, William. "Technology as Magic in the Late Middle Ages and the Renaissance." *Janus* 70 (1983), 171–212.

Points out that fact and legend, fantasy and science, were combined in Renaissance considerations of technology to a remarkable degree, to create an image of the engineer as a magus. The "technological dream" of the late Middle Ages and the Renaissance was largely a product of the magical world view. With bibliographical references.

1185. Reynolds, Terry S. *Stronger Than a Hundred Men: A History of the Vertical Water Wheel*. Baltimore, Md.: Johns Hopkins Press, 1983. 454 pp.

Chapters Two and Three cover the sixteenth and seventeenth centuries. Pp. 399–430: Bibliography.

1186. Benoit, Paul. "Technology and Crisis: the Great Depression of the Middle Ages and the Technology of the Renaissance (Fourteenth to Sixteenth Centuries)." *History and Technology*, 1984, pp. 319–34.

"An extremely violent crisis hit Western Europe towards the end of the Middle Ages: population, prices and agricultural production declined; the stretch of cultivated land shrank when, simultaneously, wars and internal troubles shook up the new states. While the technological level remained stagnant in agriculture as well as in the textile and building industries, it progressed remarkably in other sectors, particularly those of mining and metallurgy."

1187. Hart, Clive. *The Prehistory of Flight*. Berkeley: Univ. of California Press, 1985. 279 pp.

Study of the development of ideas about flying from preclassical times to the end of the eighteenth century. Pp. 56–145: How do birds fly? Leonardo's theory of bird flight and his last ornithopters; Will man fly? Burattini's flying dragon. Pp. 227–74: Bibliography.

1188. Hollister-Short, G. J. "Gunpowder and Mining in Sixteenth- and Seventeenth-Century Europe." *History of Technology* 10 (1985), 31–66.

On the invention and diffusion of the technique of gunpowder blasting.

1189. Long, Pamela O. "The Contribution of Architectural Writers to a 'Scientific' Outlook in the Fifteenth and Sixteenth Centuries." *Journal of Medieval and Renaissance Studies* 15 (1985), 265–98.

"The significance of architectural writings is not that they were the first result of the interaction of humanist and artisan cultures but that they constituted the first textual tradition which, as a result of that interaction, disseminated in writing the ideal of the unity of theory and practice."

1190. Multhauf, Lettie S. "The Light of Lamp-Lanterns: Street Lighting in Seventeenth-Century Amsterdam." *Technology and Culture* 26 (1985), 236–52.

Prosperity, city pride, and concern for security prompted the city fathers to install a system of street-lighting in 1669 (and fire-fighting equipment in 1673). Their example was followed by other cities of Holland in the 1670s.

1191. Barnadas, Josep M. *Alvaro Alonso Barba (1569–1662): investigaciónes sobre su vida y obra.* Biblioteca Minera Boliviana, 3. La Paz: Papiro, 1986. 285 pp.

Barba was a Jesuit who lived in Peru between 1604 and 1650. His *El arte de los metales* (1640) summarized the Spanish metallurgical techniques used in the New World. Includes a bibliography of editions and translations of *El arte* and of secondary literature. Reviewed in *Isis* 79 (1988), 730–31.

1192. Pedretti, Carlo. *Leonardo, Architect.* With 544 illustrations. Translated by Sue Grill, edited by I. Grafe. London: Thames and Hudson, 1986. 365 pp. [Translation of the 1981 Italian ed.]

A corpus of Leonardo da Vinci's architectural studies classified in eleven chapters (35 sections): Florence, 1469–82; Milan, 1482–99; The return to Florence, 1500–1505; Milan, 1506–13; Rome and Florence, 1513–16; France, 1517–19; Fictive architecture; Theatrical architecture; Ornaments and emblems; The *Ingegni*; Industrial design. With commentary. Comprehensive bibliography and indexes (names and places; MSS).

1193. Galluzzi, Paolo, ed. *Leonardo da Vinci: Engineer and Architect.* Montreal: Montreal Museum of Fine Arts, 1987. 356 pp.

Catalogue of an exhibition at the Museum, 22 May–8 November 1987. With articles by a team of international scholars. Pp. 1–21: Introduction by Carlo Pedretti. Pp. 23–181: Leonardo, engineer (The career of a technologist; Leonardo's impossible machines; The inventions of nature and the nature of inventions; A typology of Leonardo's mechanisms and machines; Leonardo, Brunelleschi and the machinery of the construction site). Pp. 183–314: Leonardo, architect (The problems of Leonardo's architecture in the context of his scientific theories; Leonardo and architecture; Leonardo as urban planner; Leonardo, fortified architecture, and its structural problems). Pp. 317–38: Notes and bibliographies. Pp. 339–56: List of exhibits. Reviewed in *Isis* 80 (1989), 522–24.

1194. Schneider, Gerhard. *300 Jahre Denis Papin, Naturforscher und Erfinder in Hessen. Ausstellung in der Universitätsbibliothek Mar-*

burg und den Hessischen Landesmuseum Kassel. Schriften der Universitätsbibliothek Marburg, 35. Marburg, 1987. 201 pp.

Catalogue of an exhibition illustrating the life and times of Papin (1647–1712?). Bibliography.

1195. Crossley, David. *Post-medieval Archaeology in Britain.* London: Leicester Univ. Press, 1990. 328 pp.

General survey of the present state of knowledge. Topics include: rural building in the sixteenth and seventeenth centuries; fortifications; power before the steam engine; metals; glass. Pp. 309–28: Name and subject index.

Instruments

BIBLIOGRAPHIES

1196. Baillie, G. H. *Clocks and Watches: An Historical Bibliography.* London: N.A.G. Press, 1951. 414 pp. [Reprinted by Holland Press, London, 1978]

Bibliography of publications on mechanical timepieces, from 1344 to 1800 (from Giovanni de Dondi dall'Orologio to Thomas Mudge, Jr.). Annotated, with name and subject indexes. Pp. 3–130 cover the period from Leonardo da Vinci to Huygens.

1197. Maddison, Francis. "Early Astronomical and Mathematical Instruments. A Brief Survey of Sources and Modern Studies." *History of Science* 2 (1963), 17–50.

Pp. 34–50: Select bibliography of the important modern literature.

1198. Turner, G. L'E. "The History of Optical Instruments. A Brief Survey of Sources and Modern Studies." *History of Science* 8 (1969), 53–93.

Pp. 75–93: Bibliography.

1199. International Union of the History and Philosophy of Science— Scientific Instrument Commission. *Bibliography of Books, Pamphlets, Catalogues and Articles on or Connected with Historical Studies on Scientific Instruments,* 1983–

Current bibliography published annually. No. 8 (1990) contains eighty items of which about twenty-five are relevant. Alphabetically arranged in one sequence, with no classification.

1200. Turner, G. L'E. "Scientific Instruments." *Information Sources in the History of Science and Medicine.* Pietro Corsi and Paul Weindling, eds. London: Butterworth Scientific, 1983, pp. 243–58.

Historical survey with a select bibliography.

MONOGRAPHS AND ARTICLES

1201. Burckhardt, Fritz. "Die Erfindung des Thermometers und seine Gestaltung im XVII Jahrhundert." *Bericht des Pädagogiums,* Basel 1867, pp. 1–48.

Documented study of the invention of the thermometer, from the Commandino ed. of Hero's *Pneumatica* (1575) to Halley's experiments (1688).

1202. Burckhardt, Fritz. *Zur Geschichte des Thermometers. Berichtigungen und Ergänzungen.* Wissenschaftliche Beilage zum Bericht über das Gymnasium in Basel, Schuljahr 1901/02. Basel: Fr. Reinhardt, 1902. 22 pp.

Additions and corrections to the above.

1203. Vorsterman van Oijen, G. A. "Quelques arpenteurs hollandais de la fin du XVIe et du commencement du XVIIe siècle et leurs instruments." *Bullettino di bibliografia e di storia delle scienze matematiche e fisiche* 3 (1870), 323–76.

On mathematical instruments and textbooks of Dutch surveyors.

1204. Caverni, Raffaello. "Notizie storiche intorno all'invenzione del termometro." *Bullettino di bibliografia e di storia delle scienze matematiche e fisiche* 11 (1878), 531–86.

Notes on the thermometers of Galileo, Santorio, Sagredo, Torricelli, and others.

1205. Gunther, Robert Theodor. *Early Science in Oxford.* Oxford: Oxford Univ. Press, 1920–23. 2 vols.

"Essentially a *catalogue raisonné* of the ancient instruments in Oxford." With biographical notes on the Oxford scientists and other information. Vol. 1: Chemistry, mathematics, physics, and surveying. Vol. 2: Astronomy. No chronological grouping. Reviewed in *Isis* 6 (1924), 449–53.

1206. Hellmann, G., "Beiträge zur Erfindungsgeschichte meteorologischer Instrumente." *Abhandlungen, Preussische Akademie der Wissenschaften,* phys. math. Kl., Jahrgang 1920, Nr. 1. 59 pp.

Notes on the thermometer, barometer, raingauge, compass-card. With bibliographies.

1207. Rohde, Alfred *Die Geschichte der wissenschaftlichen Instrumente vom Beginn der Renaissance bis zum Ausgang des 18. Jahrhunderts.* Leipzig: Von Klinkhardt und Biermann, 1923. 119 pp; 139 illus.

Scientific instruments as seen from an artistic viewpoint. Covers measuring instruments (time and space) and astronomical instruments. Contains illustrations of many instruments from the Mathematischen Salon in Dresden and the Museum für Kunst und Gewerbe in Hamburg.

1208. Clay, Reginald S., and Thomas H. Court. *The History of the Microscope. Compiled from Original Instruments and Documents, up to the Introduction of the Achromatic Microscope.* London: The Holland Press, 1985. 266 pp. [First published 1932]

Pp. 1–107: History of early magnifiers up to the invention of the microscope; Robert Hooke and his microscope; The simple microscope; Early compound microscopes; The Marshall microscope.

1209. Gunther, Robert T. *The Astrolabes of the World. Based upon the Series of Instruments in the Lewis Evans Collection in the Old Ashmolean Museum at Oxford.* Oxford: Oxford Univ. Press, 1932. 2 vols.

Catalogue of 336 astrolabes and astrolabe clocks, arranged by country of origin (about half are sixteenth to seventeenth century). Vol. 1, pp. 46–47: Bibliography of treatises on the astrolabe; pp. 576–98: Bibliography of the astrolabe.

1210. Gunther, R.W.T. *Early Science in Cambridge.* Oxford: printed for the author at the University Press, 1937. 513 pp.

On scientific instruments invented (or developed) by Cambridge men. With historical notes. 336 items. Reviewed in *Isis* 28 (1938), 134.

1211. Taylor, F. Sherwood. "The Origin of the Thermometer." *Annals of Science* 5 (1941–47), 129–56.

Discusses the sixteenth-century studies of Hero's *Pneumatica* and four possible inventors of the thermometer: Sanctorius, Galileo, Fludd, and Drebbel.

1212. Middleton, W.E. Knowles. "The Early History of Hygrometry, and the Controversy between de Saussure and de Luc." *Quarterly Journal of the Royal Meteorological Society* 68 (1942), 247–61.

On the development of hygrometry from the time of Leonardo da Vinci to 1800. Pp. 247–52: Developments before 1700.

1213. Defossez, L. *Les savants du XVIIe siècle et la mesure du temps.* Lausanne: Editions du Journal suisse d'horologerie et de bijouterie, 1946. 341 pp.

Study of time measurement from Jost Bürgi to Leibniz, Huygens, and Nicolas Fatio. Includes a chapter on sixteenth-century mechanics and the theory of clockwork.

1214. King, H. C. *History of the Telescope.* London: Griffin, 1955. 456 pp.

Pp. 15–110: Developments in the sixteenth and seventeenth centuries. Reviewed in *Isis* 48 (1957), 357–58.

1215. Zinner, Ernst. *Deutsche und niederländische astronomische Instrumente des 11. bis 18. Jahrhunderts.* Munich: C.H. Beck'sche Verlagsbuchhandlung, 1956. 678 pp.; pls.

Historical survey (Part 1, pp. 1–226) of some one-hundred types. References are made to Part 2 (pp. 229–601) which gives biobibliographical details of instrument makers. Chronological table and bibliography.

1216. Chilton, D. "Land Measurement in the Sixteenth Century." *Transactions of the Newcomen Society* 31 (1957–59), 111–29.

Notes on measuring instruments (plane table, chain, rod, cord, perambulator, carriage waywiser, pace counter, geometrical square, altazimuth theodolite, cross-staff, mariner's astrolabe, surveyor's cross).

1217. Hill, H. O., and E. W. Paget-Thomlinson, comps. *Instruments of Navigation. A Catalogue of Instruments at the National Maritime Museum with Notes upon Their Use.* London: HMSO, 1958. 90 pp.

A catalogue and description of the tools used by seamen to find their way. Includes sixteenth- and seventeenth-century instruments: Quadrant; astrolabe; fore-staff; universal ring dial; nocturnal; sand- or hour-glass; compass; lodestones; dividers and sectors.

1218. Donnelly, M. C. "Astronomical Observatories in the Seventeenth and Eighteenth Centuries." *Mémoires, Académie royale de Belgique, cl. sci.,* 34 (1964), (5), 1–37. pls.

Pp. 1–19 contain notes on early observatories: Leiden (1632), Ingolstadt and Copenhagen (1637), Utrecht (1642), Jena (1656), Altdorf (1657), Nuremberg (1678), Lyons (1690).

1219. Middleton, W.E. Knowles. *The History of the Barometer*. Baltimore, Md.: Johns Hopkins Press, 1964. 489 pp.

Pp. 1–82 cover the seventeenth-century experiments which led to the invention of the mercury barometer. Reviewed in *Isis* 55 (1964), 450–51.

1220. Körber, Hans Günther. *Zur Geschichte der Konstruktion von Sonnenuhren und Kompassen des 16. bis 18. Jahrhunderts. (Unter besonderer Berücksichtigung der im Geomagnetischen Institut Potsdam und im Staatl. Mathematisch-Physikalischen Salon Dresden vorhandenen Instrumente)*. Veroffentlichungen des Staatlichen Math.-Physikalischen Salons—Forschungsstelle—Dresden-Zwinger, Bd. 3. Berlin: VEB Deutscher Verlag der Wissenschaften, 1965. 206 pp.; illus.

On the history of sundials and compasses with special reference to the Potsdam and Dresden collections. Pp. 77–108: Chinese and Arabic instruments. Pp. 187–200: List of references.

1221. Middleton, W. E. Knowles. *History of the Thermometer and Its Uses in Meteorology*. Baltimore, Md.: Johns Hopkins Univ. Press, 1966. 249 pp.

Pp. 1–64 deal with the history up to 1700. Reviewed in *Isis* 59 (1968), 214–16.

1222. Bradbury, S. *Evolution of the Microscope*. Oxford: Pergamon, 1967. 357 pp.

Pp. 1–103 deal with the history up to 1700. Name and subject indexes.

1223. Michel, Henri. *Scientific Instruments in Art and History*. London: Barrie and Rockliff, 1967. 208 pp. [Translated from the French edition of 1966]

Anthology of illustrations of over one hundred and sixty scientific instruments of which one hundred belong to the period 1450–1700. Arranged in five sections: Basic elements; Measuring the earth; Measuring the heavens; Measurement of time; Physical measurements. With introduction and notes. Reviewed in *Isis* 59 (1968), 213–14.

1224. Bedini, Silvio A. "Seventeenth Century Magnetic Timepieces." *Physis* 11 (1969), 37–68.

Discusses attempts at using the loadstone and magnetic force for the measurement of time.

1225. National Maritime Museum, Greenwich. *An Inventory of the Navigation and Astronomy Collections in the National Maritime Museum, Greenwich. With a List of Instruments Used at the Royal Observatory 1676–1950 and of Other Special Collections.* Greenwich, London, 1970– .

Loose-leaf. Series of handlists in thirty-five sections (from armillary spheres to traverse boards), each arranged chronologically and with an introductory note. Index.

1226. Guye, Samuel, and Henri Michel. *Time and Space. Measuring Instruments from the Fifteenth to the Nineteenth Century.* London: Pall Mall Press, 1971. 289 pp. [Translated from the French (Fribourg, 1970)]

Pp. 41–108: Renaissance and seventeenth century (clocks and watches). Pp. 193–286: Ancient measuring instruments. Illustrated.

1227. Maddison, Francis. "Medieval Scientific Instruments and the Development of Navigational Instruments in the Fifteenth and Sixteenth Centuries." *Revista da Universidade de Coimbra* 24 (1971), 115–72.

Sixteenth-century instruments described: mariner's quadrant; mariner's or sea astrolabe; nocturnal; azimuth compass; mechanical clock as an aid to navigation; universal equinoctial ring dial; cross-staff; backstaff; sector.

1228. Daumas, Maurice. *Scientific Instruments of the Seventeenth and Eighteenth Centuries and Their Makers.* London: Batsford, 1972. 361 pp. [Translation of the 1953 French ed.]

"Internalist history of the instruments and their functions and also an externalist account of the makers and their workshops in social perspective" (De Solla Price). Review: *Isis* 64 (1974), 421–22.

1229. May, W. E., and Leonard Holder. *A History of Marine Navigation.* Henley-on-Thames, Oxon.: G. T. Foulis and Co., 1973. 280 pp.

History of navigational instruments from the compass to modern developments. Treatment is by subject.

1230. Van Helden, Albert. "The Telescope in the Seventeenth Century." *Isis* 65 (1974), 38–58.

Discusses the role played by science in the development of the telescope and assesses the influence of the telescope on scientific ideas in the seventeenth century.

1231. Goodison, Nicholas. *English Barometers 1680–1860: A History of Domestic Barometers and Their Makers and Retailers.* 2nd ed. Woodbridge, Eng.: Antique Collectors' Club, 1977. 388 pp. [Reprinted by Sotheby's, London 1969, 1985 and 1992]

History of the domestic mercurial barometer in Britain. Pp. 1–63 cover the period ending 1720. 1st ed. reviewed in *Isis* 61 (1970), 120–21.

1232. Van Helden, Albert. *The Invention of the Telescope.* Transactions of the American Philosophical Society, Vol. 67(4). Philadelphia, 1977. 67 pp.

Study of the genesis of the telescope, based on the comprehensive survey by Cornelius de Waard, *De uitvinding der verrekijkers* (The Hague, 1906). With English translations of primary sources. See also author's article in *History of Science* 13 (1975), 251–63. Reviews: *Isis* 70 (1979), 601–2; *Annals of Science* 36 (1979), 418–19; *Journal for the History of Astronomy* 10 (1979), 57–58.

1233. Wunderlich, Herbert. *Kursächsische Feldmesskunst, artilleristische Richtverfahren und Ballistik im 16. und 17. Jahrhundert. Beiträge zur Geschichte der praktischen Mathematik, der Physik und des Artilleriewesens in der Renaissance unter Zugrundelegung von Instrumenten, Karten, Hand- und Druckschriften des Staatlichen Mathematisch-Physikalischen Salons Dresden.* Veröffentlichungen des Staatlichen Mathematisch-Physikalischen Salons, Forschungsstelle, Dresden, Band 7. Berlin: VEB Deutscher Verlag der Wissenschaften, 1977. 218 pp.

Nine studies of the history of practical mathematics, physics, and military technology during the Renaissance, based on the collections of the State Hall of Mathematics and Physics of Saxony in Dresden. Topics include: Surveying, surveying instruments, and textbooks; hodometer; proportional scales; instruments for angle measurement; mathematical tables for surveyors; mathematical instruments (reducing compasses, proportional compasses, slide rule; manuscript texts); ballistics. 73 illustrations. Pp. 189–218: Bibliographical notes; name and subject index.

1234. Brown, Joyce. "Guild Organisation and Instrument-Making Trade, 1550–1830: The Grocers' and Clockmakers' Companies." *Annals of Science* 36 (1979), 1–34.

On the introduction to England of the craft of making mathematical instruments, by Flemish religious refugees in the mid-sixteenth century. Contains a list of mathematical instrument makers in the Clockmakers' Company, 1631–1730.

1235. Mackensen, Ludolph von, ed. *Die erste Sternwarte Europas mit ihren Instrumente und Uhren: 400 Jahre Jost Bürgi in Kassel.* Schriften zur Naturwissenschaft- und Technikgeschichte. 1. Munich: Callwey Verlag, 1979. 158 pp.

Pp. 12–41: On the first modern observatory and its observers. Pp. 42–60: Bürgi, the clockmaker and engineer. Pp. 61–69: Reconstruction and-testing of an early sextant. Pp. 70–88: Bürgi's two mechanical celestial globes in Kassel. Pp. 89–114: Changing conceptions of the world as shown by astronomical museum-models and planetaria. Pp. 118–54: Exhibition catalogue. Reviewed in *Annals of Science* 37 (1980), 694.

1236. Maurice, Klaus, and Otto Mayer, eds. *The Clockwork Universe: German Clocks and Automata, 1550–1650.* Washington, D.C.: Smithsonian Institution, 1980. 321 pp.

Catalogue of an exhibition portraying the invention of the mechanical clock of early modern Europe. Including fourteen essays on various aspects of the invention and its history. Reviewed in *Isis* 73 (1982), 288–89.

1237. Poulle, Emmanuel. *Les instruments de la théorie des planètes selon Ptolémée: équatoires et horlogerie planétaire du XIIIe au XVIe siècle.* Paris: Champion, 1980. 2 vols. 1,162 pp.

Detailed survey of geometrical equatoria, mathematical equatoria, and astronomical clocks. List of sources and indexes in Vol. 2. Reviews: *Historia Mathematica* 9 (1982), 97–99; *Journal for the History of Astronomy* 13 (1982), 56–60; *Technology and Culture* 23 (1982), 651–53.

1238. Randier, Jean. *Marine Navigation Instruments.* London: John Murray, 1980. 219 pp. [Translated from the French]

Illustrated history of maritime navigation viewed through the pilot's "tools." In six chapters. Two hundred and forty illustrations with notes and connecting narrative. Index of instruments and bibliography.

1239. Doggett, Rachel, ed. *Times the Greatest Innovator: Timekeeping and Time Consciousness in Early Modern Europe.* (Exhibition at the Folger Shakespeare Library, October 1986–March 1987.) Washington, D.C.: Folger Shakespeare Library, 1986. 116 pp.

Catalogue of an exhibition documenting the revolution in Western culture between 1300 and 1700. Two introductory essays: Development of timepieces from the sundial to the pendulum clock (Silvio Bedini); Changing concept of time in Renaissance literature (Ricardo Quinones). Bibliography. NOT SEEN. Reviews: *Papers, Bibliographical Society of America* 82 (1988), 114; *Isis* 80 (1989), 524–25.

1240. Howse, Derek, ed. *The Greenwich List of Observatories. A World List of Astronomical Observatories, Instruments and Clocks, 1670–1850.* Chalfont St. Giles, Bucks.: Science History Publications, 1988. 100 pp. [First published in *Journal for the History of Astronomy* 17 (1986), (4), 1–100]

Report of data collected by the International Astronomical Union (Commission 41, History of Astronomy) for a proposed *General History of Astronomy.* Includes information on observatories in the following cities (seventeenth century): Peking, Copenhagen, Paris, Kassel, Nürnberg, Kyoto, Tokyo, Seoul, Leiden, Utrecht, Lund, Greenwich, Oxford.

1241. Pinto, John. "The Renaissance City Image." *The Rational Arts of Living: Ruth and Clarence Kennedy Conference in the Renaissance, 1982.* Edited by A. C. Crombie and Nancy Siraisi. Smith College Studies in History, Vol. 50. Northampton, Mass.: Department of History of Smith College, 1987, pp. 205–54.

On the ichnographic plan (used during 1450–1550), a new conceptual approach to the depiction of cities which stimulated the development of scientific instruments designed to bridge the gap between the direct observation of the environment and its subsequent graphic abstraction.

1242. Turner, Anthony. *Early Scientific Instruments: Europe 1400–1800.* London: Sotheby's, 1987. 320 pp.

Describes and illustrates the main types of scientific instruments used in Europe during the years 1400–1800. Pp. 280–320: Notes, bibliography and index. Reviewed in *Technology and Culture* 32 (1991), 169–70.

1243. Bennet, J. A. *The Divided Circle: A History of Instruments for Astronomy, Navigation and Surveying.* Oxford: Phaidon Christie's, 1988. 224 pp.

Pp. 7–82 cover the history from ancient to early modern times—from Ptolemy's instruments to the equinoctial ring. Reviews: *Annals of Science* 45 (1988), 656–57; *Isis* 80 (1989), 299–300.

1244. Carl Zeiss-Stiftung. *A Spectacle of Spectacles. Exhibition Catalogue.* Jena: Edition Leipzig, 1988. 178 pp.

Pp. 9–16: Spectacles over seven hundred years (Introductory essay by Gerard L'E. Turner). Pp. 29–162: Catalogue (pp. 28–69 relate to the period 1450–1700).

1245. Hambly, Maya. *Drawing Instruments 1580–1980.* London: Sotheby Publications, 1988. 206 pp.

Guide to the development of European drawing instruments and aids (classified by type under eight headings, with illustrations). Pp. 19–54: History of drawing instruments and their makers; Literary evidence. Reviewed in *Technology and Culture* 31 (1990), 548–50.

1246. Pares, Jean. *La gnomonique de Desargues à Pardies.* Cahiers d'histoire et de philosophie des sciences, nouv. sér., 17. Paris: Société Française d'Histoire des Sciences et des Techniques, 1988. 181 pp.

Survey, with twenty-nine illustrations and a bibliography.

1247. Stimson, Alan. *The Mariner's Astrolabe. A Survey of Known, Surviving Sea Astrolabes.* Utrecht: HES Publishers, 1988. 190 pp.

Descriptions of sixty-five astrolabes of which fifty-nine are pre-1700. With historical introduction (pp. 13–42) and bibliography. Reviewed in *Journal for the History of Astronomy* 21 (1990), 298–300 and *TLS* 16 June 1989, p. 662.

Natural History

BIBLIOGRAPHIES

1248. Haller, Albrecht von. *Bibliotheca botanica, qua scripta ad rem herbariam facientia a rerum initiis recensentur.* Zurich: Orell, Gessner, Fuessli, et Socc., 1771–72. 2 vols. [Reprinted in 1967 by Johnson Reprint Corporation, New York with the *Index* emendatus (Berne, 1908) by J. Christian Bay]

Classified annotated bibliography of botanical literature from the beginnings to the present. Arrangement chronological. Vol. 1 (before Tournefort) is divided into eight chapters (Graeci; Arabes; Arabistae; Instauratores; Inventores; Collectores; A Bauhinis ad Rajum; A Rajo ad Tournefortium) and subdivided (830 sections) by authors and groups of authors. Annotations include bio-bibliographical information on authors, résumé of contents of books and critical comments. Vol. 2 (from Tournefort to the present) includes addenda, index of names, and classified index of anonyma. Some 15,000 entries in both vols. (A few seventeenth-century titles in Vol. 2.)

1249. Pritzel, G. A. *Thesaurus literaturae botanicae omnium gentium, inde a rerum botanicarum initiis ad nostra usque tempora, quindecim millia operum recensens.* 2nd ed. Leipzig: Brockhaus, 1872–77. 576 pp.

Bibliography of 10,874 titles in alphabetical order. These are re-arranged on pp. 374–540 in 46 categories (each chronologically ordered). With index of authors and anonyma.

1250. Jackson, B. D., ed. *Guide to the Literature of Botany: Being a Classified Selection of Botanical Works, Including Nearly Six Thousand Titles Not Given in Pritzel's* Thesaurus. London: Longman, for the Index Society, 1881. 626 pp. [Reprinted in 1964 by Hafner]

Classified bibliography in twenty chapters (nine thousand entries arranged chronologically under 124 subjects). Pp. 513–623: Index (names and subjects).

1251. Mullens, W. H., and E. Kirke Swann. *A Bibliography of British Ornithology from the Earliest Times to the End of 1912 including Biographical Accounts of the Principal Writers and Bibliographies of Their Published Works.* London: Macmillan, 1917. 691 pp.

Bio-bibliography, arranged alphabetically by author. Includes some forty writers who lived between 1500 and 1700.

1252. Rohde, Eleanor Sinclair. *The Old English Herbals.* London: Longmans, Green and Co., 1922. 243 pp.

Survey, from Anglo-Saxon herbals to the printed herbals of the seventeenth century. Pp. 189–235: Bibliographies (MS herbals, excluding Latin MSS written after 1400; printed English herbals to 1838; printed foreign herbals to 1678). Reviewed in *Isis* 5 (1923), 457–61.

1253. Schreiber, W. L. "Die Kräuterbücher des XV. und XVI. Jahrhunderts." *Hortus sanitatis.* By Johann von Cube. (Faksimileausgabe Peter Schöffer Mainz 1485). Munich: Mandruck, 1924, pp. i–lxiii.

Contains a list of 154 herbals published in Europe (Mainz, Strasburg, Oberrhein, Venice, Prague, Frankfurt, Basel, Zurich).

1254. Arents, George, Jr. *Tobacco: Its History Illustrated by the Books, Manuscripts and Engravings in the Library of George Arents, Jr., Together with an Introductory Essay, a Glossary and Bibliographic Notes* by Jerome E. Brooks. New York: The Rosenbach Co., 1937–1952. 5 vols.

Catalogue of the Arents Tobacco Collection now in the New York Public Library. Full bibliographical entries (3,386 items) arranged chronologically, from 1507 to 1942, with annotations and references to standard bibliographical works. Items 1–447 cover books published up to 1700. Vol. 1, pp. 3–173: General survey of the history of tobacco. Vol. 5: Index by Anne M. Nill (pp. 3–145, names; pp. 148–326, subjects).

1255. Schmid, Alfred. "Über alte Kräuterbücher." *Contributions bibliophiliques 1939, Schweizer Beiträge zur Buchkunde.* Berne: Société Suisse des Bibliophiles, 1939, pp. 65–140.

Outline history of old herbals and the genetical relationship between them. Includes a select, annotated bibliography of secondary sources (pp. 133–38) and a *Stammtafel* of herbals.

1256. Bourlière, François, ed. *Eléments d'un guide bibliographique du naturaliste.* Mâcon: Protat, 1940. 302 pp. Two supplements by P. Lechevalier, 1941. Pp. 303–68.

Main work contains 6,357 notices arranged in seven chapters: General; Europe; Asia; Africa; The Americas; South Asia (from Malaysia to New Zealand); Arctic and Antarctic. Each chapter is subdivided by subject (biology, botany, geology, and ethnology). No index. Pp. 335–53 of supplements (NOT SEEN) deal with the major zoological expeditions (arranged by country).

1257. Hunt Botanical Library. *Catalogue of Botanical Books in the Collection of Rachel McMasters Miller Hunt. Vol. 1: Printed Books 1477–1700. With Several Manuscripts of the Twelfth, Fifteenth, Sixteenth and Seventeenth Centuries.*, Compiled by Jane Quinby. Pittsburgh, Penn., 1958. 517 pp.

Pp. xxiii–xlvii: Introduction. Pp. xlix–lxxv: Bibliographical survey (Printing in the fifteenth and sixteenth centuries as represented in the Hunt Collection, with a note on the seventeenth century) by Margaret Bingham Stillwell. Pp. 1–424: Catlogue (405 items). Full bibliographical descriptions with references to standard bibliographical sources and annotations. Pp. 461–517: Author, title and subject index. Reviewed in *Sudhoffs Archiv* 47 (1963), 191–92.

1258. Lawrence, George H. M. "Herbals, Their History and Significance." *History of Botany: Two Papers Presented at a Symposium Held at the William Andrews Clark Memorial Library, December 7, 1963.* Los Angeles: University of California, Clark Memorial Library, and Pittsburgh: Hunt Botanical Library, 1965, pp. 1–35.

Includes a select bibliography of some 250 titles.

1259. Stearn, W. T. "Sources of Information about Botanic Gardens and Herbaria." *Biological Journal of the Linnean Society* 3 (1971), 225–33.

A rapid survey of the development of botanic gardens and herbaria from the sixteenth century to the present day. Bibliography of some sixty items classified under: botanical and horticultural collections; botanic gardens and plant introduction; herbaria.

1260. Smit, Pieter. *History of the Life Sciences: An Annotated Bibliography.* Amsterdam: Asher, 1974. 1,036 col. + 35 pp.

Lists some three thousand titles (120 bibliographies, 1,500 histories of biological sciences, 450 histories of medicine, 60 biographical dictionaries, 850 biographies), classified under 130 headings and arranged alphabetically under each. Detailed annotations include references to less important works. Col. 1–4: philosophy and methodology; 5–28: general histories of science; 40–61: bibliographies (general biology, plant sciences, animal sciences); 79–82: biographical dictionaries; 101–33: other reference works; 559–715: histories of the life sciences in general, animal and plant sciences, during the Renaissance and later periods; 863–1035: select list of biographies (arranged alphabetically by biographee). Pp. 1037–71: index of personal names. No subject index. Reviewed in *Isis* 68 (1977), 310–11 and in *Annals of Science* 34 (1977), 84–85.

1261. Bourdier, Georgette Légée, Michel Guédès, Jean Théodorides, and Yves Laissus, eds. *Introduction bibliographique à l'histoire de la biologie.* Publié avec le concours de l'Académie des Sciences de l'Institut de France. (*Histoire et nature*, nouv. série, numéro 5–6, fasc. 3–4, 1974–1975). Paris: Museum Nationale d'histoire Naturelle, 1975. 195 pp.

Guide for study and research in the history of biology. In three parts: 1. Generalities. 2. The development of biology—by period (pp. 55–67 cover the years 1453–1715). 3. Subject bibliography (general biology, botany, microbiology, zoology, physiology, ethology, evolution, natural history collections); includes studies of individual biologists. Many titles annotated. Reviews: *Revue de synthèse* 98 (1977), 390–91; *Annals of Science* 35 (1978), 210–11; *Isis* 69 (1978), 605–11.

1262. Henrey, Blanche. *British Botanical and Horticultural Literature before 1800, Comprising a History and Bibliography of Botanical and Horticultural Books Printed in England, Scotland, and Ireland from the Earliest Times until 1800.* Vol. 1. *The Sixteenth and Seventeenth Centuries. History and Bibliography.* London: Oxford Univ. Press, 1975. 290 pp.

Surveys the history of herbals, gardening books, and works on cultivation, and early scientific and floristic works. Pp. 229–76: Bibliography of 376 titles. Full bibliographic descriptions, with locations. Not annotated. Reviewed in *Isis* 68 (1977), 467–69 and *The Library* (6), 2 (1980), 235–40.

1263. Keynes, Sir Geoffrey. *John Ray, 1627–1705: A Bibliography, 1660–1970. A Descriptive Bibliography of the Works of John Ray, English Naturalist, Philologist and Theologian.* With introductions, annotations, various indexes, and a supplement of new entries, additions and corrections by the author. Amsterdam: Gerard Th. van Heusden, 1976. 184 pp. [First published in 1951].

Bibliography was originally compiled following the publication of Raven's biography of Ray (item 1299). The description of each work is preceded by a historical note. Reviewed in *Isis* 43 (1952), 276–77.

1264. Freeman, R. B. *British Natural History Banks, 1495–1900. A Handlist.* Folkestone: Dawson, 1980. 437 pp.

Pp. 25–384: alphabetical list (by author); pp. 385–400: chronological list; pp. 403–37: subject index. No annotations. Includes some one-hundred titles published before 1700. Reviewed in *Isis* 72 (1981), 651.

1265. Creasey, J. S. "Agrarian and Food History." *Information Sources in Agriculture and Food Sciences* G. P. Lilley, ed. London: Butterworth, 1981, pp. 526–83.

Bibliographical essay. Topics include: World agrarian history; Study of agrarian history in Great Britain; Bibliographies and library catalogues; Sources for British agrarian history; Social groups in agriculture; Agricultural science; Agricultural history in other countries; Agricultural technology; Farm crops; Horticulture and gardening; Forestry and woodland; Livestock husbandry; Food.

1266. Museum of Fine Arts, Houston. *Leonardo da Vinci, Nature Studies from the Royal Library at Windsor Castle.* [Exhibition] *7 February–4 April 1982. Catalogue* by Carlo Pedretti. Introduction by Kenneth Clark. New York: Johnson Reprint Corporation, 1981. 95 pp.; pls.

Catalogue of some fifty drawings (concerned with landscapes, botanical drawings, and studies of the flow of water) arranged roughly in chronological order—an almost complete corpus of Leonardo's nature studies exhibited at Windsor Castle—subdivided into ten categories with introductory notes. Pp. 94–95: Bibliographical note.

1267. Allen, David Elliston. "Life Sciences: Natural History." *Information Sources in the History of Science and Medicine.* Editors, Pietro Corsi and Paul Weindling. Butterworth's Guides to Information Science. London: Butterworth Scientific, 1983, pp. 349–60.

Bibliographical survey of the history of natural history. Some 80 references.

1268. Hocker, Sally Haines. *Herbals and Closely Related Medico-Botanical Works, 1472–1753 in the Department of Special Collections, Kenneth Spencer Research Library, and the History of Medicine Collection, Clendening Medical Library.* University of Kansas publications, Library series, 50. Lawrence: Univ. of Kansas Libraries, 1985. 94 pp.

Catalogue of some 260 herbals, of which 231 were published between 1472 and 1700. Full bibliographical descriptions, no annotations.

1269. Guerrini, Anita. *Natural History and the New World, 1524–1770. An Annotated Bibliography of Printed Materials in the Library of the American Philosophical Society.* American Philosophical Society Library Publication, 11. Philadelphia, 1986. 83 pp.

NOT SEEN.

1270. Bridson, Gavin D. R., and James J. White, comps. *Plant, Animal and Anatomical Illustration in Art and Science: a Bibliographical Guide from the Sixteenth Century to the Present Day.* Winchester: St. Paul's Bibliographies (in association with Hunt Institute for Botanical Documentation), 1990. 450 pp.

Some 4,500 entries arranged in six main sections: A. Bibliographies; B. Nature in general; C. Plants; D. Animals; E. The Human body; F. Artist biographies. Sections B–E each include four categories of literature: primary literature on hand-created illustration methods from the sixteenth century to the present day; secondary literature on the history and criticism of natural history illustration; selected literature on the history of plant, animal, and anatomical representation in art; and primary and secondary literature on illustrative photography from the 1840s to the present day. Section F includes secondary literature on the biography and work of individual artists, illustrators, and photographers. Pp. 325–450: Indexes (title; subject; name).

1270A. Bridson, Gavin. *The History of Natural History. An Annotated Bibliography.* Bibliographies of the History of Science and Technology, 24. New York: Garland Publishing, 1994. 740 pp.

Some 7,000 entries arranged in eight sections. With indexes. Covers natural history from the end of the Middle Ages to the present day. Includes bibliographies.

MONOGRAPHS AND ARTICLES

1271. Cuvier, Georges. *Histoire des sciences naturelles, depuis leur origine, jusqu'à nos jours, chez tous les peuples connus, professée au Col-*

lège de France. Tome 2. Paris: chez Fortin, Masson et Cie., 1841. 558 pp.

Nineteen lectures dealing with the sixteenth and seventeenth centuries. With index.

1272. Saint-Lager, J. B. "Histoire des herbiers." *Annales de la Société Botanique de Lyon* 13 (1885), 1–120.

Study of sixteenth-century herbaria from Luca Ghini (c. 1490–1556) to Jean Bauhin (1541–1613). Includes descriptions of the herbaria of Ghini's pupils Aldrovandi and Cesalpino and also of Jean Girault, Rauwolf, and the Duke of Ferrara. No bibliography.

1273. Kraus, Gregor. *Geschichte der Pflanzeneinführungen in die europäischen botanischen Gärten*. Leipzig: Engelmann, 1894. 73 pp.

Study of the introduction of plants into some fifty botanical gardens of Continental Europe, from the time of Gesner to the nineteenth century. In seven stages, four of which fall in the sixteenth and seventeenth centuries: Europe, The Orient, Canada-Virginia, Cape. For each stage the following details are given in tabulated form: name of garden; name of director, year, number of plants; source of information; comments. Twenty-one gardens are listed for the period 1560–1700.

1274. Hamy, E. T. "Vespasien Robin, arboriste du Roy, premier sous-démonstrateur de botanique du Jardin Royal des Plantes (1635–1662)." *Nouvelles archives du Muséum d'Histoire Naturelle* (3), 8 (1896), 1–24.

Robin (1579–1662), son of Jean Robin (1550–1629) apothecary and botanist, was a plant collector from childhood. The Robins introduced into Europe the first acacia tree which Vespasien planted in Guy de la Brosse's garden in 1636 and which still survives. Pp. 15–24 contain extracts of correspondence with Peiresc and Valavez.

1275. Bretschneider, B. *History of European Botanical Discoveries in China*. Vol. 1. London: Sampson Low, Marston and Co., 1898. 1,167 pp.

Pp. 5–44: Early sea trade of the Portuguese with China; early botanical information with respect to China supplied by Catholic missionaries, especially the Jesuits; sea trade of the Dutch with Eastern Asia in the seventeenth century; early English intercourse with China and first botanical collections made there.

1276. Legré, Ludovic. *La botanique en Provence au XVIe siècle. [1] Pierre Pena et Mathias Lobel. [2] Hugues de Solier. [3] Felix et Thomas Plat-*

ter. *[4] Léonard Rauwolff-Jacques Raynaudet. [5] Louis Anguillara, Pierre Belon, Charles de l'Escluse, Antoine Constantin. [6] Les deux Bauhin, Jean-Henri Cheler et Valerand Dourez.* Marseilles: Aubertin et Rolle, 1899–1904. 6 vols.

Studies of plant explorers in Provence. Many of them studied or taught at the University of Montpellier. The famous herbarium assembled by Rauwolff is now at the University of Leiden.

1277. Sachs, Julius von. *History of Botany (1530–1860).* 2nd ed. Oxford: Clarendon Press, 1906. 865 pp.

Pp. 13–36: Botanists of Germany and the Netherlands from Brunfels to Caspar Bauhin, 1530–1623. Pp. 37–107: Artificial systems and terminology from Cesalpino to Linnaeus, 1583–1760. Pp. 229–45: Phytotomy founded by Malpighi and Grew, 1671–1682. Pp. 376–90: From Aristotle to Camerarius; establishment of doctrine of sexuality in plants. Pp. 445–53: History of the theory of nutrition of plants.

1278. Giraud, Louise. *Le premier Jardin des Plantes français. Création et restauration du Jardin du Roy à Montpellier par Pierre Richer de Bellaval (1593–1632).* Montpellier: Impr. Roumégous et Déhan, 1911. 396 pp.

NOT SEEN.

1279. Miall, L. C. *The Early Naturalists, Their Lives and Work (1530–1789).* London: Macmillan, 1912. 396 pp.

Pp. 12–239: From Otto Brunfels to Claude Perrault and Francesco Redi (topics: The new biology; the natural history of distant lands; early English naturalists; Ray and his fellow workers; the minute anatomists; early studies in comparative anatomy). No bibliography.

1280. Britten, James. "Some Early Cape Botanists and Collectors." *Journal of the Linnean Society* (Botany) 45 (1920), 29–51.

Notes on early botanists and collectors whose work is represented in the Department of Botany of the British Museum (*now* Natural History Museum). Includes: Justus Heurnius, Paul Hermann (1640–1698), Thos. Bartholinus (1616–1680), Henric Bern, Oldenlandus (d. 1699), John Starrenburgh (fl. 1700–09), John Foxe, Franz Kiggelaer (d. 1722).

1281. Gunther, R. T. *Early British Botanists and Their Gardens. Based on Unpublished Writings of Goodyear, Tradescant, and Others.* Oxford: printed by Frederick Hall for the author at the University Press, 1922. 417 pp.

Pp. 1–99: Life of John Goodyear (1592–1664). Pp. 100–96: Descriptions of plants by Goodyear. Pp. 197–232: Goodyear's library (with catalogue). Pp. 233–302: Notes on contemporary botanists. Pp. 303–57: Lists of plants grown in English gardens. Pp. 358–71: Lists of exotic plants.

1282. Savage, F. G. *The Flora and Folk Lore of Shakespeare.* London: Burrow, 1923. 420 pp.

Revised version of a series of 155 articles which appeared in the Stratford-upon-Avon *Herald* during the years 1909 to 1916. Every plant mentioned by the poet is described with allusions. With an index of plants under their Shakespearian names, botanical names, and natural orders.

For an earlier study see Henry N. Ellacombe, *The Plant-Lore and Garden-Craft of Shakespeare*, new ed. (London: Edwin Arnold, 1896; 384 pp.).

1283. Gunther, R. T. *Early Science in Oxford.* Vol. 3. Part 1. *The Biological Sciences.* Part 2. *The Biological Collections.* Oxford: Privately printed, 1925. 564 pp.

The result of "stock-taking the scientific resources of the University of Oxford since the earliest times." Historical introduction and study of the biological collections with a *catalogue raisonné.* Subjects include medicine, anatomy and physiology. Reviewed in *Isis* 8 (1926), 375–77.

1284. Delaunay, Paul. *L'aventureuse existence de Pierre Belon du Mans.* Paris: Champion, 1926. 177 pp.

Study of the life and work of Belon, physician and explorer-naturalist (1517–1564).

1285. Hunger, F. W. T. *Charles de l'Escluse (Carolus Clusius) Nederlandsch kruidkundige, 1526–1609.* The Hague: Martinus Nijhoff, 1927–42. 2 vols.

Biography of one of the most famous botanists of the sixteenth century, a rich source for the history of botany (see Smit, item 1260, col. 898). Vol. 2 contains his letters to Joachim Camerarius (part of the text is in German). Clusius was a student and collaborator of Rondelet; he spent the last sixteen years of his life as professor of botany and *horti praefectus* at the University of Leiden.

1286. Hawks, Allison. *Pioneers of Plant Study.* London: Sheldon Press, 1928. 288 pp.

Traces the work of the pioneers who opened up to the world some of the treasures of the "vegetable kingdom." Pp. 126–219 cover the period from the earliest printed books on plants to Hans Sloane.

1287. Nordenskiold, Erik. *The History of Biology: A Survey.* London: Kegan Paul, Trench, Trubner and Co., 1929. 629 pp.

Based on a course of lectures, 1916–17. Pp. 82–233 relevant.

1288. Rytz, Walther. "Das Herbarium Felix Platters. Ein Beitrag zur Geschichte der Botanik des XVI Jahrhunderts." *Verhandlungen der Naturforschenden Gesellschaft in Basel* 44(1), (1932–33), 1–222.

Pp. 1–126: Study of a herbarium identified as that of Felix Platter of Basel (1536–1614), collected when he was a medical student at Montpellier. With a biography and a brief history of herbaria. References (pp. 31–83) include a bibliography of herbals, herbaria, and their history. Pp. 126–222: Appendices (Descriptions of: contents of herbarium; illustrations; water colours).

1289. Fournier, P. *Voyages et découvertes scientifiques des missionaires naturalistes français à travers le monde pendant cinq siècles, XVᵉ à XXᵉ siècles.* Iᵉ Partie: *Les voyages naturalistes du clergé français avant la Révolution.* Encyclopédie biologique, 10. Paris: Paul Lechevalier, 1932. 108 pp.

Pp. 1–45 cover the sixteenth and seventeenth centuries: Brazil (André Thevet); Canada (Jesuits, Récollets); Equatorial France (Claude d'Abbéville and Yves d'Evreux); West Indies (Du Tertre); Far East. With bibliographies.

1290. Rohde, Eleanour Sinclair. *The Story of the Garden.* With a chapter on American gardens by Mrs. Francis King. London: The Medici Society, 1932. 326 pp.

Pp. 61–171: The Tudor Age; Stuart times; French and Dutch influences. Pp. 273–82 of the bibliography cover the years 1495–1706 (some 200 titles).

1291. Cowan, John Macqueen. "The History of the Royal Botanic Garden, Edinburgh." *Notes from the Royal Botanic Garden, Edinburgh* 19 (1933), 1–134.

Pp. 1–62 contain the early history of the garden (estd. 1670) under James Sutherland, the first Keeper and compiler of *Hortus medicus Edinburgensis* (Edinburgh, 1683).

1292. Wright, Richardson. *The Story of Gardening: from the Hanging Gardens of Babylon to the Hanging Gardens of New York.* London: George Routledge and Sons, 1934. 475 pp.

Pp. 178–296 (Chapters 10, 11, 12) relate to the Renaissance and seventeenth century. Ch. 10: The garden flowering of the Renaissance and afterward (includes: Italian Renaissance botanists; new plants from the new world; gardens the Conquistadores made; French, German, and Dutch botanists). Ch. 11: Two centuries of English gardening (includes: gardening books, herbals; physic gardens and botanists; seventeenth-century colonial gardens in America). Ch. 12: André Le Notre (1613–1700) and the spread of his gardens. Pp. 433–46: Bibliography.

1293. Veendorp, H., and L. G. M. Baas Becking. *Hortus Academicus Lugduno-Batavus, 1587–1937: The Development of the Gardens of Leyden University.* Haarlem: Enschedé, 1938. 218 pp.

Pp. 9–114: The gardens from the beginnings through Clusius and Pauw to Boerhaave. With a bibliography and indexes (plants, personal and geographical names).

1294. Alvarez Lopez, Enrique. "Las plantas de América en la botánica europea del siglo XVI." *Revista de Indias, Madrid* 6 (1945), 221–88.

A study of American plants and their place in the botanical works of sixteenth-century Europe.

1295. Bouvet, M. "Les anciens jardins botaniques médicaux de Paris." *Revue d'histoire de pharmacie* [35e année], 9 (1947), 221–28.

Notes on twelve botanical gardens of the sixteenth and seventeenth centuries.

1296. Raven, C. E. *English Naturalists from Neckham to Ray. A Study of the Making of the Modern World.* Cambridge: University Press, 1947. 379 pp.

Survey of the exploration and study of nature in England. Reviewed in *Isis* 39 (1948), 196–97.

1297. Jaramillo-Arango, Jaime. "A Critical Review of the Basic Facts in the History of Cinchona." *Journal of the Linnean Society (Botany)* 53 (1949), 272–311.

Throws new light on the early history of cinchona (introduced to Europe from the New World and, until recently, the only effective remedy against malaria).

1298. Blunt, Wilfrid. *The Art of Botanical Illustration.* London: Collins, 1950. 304 pp.

General survey of the development of botanical illustration. Pp. 45–131: from the herbals of Brunfels and Fuchs to the Dutch flower-pieces.

1299. Raven, Charles E. *John Ray, Naturalist: His Life and Works.* 2nd ed. Cambridge: Cambridge Univ. Press, 1950. 506 pp. [1st ed. 1942]

Comprehensive study of the father of English botany (1627–1705). Reviewed by Agnes Arber in *Isis* 34 (1943), 319–24.

1300. Callot, Emile. *La renaissance des sciences de la vie au XVIe siècle.* Paris: Presses Universitaires de France, 1951. 204 pp.

Reviews the history of biology of the sixteenth century against the philosophical and cultural background of the period. In five chapters: methodology; morphology; anatomy; physiology; taxology. Topics discussed include: the discovery of new plants and animals; herbals and botanical gardens; cultivation of plants; breeding of animals. With a bibliography. Reviewed in *Archives internationales d'histoire des sciences* 4 (1951), 1001–4.

1300A. Karsten, Mia C. *The Old Company's Garden at the Cape and Its Superintendents, Involving an Historical Account of Early Cape Botany.* Cape Town: Maskew Miller, 1951. 188 pp.

An account of the garden colony established (1652) by the Dutch East India Co.—the forerunner of the Botanic Gardens at Kirstenbosch—to supply vegetables and fruit "essential for the scurvy-ridden sailors on their long voyages between Europe and the Indies." Pp. 1–105 cover the history of the garden from Jan van Riebeeck to Simon van der Stel. Pp. 169–76: Bibliography.

1301. Stearns, Raymond Phineas. "James Petiver, Promoter of Natural Science, *c.* 1663–1718." *Proceedings of the American Antiquarian Society* 62 (1952), 243–365.

For more than thirty years (after 1685), Petiver was the proprietor of an apothecary shop in Aldersgate Street, London. His correspondents and customers extended "from Moscow to the Cape of Good Hope and from the British Colonies in the New World to the Spanish settlements in the Philippines . . . . Petiver's vigilant, even nagging, correspondence served to introduce the study of natural science, to underscore its importance, and to suggest something of its methodology to scores of colonials."

1302. Chiarugi, Alberto. "Le date di fondazione dei primi orti botanici del mondo: Pisa (estate 1543), Padova (7 luglio 1545), Firenze (1 dicembre 1545)." *Nuovo giornale botanico italiano,* n.s., 60 (1953), 785–839.

Notes on the foundation of the early university botanical gardens and the teaching of *materia medica*. Establishes the correct chronology of the oldest university botanical gardens in the world. Documented. Also gives the dates of foundation of the other *orti*: Bologna 1568, Leipzig 1580, Koenigsberg 1581, Breslau 1587, Paris (Jardin Royale de Louvre) 1590, Montpellier 1598, Oxford 1621, Paris (Jardin des Plantes) 1626, Jena 1629, Messina 1639.

1303. Crestois, Paul. *L'enseignement de la botanique au Jardin royal des plantes.* Contribution à l'histoire de l'enseignement de la pharmacie. Thèse, Strasbourg, 1953. Cahors: A. Coulesant, 1953. 132 pp.

From the beginnings to the end of the eighteenth century. Bibliography.

1304. De Virville, Ad. Davy. *Histoire de la botanique en France.* Paris: Société d'Edition d'Enseignement Supérieur, 1954. 394 pp.

Pp. 21–61: From Jean de Ruel (1474–1537) to Tournefort (1656–1708). Includes botanical expeditions.

1305. Jacquot, Jean. *Le naturaliste Sir Hans Sloane (1660–1753) et les échanges scientifiques entre la France et l'Angleterre.* Conférence faite au Palais de la Découverte, Université de Paris, série D, no. 29. Paris, 1954. 25 pp.

Sloane's relations with French scientists (Chirac, Magnol, Tournefort) began during his sojourns as a student in Paris and Montpellier.

1306. Museum National d'Histoire Naturelle. *Tournefort.* By G. Becker and others. Les grands naturalistes français. Paris, 1957. 321 pp.

Fifteen essays on the life and work of Joseph Pitton de Tournefort (1656–1708).

1307. British Museum (Natural History). *The Sloane Herbarium. An Annotated List of the* Horti Sicci *Composing It; With Biographical Accounts of the Principal Contributors.* Based on records compiled by the late James Brown. Revised and edited by J. E. Dandy. London, 1958. 246 pp.

Pp. 78–231 contain biographical notes on some sixty plant collectors of the seventeenth century.

1308. Stearn, W.T. "Botanical Exploration to the Time of Linnaeus." *Proceedings of the Linnean Society* 169 (1958), 173–96.

On the early collectors whose work provided a foundation for modern taxonomy and plant geography. With bibliography.

1309. Petit, Georges, and Jean Théodoridès. *Histoire de la zoologie des origines à Linné*. Paris: Hermann, 1962. 360 pp.

Pp. 191–348: Beginnings of modern zoology. Reviewed in *Isis* 54 (1963), 293–95.

1310. Karsten, Mia C. "Heurnius and Hermann, the Earliest Known Plant Collectors at the Cape." *Journal of South African Botany* 29 (1963), 25–32; 33 (1967), 117–32, 161–75.

In two parts; 1. Justus Heurnius (Van Heurne or Van Horne, 1587–1651?), Dutch missionary, his travel to the Cape (en route to Batavia) and the plants drawn and described by him. 2. Paul Hermann, his life, his plant collecting, and his descriptions of Cape plants.

1311. Boxer, C. R. *Two Pioneers of Tropical Medicine. Garcia d'Orta and Nicholás Monardes*. Lecture series, no. 1. London: Wellcome Historical Medical Library, 1963. 36 pp.

Lecture delivered on the fourth centenary of the publication in Goa of d'Orta's *Coloquios dos simples e drogas e consas medicinais da India*. D'Orta (1501–1568) and Monardes (1493–1588) respectively gave to the learned world of the West the first accurate accounts of Asian and American plants.

1312. Delaunay, Paul. *La zoologie au seizième siècle*. Paris: Hermann, 1963. 338 pp.

Topics covered include: use of animals (for food, clothing, etc.); bestiaries; voyages of exploration; collections and cabinets; zoological iconography; nomenclature; bloodless animals.

1313. *Garcia de Orta. Revista da Junta de Investigações do Ultramar*, 11 (1963), (4), 607–876.

Special issue commemorating the fourth centenary of the publication of Garcia de Orta's *Coloquios dos simples*. Includes thirteen papers on his life and work and a bibliography (pp. 857–75).

1314. Stearn, W. T. "The Influence of Leyden on Botany in the Seventeenth and Eighteenth Centuries." *British Journal of the History of Science* 1 (1963), 137–58.

Founded in 1575, the University of Leiden became well known for its religious tolerance and drew students from all over Europe (and the British American colonies). Successor to Montpellier and Padua, Leiden made its greatest contribution to botany and medicine through the work of Herman Boerhaave (1668–1738). The article is a revised version of a lecture delivered at Leiden University (1960) on the occasion of the award of the honorary Doctorate to the author.

1315. Allan, Mea. *The Tradescants, Their Plants, Gardens and Museum, 1570–1662.* London: Michael Joseph, 1964. 345 pp.

The story of the two John Tradescants, father and son, who "formed the collection of curiosities which became Britain's first public museum, and in the world of horticulture gave us our first lilac, our first acacia trees, our first occidental plane which, crossed with the oriental plane, made possible the dapple-boled *Platanus acerifolia*, joy of our London squares and parks and of all droughty and smoky places, in Ohio, Pennsylvania and New England as well as in Old England. To them we owe scores of the trees, shrubs, fruits and flowers we expect to find in a well-stocked garden today."

1316. Dulieu, L. "Guillaume Rondelet." *Clio medica* 1 (1966), 89–111.

On Rondelet (1507–1566), his work, and his students at Montpellier. With notes and references.

1317. Dannenfeldt, Karl H. *Leonard Rauwolf: Sixteenth-Century Physician, Botanist, and Trader.* Harvard Monographs in the History of Science. Cambridge, Mass.: Harvard Univ. Press, 1968. 321 pp.

Life and work of Rauwolf, the first modern botanist to collect and describe the flora of the Near East. Based on his travel book (pp. 235–49: selections). Pp. 251–79: Plants named or described by him. Pp. 283–313: Bibliography and notes.

1318. Coats, Alice M. *The Quest for Plants: A History of the Horticultural Explorers.* London: Studio Vista, 1969. 400 pp.

Book is chiefly concerned with the exploits of professional gardeners who travelled abroad for the purpose of collecting hardy horticultural plants. Includes: Belon, Rauwolf, George Wheler (Turkish Empire); Guillaume Boel (Iberian Peninsula); John Tradescant (Russia); Engelbert Kaempfer (Japan); James Cunninghame (China); Samuel Browne (India, Burma, Tibet); Georg Eberhard Rumpf, Joseph Kamel (East Indies); John Tradescant II, John Banister (North America); Francisco Hernandez, Thomas Willisel, Hans Sloane, James Barlow, Charles Plumier, James Reed (Mexico and the Spanish Main); William Piso, Georg Marcgrave (The Amazon).

1319. Laissus, Yves, and Anne-Marie Monseigny. "Les plantes du roi. Note sur un grand ouvrage de botanique préparé au XVIIe siècle par l'Académie royale des sciences." *Revue d'histoire des sciences* 22 (1969), 193–236.

On a projected *Histoire des plantes* to be directed by Denis Dodart (1634–1709), not completed. All that was published was *Mémoires pour servir à l'histoire des plantes* (Paris, 1676).

1320. Rauschert, Stephan. "Das Herbarium von Paul Hermann (1646–1695) in den Forschungsbibliothek Gotha." *Hercynia* 7 (1970), (4), 301–28.

Describes the herbarium collected by Hermann in Ceylon and the Cape of Good Hope, now in Gotha. Hermann served the Dutch East India Co. as Chief Medical Officer in Colombo from 1672 to 1677. He was a keen plant collector, and on his way to Ceylon he stopped at the Cape to study the flora. In 1679 he was called to the Chair of Botany at Leiden. His other herbaria can be found in London, Paris, and Leiden. Many plants collected by him are in the Sloane herbarium (London).

Pp. 301–9 contain a brief account of his life and career and a discussion of his place as the "discoverer" of Cape flora and as the first botanist of Ceylon. Pp. 325–28: Bibliography. See also D. C. Gunawardene, "Medicinal and Economic Plants of the Musaeum Zeylanicum of Paul Hermann," *Journal of the Royal Asiatic Society* (Sri Lanka Branch), NS, 19 (1975), 33–48. See also items 1280, 1310, and 1330.

1321. Crosby, Alfred W., Jr. *The Columbian Exchange: Biological and Cultural Consequences of 1492.* Westport, Conn.: Greenwood Press, 1972. 268 pp.

An examination of the impact of Columbus's voyages on the global ecosystem. Discusses: the role of smallpox in the Conquistadors' exploitation of New World Indians; Old World plants and animals in the Americas; the biohistory of syphilis; influence of New World foods (maize, manioc, . . .) on Old World demography. Reviews: *Choice* 10 (1973), 128; *Economic Botany* 27 (1973), 348–49; *William and Mary Quarterly* (3), 30 (1973), 542.

1322. O'Brien, Patricia J. "The Sweet Potato: Its Origin and Dispersal." *American Anthropologist* 74 (1972), 342–65.

Reviews the question of the plant's dispersal from its New World home to the Old World and examines its spread into Polynesia before European contact. ("When Columbus discovered the New World, the sweet potato made its first appearance in European history and began its movement throughout the Old World.")

1323. Smit, P. "Carolus Clusius [1526–1609] and the Beginning of Botany in Leiden University." *Janus* 60 (1973), 87–92.

Clusius founded a new botanical tradition and made Leiden the botanical center of Europe.

1324. Van der Pas, Peter. "The Earliest European Descriptions of Japan's Flora." *Janus* 61 (1974), 281–95.

Notes on Georg Meister (gardener) and Andreas Cleyer (Chief Physician, Dutch East India Co. in Batavia) and their attempts at describing Japanese flora in the seventeenth century.

1325. Streseman, Erwin. *Ornithology, from Aristotle to the Present.* Cambridge, Mass.: Harvard Univ. Press, 1975. 432 pp. [Translation of the 1951 German ed.]

Pp. 3–48 cover the years before 1700: from classical times to the Renaissance; the beginnings of exotic ornithology; the influence of methodology.

1326. Andrade-Lima, Dárdano de, Anne Fox Maule, Troels Myndel Pedersen, and Knud Rahn. "Marcgrave's Brazilian Herbarium Collected 1638–44." *Botanisk Tidsskrift* 71 (1976), 121–60.

Study, with notes on the historical background, of the herbarium collected by Georg Marcgrave (1610–1644) when he served the Dutch West India Co. in Brazil, now (since 1653) in the Botanical Museum of the University of Copenhagen.

1327. Reeds, K. M. "Renaissance Humanism and Botany." *Annals of Science* 33 (1976), 519–42.

On the influence of the Renaissance eds. of the works of Theophrastus, Pliny, Galen, and Dioscorides.

1328. Fastlicht, Samuel. "Spain's First Scientific Expedition to Mexico in the XVI [*sic*] Century: The Tragic Story of Francisco Hernandez." *Bulletin of the History of Dentistry* 28 (1), (1980), 3–10.

On the "prolific work" of Hernández, which, because of his desire for perfection, was not published See also items 1490, 1494.

1329. George, Wilma. "Sources and Background to Discoveries of New Animals in the Sixteenth and Seventeenth Centuries." *History of Science* 18 (1980), 79–104.

On the explosive increase in knowledge of far-away faunas following the discovery of the New World and the extension of the spice-trade routes in the East. Documented (113 references).

1330. Gunn, Mary, and L. E. Codd. *Botanical Exploration of Southern Africa: An Illustrated History of Early Botanical Literature on the Cape Flora. Biographical Accounts of the Leading Plant Collectors and Their Activities in Southern Africa from the Days of the East India*

Company until Modern Times. Cape Town: A.A. Balkema (for the Botanical Research Institute), 1981. 400 pp.

Part 1: Historical outline up to about 1750 (pp. 1–74). Part 2: Dictionary of plant collectors (pp. 77–387). (Pp. 1–54 relate to the period from the voyages of discovery to 1700. Biographical notes on some twenty-four collectors of the seventeenth century in Part 2.)

1331. Howard, Rio C. "Guy de la Brosse [ca. 1586–1641] and the Jardin des Plantes in Paris." *The Analytic Spirit: Essays in the History of Science in Honor of Henry Guerlac,* edited by Harry Woolf. Ithaca, N.Y.: Cornell Univ. Press, 1981, pp. 195–224.

On the early history of the Jardin des Plantes (*now* Musée d'Histoire Naturelle) and its role in the botanical renaissance of the sixteenth and seventeenth centuries. See also items 1001, 1337.

1332. Morton, A. G. *History of Botanical Science: An Account of the Development of Botany from Ancient Times to the Present Day.* London: Academic Press, 1981. 474 pp.

Pp. 115–64: Renascence of botany in Europe; from herbal to flora (1483 to 1623). Pp. 165–231: From Jungius to Camerarius.

1333. Duval, Marguerite. *The King's Garden.* Charlottesville, Va.: Univ. Press of Virginia, 1982. 214 pp. [Translation of *La planète des fleurs* (1977)]

"During more than three exciting centuries, France and Frenchmen dominated the field of plant hunting and international botanic exchanges. Most of the voyages of botanical discovery were French-sponsored until Captain Cook's first voyage to the Pacific." Pp. 9–59: Pierre Belon, first botanist-traveler; creation of the Jardin des Plantes; seventeenth-century missionaries; Tournefort in the Levant; Etienne de Flacourt (Madagascar) and Robin Cavelier de la Salle (Canada).

1334. Marcus, Joyce. "The Plant World of the Sixteenth and Seventeenth Century Lowland Maya." *Maya Subsistence: Studies in Memory of Dennis E. Puleston.* Kent V. Flannery, ed. New York: Academic Press, 1982, pp. 239–73.

An attempt to study the plants used by the ancient Maya during the period A.D. 1566–1696, when "European eyewitnesses gave us our first documentary evidence of how the lowland Maya classified, named, grew, and used the rich variety of plant species known to them."

1335. Palm, L. C., and H. A. M. Snelders, eds. *Antoni van Leeuwenhoek 1632–1723. Studies on the Life and Work of the Delft Scientist Commemorating the 350th Anniversary of His Birthday.* Amsterdam: Rodopi, 1982. 212 pp.

Subjects covered: Schooling; use of language; microscopes; mechanistic view of the world; contributions to plant anatomy; studies of fleas, lice, and mites (with Swammerdam); studies of sexual reproduction; malacological researches; ideas on spontaneous generation; reactions of the scientific establishment. Reviewed in *Annals of Science* 40 (1983), 300.

1336. Greene, Edward Lee. *Landmarks of Botanical History.* Frank N. Egerton, ed. Publication of the Hunt Institute for Botanical Documentation, Carnegie-Mellon University. Stanford, Calif.: Stanford Univ. Press, 1983. 2 vols.

Twenty-six essays, mostly on botanists of the sixteenth and seventeenth centuries. In Appendix: Pre-Grecian knowledge of plants; medieval knowledge of plants; botany in the seventeenth century. Notes and bibliography. Pp. 18–84: Bio-bibliographical note and appreciation of the work of E. L. Greene.

1337. Howard, Rio. *La bibliothèque et le laboratoire de Guy de la Brosse au Jardin des Plantes à Paris.* Histoire et civilisation du livre, 13. Geneva: Droz, 1983. 133 pp.

Transcript of the 1641 inventory with introduction (pp. 1–20), notes (pp. 49–123) and bibliography (pp. 125–33). Of the 1,447 volumes, 531 have been identified. La Brosse was the founder and first director of the *Jardin des Plantes.* See also item 1331.

1338. Thomas, Keith. *Man and the Natural World. Changing Attitudes in England 1500–1800.* Harmondsworth, England: Penguin Books, 1984. 425 pp. [First published 1983]

Expanded version of the George Macaulay Trevelyan lectures, Cambridge, 1979. Documents the changes in English views of man's relationship to the natural world. Reviews: *Observer* (London), 4 Dec. 1983, p. 25; *American Historical Review* 89 (1984), 733; *English Historical Review* 99 (1984), 571–74; *Historical Journal* 28 (1985), 225–29; *Isis* 75 (1984), 588–89.

1339. Leith-Ross, Prudence. *The John Tradescants: Gardeners to the Rose and Lily Queen.* London: Owen, 1984. 320 pp.

"New" biography of the Tradescants, father and son. Pp. 181–95: The Tradescants' plant introductions. For a critical review see *Annals of Science* 42 (1985), 167–69 (A. V. Simlock).

1340. Rhodes, D. E. "The Botanical Garden of Padua: the First Hundred Years." *Journal of Garden History*, (4), 4 (1984), 327–31.

On the *orto botanico* (founded 1545) and its curators.

1341. Struik, Dirk J. "Early Colonial Science in North America and Mexico." *Quipu: Revista latinoamericana de historia de las ciencias y la tecnologia* 1 (1984), 25–54, 323–25.

Survey, from the Vespucci pamphlets to the book on medicinal properties of American plants by Monardes and the work of the engineer-astronomer Enrico Martinez. Bibliography.

1342. *Le* Traité de matière médicale *de Dioscoride à la Renaissance. Exposition d'exemplaires des XVe et XVIe siècles, de 11 au 24 Février 1984, à Louvain-en-Woluwe à la Bibliothèque de la Faculté de Médecine de l'U.C.L.* Louvain: Université Catholique de Louvain, Centre d'Histoire de la Pharmacie et du Médicament, 1984. 40 pp.

Catalogue of an exhibition illustrating the influence and diffusion of the *Materia medica* of Dioscorides during the Renaissance. Manuscript copies, eds. and translations, and commentaries.

1343. Koreny, Fritz. *Albrecht Dürer und die Tier- und Pflanzenstudien der Renaissance.* Munich: Prestel Verlag, 1985. 278 pp. [English ed. published in Boston, 1988]

Exhibition catalogue of 93 drawings by Dürer, his predecessors and contemporaries. Full catalogue entries. With introduction and bibliography.

1344. Arber, Agnes. *Herbals, Their Origin and Evolution. A Chapter in the History of Botany, 1470–1670.* 3rd ed., with an introduction and annotations by William T. Stearn. Cambridge: Cambridge Univ. Press, 1986. 358 pp. [First published 1912]

Outlines the evolution of the printed herbal in Europe. Appendix 1 (pp. 271–85): Principal herbals and related botanical works, listed chronologically. Appendix 2 (pp. 286–302): Historical and critical works consulted. Appendix 3 (pp. 303–7): Subject index to Appendix 2. Pp. 319–38: Reprint of author's essay "From Medieval Herbalism to the Birth of Modern Botany."

1345. Baldwin, Stuart A. *John Ray (1627–1705), Essex Naturalist. A Summary of His Life, Work and Scientific Significance.* Witham, Essex: privately published, 1986. 80 pp.

Written for the John Ray celebrations held to commemorate the third centenary of the publication of his *Historia plantarum.*

1346. Heniger, J. *Hendrik Adriaan van Reede tot Drankenstein (1636–1691) and Hortus Malabaricus. A Contribution to the History of Dutch Colonial Botany.* Rotterdam: A. A. Balkama, 1986. 295 pp.

Life of Van Reede, servant of the Dutch East India Co. and organizer of the study of colonial botany. Analyses the previous history and genesis of his *Hortus Malabaricus* (1678–93), dealing with the flora of Malabar. Surveys the process of its insertion in the botanical literature.

1347. Emboden, William A. *Leonardo da Vinci on Plants and Gardens.* Foreword by Carlo Pedretti. London: Christopher Helm, 1987. 234 pp.

"Here, for the first time, Leonardo's plant studies are presented in their historical and cultural context with a wealth of information and interpretation that has the vividness and precision of a Leonardo drawing" (Carlo Pedretti). A comprehensive study of Leonardo in the dual role of artist and botanist, with illustrations from his notebooks. Includes a list of plants mentioned in his notebooks, a bibliography, and index. Reviewed in *Archives of Natural History* 16 (1989), 355.

1348. Scarborough, John. "Botany, Pharmacy, and the Culinary Arts." *The Rational Arts of Living.* A. C. Crombie and Nancy Siraisi, eds. Ruth and Clarence Kennedy Conference in the Renaissance, 1982. Smith College Studies in History, Vol. 50. Northampton, Mass.: Department of History of Smith College, 1987, pp. 161–202.

"Reconstruction of the transformation of the European biological environment and its potentialities in the early modern period by the introduction of so many new plants" from the voyages to different regions of Asia, Africa, and the Americas.

1349. George, Wilma. *Thomas Harriot and the Fauna of North America.* The Durham Thomas Harriot Seminar, Occasional Paper no. 5. Durham, 1988. 28 pp.

On Harriot's *A Briefe and True Report of the New Found Land of Virginia* (1588).

1350. Jacques, David, and Arend Jan Van der Horst, eds. *The Gardens of William and Mary.* William and Mary: 1688–1988, Anglo-Dutch Celebration. London: Christopher Helm, 1988. 224 pp.

Study of Dutch and English garden history, with descriptions of Dutch and English gardens during the reign of William and Mary. Pp. 167–80: Exotics. Pp. 211–12: Selected list of exotics imported to the Netherlands between 1685 and 1700.

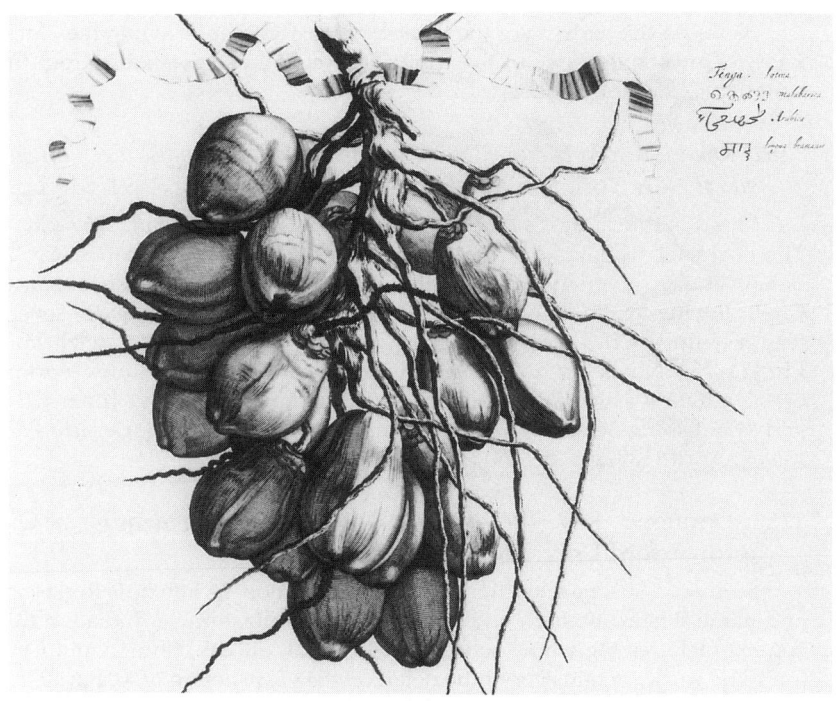

Bunch of coconuts. From the *Hortus (Indicus) Malabaricus*, vol. 1, fig. 3 (Bodleian Library, Oxford, 19163.b.l). **Entry 1346.**

1351. Heniger, J. and M. S. M. Sosef. "Antoni Gaymans (*ca.* 1630–1680) and His Herbaria." *Archives of Natural History* 16 (1989), (2), 147–68.

Notes on the herbaria of the Leiden pharmacist, representing the oldest known Dutch collections of this kind. They contain some exotics from the Americas, Asia, and South Africa.

1352. Reeds, Karen Meier. *Botany in Medieval and Renaissance Universities*. New York and London: Garland Publishing, 1991. 313 pp.

Based on the author's Ph.D. thesis, University of Michigan, 1969. Ch. 1: The character of botany in the Middle Ages and the Renaissance. Ch. 2: Botany at the University of Montpellier in the Middle Ages and Renaissance. Ch. 3: Botany at the University of Basle in the sixteenth and early seventeenth century. Ch. 4: Learning botany from books. Pp. 167–70: Conclusion. Pp. 171–259: Notes. Pp. 261–313: Bibliography and index. Includes reprints of the author's articles: "Renaissance Humanism and Botany" (item 1327) and "Publishing Scholarly Books in the Sixteenth Century" (*Scholarly Publishing*, April 1983, pp. 259–74)

1353. Desmond, Ray. *The European Discovery of the Indian Flora*. Oxford: Oxford Univ. Press, 1992. 355 pp.

Study of "Europe's reaction to India's vegetation, its attempts to classify and name this exotic flora and its subsequent utilization." ("India" in this context includes the whole subcontinent, Nepal, Sikkim, Bhutan and Ceylon.) The work contains very little detail on the years before 1700. Pp. 12–38: The first European settlements; Arrival of the British.

Medicine

GENERAL

BIBLIOGRAPHICAL GUIDES

1354. Artelt, Walter. *Einführung in die Medizinhistorik, ihr Wesen, ihre Arbeitweise und ihre Hilfsmittel.* Stuttgart: Enke, 1949, 240 pp.

A manual of research for medical historians. Pp. 8–90: Research methods and sources. Pp. 91–172: Criticism and interpretation. Pp. 173–201: Presentation of results. Pp. 202–37: History of the historiography of medicine (by Edith Heischkel). A 70–page author/title index was published in 1972 by Dagmar Müller and Rolf Winau (Institute for the History of Medicine, Mainz). Reviews: *Centaurus* 1 (1950), 189–90; *Bulletin of the History of Medicine* 26 (1952), 489–91.

1355. Blake, John B., and Charles Roos, eds. *Medical Reference Works, 1679–1966: A Selected Bibliography.* Chicago: Medical Library Association, 1967. 343 pp.

Annotated bibliography in three main sections: General reference (413 items); History of medicine (419); Special subjects (1,871). Pp. 280–343: Index of authors, titles, and subjects. Reviewed in *Isis* 59 (1968), 335–36 ("Impressive is the large number of bibliographies, indexes, abstracts, and histories which abound throughout the work.") Three supplements issued (1970–75) covering the period 1967–74 (1,059 items), but these do not include histories of medicine (for which see item 1363). The task of updating this work has been assumed by the *National Library of Medicine Current Catalog*.

1356. Pelling, Margaret. "Medicine since 1500. " *Information Sources in the History of Science and Medicine.* Pietro Corsi and Paul Weindling, eds. London: Butterworth Scientific, 1983, pp. 379–407.

Bibliographical essay. Pp. 393–407: Bibliography. See item 21.

1357. Webster, Charles. "The Historiography of Medicine." *Information Sources in the History of Science and Medicine.* Pietro Corsi and Paul Weindling, eds. London: Butterworth Scientific, 1983, pp. 29–43.

Survey of the major standard histories of medicine. With bibliography (57 references). See item 21.

1358. Symons, John. "Medical Bibliographies and Bibliographers." *Thornton's Medical Books, Libraries and Collectors.* 3rd ed. by Alain Besson. Aldershot (Hants): Gower Publishing, 1990, pp. 239–66.

A survey of bibliographies and bibliographical guides. Contains informative notes on bibliographies of pre-1700 literature and their compilers.

1359. Morton, L. T., and Shane Godbolt, eds. *Information Sources in the Medical Sciences.* 4th ed. London: Bowker-Saur, 1992. 608 pp. [First published in 1974 as *The Uses of Medical Literature*]

Series of bibliographical essays in 24 chapters. Ch. 24, pp. 557–91: Historical, biographical, and bibliographical sources (by E. J. Freeman). The chapters devoted to the various medical and surgical specialities contain sections briefly dealing with their respective histories.

BIBLIOGRAPHIES

1360. Haller, Albrecht von. *Bibliotheca medicinae praticae, qua scripta ad partem medicinae praticam facientia a rerum initiis recensentur.* Berne: Em. Haller, 1776–88. 4 vols. [Vol. 4 edited by J. D. Brandis. Reprinted in 1986 by Georg Olms, Hildesheim]

Classified annotated bibliography of the literature of medicine from the beginnings to 1707. Arrangement chronological in 12 chapters (Graeci; Arabes; Arabistae; Instauratores; Chemici; Schola Hippocratica; Seculum XVII; Van Helmont; Sylvius de le Boë; Thomas Sydenham; Stahl; Boerhaave), subdivided (1,041 sections) by authors, groups of authors, articles in journals, disputations. Annotations include bio-bibliographical information on authors, résumé of contents of books and critical comments. Some 23,500 entries. Vol. 4, pp. 465–598: Author index.

1361. Meissner, Friedrich Ludwig. *Grundlage der Literatur der Pädiatrik, enthaltend die Monographien über die Kinderkrankheiten.* Leipzig: Fest'sche Verlagsbuchhandlung, 1850. 246 pp.

Bibliography of some 7,000 works on pediatrics classified under ninety headings. Includes 503 pre-1700 titles.

1362. Pazzini, Adalberto. *Bibliografia di storia della medicina italiana.* Rome: Tosi, 1946. 455 pp.

Classified list (21 sections) of 7,451 books and papers, some annotated, on Italian medical history and biography. Reprint of the 1940 ed.

1363. National Library of Medicine, Bethesda, Md. *Bibliography of the History of Medicine,* no. 1 (1965)– . Bethesda, Md., 1965– .

Annual bibliography, cumulated every five years. In three parts: 1. Biographies (with chronological and geographical summary). 2. Subjects other than biographies. (Some 300 divisions, subdivided chronologically and geographically.) 3. Author/title index. Review: *Isis* 59 (1968), 107–8.

1364. O'Malley, C. D. "Tudor Medicine and Biology." *Huntington Library Quarterly* 32 (1968–69), 1–27.

A literature survey of some 120 works, written for a projected history of Tudor prose literature.

1365. Smit, Pieter. *History of the Life Sciences: An Annotated Bibliography.* Amsterdam: Asher, 1974. 1,036 col. + 35 pp.

Coverage includes medicine. Col. 61–73: Bibliographies. Col. 83–90: Biographical dictionaries. Col. 133–48: Other reference works. Col. 715–862: History of the medical sciences during the Renaissance and later periods; history of therapy. Detailed annotations include references to less important works. See main entry—item 1260.

1366. Lindeboom, C. A., comp. *A Classified Bibliography of the History of Dutch Medicine, 1900–1974.* The Hague: Nijhoff, 1975. 663 pp.

Bibliography of some 5,500 books and articles covering mainly medicine in the Northern Netherlands from early times to the present. Includes *all* work on medical historical topics by Netherlands authors. All chapters excluding III, IV and V contain material relevant to the sixteenth and seventeenth centuries. Minute subdivision of chapters compensates for the lack of a subject index. Each chapter has an introduction. Indexes: authors; historical personalities and places.

1367. Maloney, G., and R. Savoie, eds. *Cinq cents ans de bibliographie hippocratique, 1473–1982.* St. Jean-Chrysostome, Québec: Les Editions du Sphinx, 1982. 291 pp.

Chronological list of printed material on Hippocrates and the Hippocratic corpus. Includes eds., translations, commentaries, books, and articles

(3,332 items). Confined to publications in Europe and the Americas. Author index. Reviews: *Medical History* 27 (1983), 318–19; *Revue d'histoire des sciences* 37 (1984), 179–80; *Annals of Science* 41 (1984), 188.

1368. Bruni Celli, Blas, ed. *Bibliografía hippocrática*. Caracas: Universidad Central de Venezuela, Ediciones del Rectorado, 1984. 507 pp.

List of 4,496 titles arranged alphabetically by author and ed. Indexes (name; subject; Hippocratic corpus; place of publication). Reviews: *Medical History* 29 (1985), 329–30; *Journal of the History of Medicine* 42 (1987), 356–57.

1369. Norman, Jeremy M., ed. *Morton's Medical Bibliography: An Annotated Check-list of Texts Illustrating the History of Medicine (Garrison and Morton)*. 5th ed. Aldershot (Hants): Scholar Press, 1991. 1,243 pp. [1st ed. 1943]

The publishing history of this monumental work, which had its beginnings in the *Index-Catalogue of the Surgeon General's Office*, Vol. 17 (1912), 2nd ser., pp. 89–178, is recounted by Leslie T. Morton in the Introduction (pp. v–xi). The work consists of a classified, annotated bibliography (8,927 entries in 56 sections) of the most important contributions to the development of medicine from antiquity to the present day. For each subject, the items are arranged "chronologically" so as to show its historical development. Pp. 983–1026: History of medicine (Histories of special subjects are given separately under the subjects). Pp. 1026–31: Medical biography. Pp. 1032–43: Medical bibliography. Pp. 1044–46: Medical lexicography. Name and subject indexes.

CATALOGUES

1370. Bachmann, Augustus Rivinus. *Bibliotheca Riviniana sive Catalogus librorum philologico-philosophico-historicorum, itinerariorum, inprimis autem medicorum, botanicorum et historiae naturalis scriptorum etc. rariorum, quam magno studio et sumptu sibi comparavit D. Avg. Rivinus . . . vendenda in vaporario collegii rubri a die XXVII Octobr. MDCCXXVII*. Leipzig, 1727. 740 pp. + index. [Reprinted in 1966 by A. Asher and Co., Amsterdam]

Sale catalogue of the library of Bachmann (also known as Rivinus, 1652–1723), botanist, professor of poetry (and later, of physiology) at the University of Leipzig. Some 8,000 vols., of which about half are on medical subjects. Reprint includes a brief bio-bibliographical note by Rudolf Zaunick.

1371. Sallander, Hans, comp. *Bibliotheca Walleriana. The Books Illustrating the History of Medicine and Science, Collected by Dr. Erik Waller and Bequeathed to the Library of the Royal University of Uppsala.* Stockholm: Almqvist and Wiksell, 1955. 2 vols.

Catalogue of some 20,000 books arranged in 18 sections, including: medicine (10,000 items); history of medicine (3,000); biography (1,600); bibliography (1,200). Arrangement within each section is alphabetical by author (except for biographies). Vol. 2, pp. 401–94: Index of authors and subjects. Bibliographies and histories of special topics are entered under the topics in the index. Reviews: *Book Collector* 5 (1956), 176–80; *Isis* 48 (1957), 353–55.

1372. Wellcome Historical Medical Library (*now* Wellcome Institute for the History of Medicine), London. *Catalogue of Western Manuscripts on Medicine and Science in the Library.* By S. A. J. Moorat. London, 1962–76. 3 vols.

Vol. 1: MSS written before 1650. (Some 1,500 individual works.) Vol. 2 (2 pts.): MSS written after 1650. (Some 4,000 MSS of which about 500 are seventeenth-century.) Entries are annotated. Indexes include subject indexes. Review: *Isis* 66 (1975), 105–8.

1373. Wellcome Historical Medical Library (*now* Wellcome Institute for the History of Medicine), London. *Catalogue of the Wellcome Historical Medical Library. I. Books Printed before 1641.* London, 1962. 407 pp.

Some 7,000 titles from the world's largest historical medical library. With indexes (places of publication; printers and publishers) and concordance of English books with the STC (item 109). Vols. II and III (1966–76) contain books (authors A–L) published from 1641 to 1850. Reviews: *Isis* 55 (1964), 107–8; *The Library* (5), 18 (1963), 315–16.

1374. Durling, Richard J., comp. *A Catalogue of Sixteenth Century Printed Books in the National Library of Medicine.* Bethesda, Md.: National Library of Medicine, 1967. 698 pp.

4,808 titles, arranged alphabetically by author. Indexes: Printers and publishers; vernacular imprints. For supplement, see Krivatsy, Peter, comp. *A Catalogue of Incunabula and Sixteenth Century Printed Books in the National Library of Medicine, First Supplement.* Bethesda, Md., 1971.

1375. Eales, Nellie B. *The Cole Library of Early Medicine and Zoology. Catalogue of Books and Pamphlets. Part I. 1472 to 1800.* Oxford: The Alden Press, for The Library, University of Reading, 1969. 425 pp.

Catalogue includes some one thousand titles by authors who flourished during the period 1450 to 1700. Reviews: *Journal of the History of Medicine* 25 (1970), 358–59; *Annals of Science* 30 (1973), 124.

1376. Francis, W. W., R. H. Hill, and Archibald Malloch, eds. *Bibliotheca Osleriana: A Catalogue of Books Illustrating the History of Medicine and Science, Collected, Arranged, and Annotated by Sir William Osler and Bequeathed to McGill University.* 2nd ed. Montreal: McGill-Queen's Univ. Press, 1969. 792 pp. [First published 1929. Reprinted with additions, corrections, and new prologue.]

Catalogue of the library of Sir William Osler, clinician, bibliophile, and medical historian. After thirty years of medical and clinical teaching in North America (McGill, Philadelphia, Johns Hopkins) was Regius Professor of Medicine at Oxford (from 1909 until his death). Consists of 7,787 items arranged as follows: Essential literature of science and medicine (in chronological order) (1–1702); Works of secondary importance (1703–4298); Literary works of physicians, including medical and philosophical poems (4299–5609); Historical works including literature on medical institutions (5610–6565); Biographies (6863–7400); Incunables (7401–7505); Manuscripts (7506–7668); Addenda (7669–7787). Pp. 705–85: Subject and name index. Reviews: *The Library* (4), 10 (1929–30), 340–44; (5), 25 (1970), 271–72.

1377. Manchester University Library. *Catalogue of Medical Books, 1480–1700.* Compiled by Ethel M. Parkinson. University of Manchester Library, Bibliographical Series, no. 3. Manchester: Manchester Univ. Press, 1972. 399 pp.

Alphabetical catalogue of 2,685 items, with a subject index. References made to catalogues of other collections (Wellcome, British Library, Osler Library, etc.). Reviewed in *Isis* 66 (1975), 105–8.

1378. Savelli, Rodolfo, ed. *Catalogo del fondo Demetrio Canevari della Biblioteca Civica Berio di Genova.* Florence: La Nuova Italia Editrice, 1974. 476 pp.

Canevari, Genoese bibliophile and physician, assembled this library during his years in Rome (1584–1625). The catalogue (2,450 titles) is arranged alphabetically in four sequences: incunables, sixteenth century, seventeenth century, and MSS. Indexes: authors, editors, translators, and publishers. For biography, see *Dizionario biografico degli italiani*, Vol. 18 (1975), pp. 59–61.

1379. Herzog August Bibliothek, Wolfenbüttel. *Verzeichnis medizinischer und naturwissenschaftlicher Drucke, 1472–1830.* Bearbeitet von Ursula Zachert. Nendeln, Liechtenstein: Kraus-Thomson, 1976–82; Munich: K. G. Saur, 1987. 14 vols.

Card catalogue of early printed books in medicine and science in the Library (founded in 1572 and named after its principal benefactor). Some 20,000 titles arranged in four separate sequences: alphabetical (Vols. 1–4); chronological (5–7); place of publication (8–10); systematic (11–14). Vols. 5–6 (pp. 1–575) contain pre-1701 titles. Vols. 11–13 (medicine and veterinary science) and 14 (botany, zoology, chemistry, physics, mathematics, and astronomy) have a subject index.

1380. Benetti, Francesca Zen. "La libreria di Girolamo Fabrici d'Acquapendente." *Quaderni per la storia dell'Università di Padova* 9/10 (1976–77), 161–83.

Catalogue of the library of Fabricius, based on the inventory of 1621 (121 titles, of which some eighty are scientific/medical). With an introduction (pp. 161–71).

1381. Wellcome Institute for the History of Medicine. *Subject Catalogue of the History of Medicine and Related Sciences*. Munich: Kraus International Publications, 1980. 18 vols.

Collection consists of some 400,000 printed books (including 650 incunables), 10,000 MSS, 10,000 autograph letters, and 50,000 pamphlets, besides 2,669 periodicals. Vols. 1–9: Subjects (c. 20,000 subjects, with chronological subdivisions where relevant). Vols. 10–13: Topographical (entries under country, subdivided by subject). Vols. 14–18: Biographical. Includes articles in periodicals and off-prints.

1382. Bird, D. T., comp. *A Catalogue of Sixteenth-Century Medical Books in Edinburgh Libraries*. Edinburgh: Royal College of Physicians, 1982. 298 pp.

Catalogue of some 2,500 books. Includes a large number of titles not found in the Wellcome and National Library of Medicine catalogues. Indexes include a select subject index. Reviews: *The Library* (6), 5 (1983), 420–25; *Archives interrnationales d'histoire des sciences* 33 (1983), 397–98.

1383. Monti, Maria Teresa, comp. *Catalogo del fondo Haller della Biblioteca Nazionale Braidense di Milano*. Series: Filosofia e Scienza nel Cinquecento e nel Seicento. Milan: Franco Angeli, 1983–92. 10 vols. in 11. *In progress*.

Catalogue of the Haller collection—library of Albrecht von Haller (1708–77), anatomist, physiologist, botanist, and bibliographer—in the Brera library (Milan) and other libraries of Lombardy (see item 1358, pp. 246–47). The collection contains some 21,000 printed books and dissertations.

1384. Durling, Richard J. "A Guide to the Medical Manuscripts Mentioned in Kristeller's *Iter Italicum* I–IV." *Traditio* 41 (1985), 341–65; 44 (1988), 485–536; 46 (1991), 347–79.

A summary of the medical items which are only "sparsely indicated" in Kristeller's *Iter Italicum* (item 102).

1385. Krivatsy, Peter, comp. *A Catalogue of Seventeenth Century Printed Books in the National Library of Medicine.* Bethesda, Md.: National Library of Medicine, 1989. 1,315 pp.

Catalogue of some 13,300 books arranged alphabetically by author. Includes dissertations. No subject index. Review: *Book Collector* 39 (1990), 416–18.

MONOGRAPHS AND ARTICLES

1386. Kruger, M. S. *Synchronistische Tabellen zur Geschichte der Medizin. Ein Leitfaden zu akademischen Vorlesungen so wie zum Privatgebrauche.* Berlin: A. Hirschwald, 1840. 66 pp.

Chronological tables. Pp. 22–39 cover the years 1450 to 1700.

1387. Daremberg, Charles. *Histoire des sciences médicales comprenant l'anatomie, la physiologie, la médecine, la chirurgie et les doctrines de pathologie générales.* Tome 1: *Depuis les temps historiques jusqu'à Harvey.* Tome 2: *Depuis Harvey jusqu'au XIXe siècle.* Paris: J. B. Bailliere et Fils, 1870. 2 vols.; 1,303 pp.

Summary in 34 chapters of 175 lectures given at the Collège de France from 1864 to 1867. Pp. 311–1000 (Ch. XV–XXVII) cover the fifteenth to seventeenth centuries.

1388. Gautier, Léon. *La médecine à Genève jusqu'à la fin du dix-huitième siècle.* Geneva: Jullien, 1906. 696 pp.

Pp. 25–314: History of medicine in Geneva from the time of Calvin to 1700. Pp. 417–555: List of doctors and bibliography.

1389. Hirschberg, J. *Geschichte der Augenheilkunde in Mittelalter und in der Neuzeit.* Graefe-Saemisch Handbuch der gesamten Augenheilkunde, hrsg. von Theodor Saemisch, 2nd ed., Vol. 13. Leipzig: Engelmann, 1908. 546 pp.

Pp. 265–85: History of spectacles (with bibliography). Pp. 285–357: History of ophthalmology in the sixteenth and seventeenth centuries.

1390. Pagel, Julius Leopold. *Zeittafeln zur Geschichte der Medizin.* Berlin: A. Hirschwald, 1908. 16 pp.

Chronological tables. Tables VIII to IX^c cover the sixteenth and seventeenth centuries.

1391. Smith, Sir Frederick. *The Early History of Veterinary Literature and Its British Development.* Vol. 1. London: J. A. Allen, 1976. 373 pp. [Reprinted from the *Journal of Comparative Pathology and Therapeutics*, 1912–1918]

Pp. 103–369 contain a survey of the literature of the fifteenth, sixteenth and seventeenth centuries. Review: *Isis* 3 (1920), 307–8.

1392. Holländer, Eugen. *Die Medizin in der klassischen Malerei.* 3rd ed. Stuttgart: Enke, 1923. 488 pp.; 307 illus. [Reprinted 1950]

Medicine and its practice seen through classical paintings. Subjects include: anatomical drawings; anatomy lessons; pregnancy; childbirth; the plague; surgery. The work has been considered a classic of medico-artistic documentation. Review: *Archives internationales d'histoire des sciences* 4 (1951), 1074–79.

1393. Garrison, Fielding H. *An Introduction to the History of Medicine. With Medical Chronology, Suggestions for Study and Bibliographic Data.* 4th ed., revised and enlarged. Philadelphia: W. B. Saunders Co., 1929. 996 pp. [First published in 1913]

Pp. 195–309: The period of the Renaissance, the revival of learning, and the Reformation (1453–1600); The seventeenth century. Pp. 809–79: Chronology of medicine and public hygiene (pp. 816–29 relevant). Reviewed in: *Bulletin, Johns Hopkins Hospital* 25 (1914), 284–85; *Isis* 4 (1922), 554–56; *Isis* 13 (1929), 137–38.

1394. Still, George Frederic. *The History of Pediatrics. The Progress of the Study of Diseases of Children up to the End of the XVIIIth Century.* London: Oxford Univ. Press, 1931. 526 pp. [Reprinted in 1965 by Dawsons of Pall Mall]

Pp. 58–304: From Paolo Bagellardo's *Libellus de egritudinibus infantium* (1472) to John Pechey's *A General Treatise of the Diseases of Infants and Children* (1697). Pp. 305–22: Medicine and pedagogy (Sir Thomas Elyot, George Maler, Sir William Petty). Appendix I (pp. 505–13): List of 213 inaugural dissertations by students at Continental universities. Pp. 517–26: Indexes: name and subject).

1395. Pagel, Walter. "The Reaction to Aristotle in Seventeenth-Century Biological Thought: Campanella, van Helmont, Glanvill, Charleton, Harvey, Glisson, Descartes." *Science, Medicine and History: Essays in Honour of Charles Singer.* Edited by E. Ashworth Underwood. London: Oxford Univ. Press, 1953, Vol. 1, pp. 489–509.

Anti-Aristotelian currents are compared with the Aristotelian biological philosophies of Harvey and Glisson.

1396. Keevil, John J. *Medicine and the Navy, 1200–1900.* Edinburgh: Livingstone, 1957–58. 4 vols.

Vol. 1, pp. 44–235: The Tudor period; The early Stuart period. Vol. 2: The Commonwealth and Protectorate period; The late Stuart period. Each volume has a chronology and Vol. 2 has lists of sources. Reviews: *Isis* 49 (1958), 354–55; 50 (1959), 493–94.

1397. Meyer-Steineg, Th., and K. Sudhoff. *Illustrierte Geschichte der Medizin.* Fünfte . . . Auflage, hrsg. von Robert Herrlinger und Fridholf Kudlien. Stuttgart: Gustav Fischer, 1965. 349 pp.; 227 illus.

Pp. 171–257 cover the years 1450–1700 (65 illus.). Reviews: *Archives internationales d'histoire des sciences* 18 (1965), 343–44; *Clio Medica* 1 (1966), 376–77.

1398. Herrlinger, Robert. *Geschichte der medizinischen Abbildung.* I. *Von der Antike bis um 1600.* Munich: Heinz Moos Verlag, 1967. 180 pp.; 348 illus., 23 pls. [English translation published in 1970 by Pitman Medical and Scientific Publishing Co., London]

A history of medical illustration from antiquity to about 1600, pictorially documented. List of MSS on p. 180. No bibliography of printed works, but illustrations are documented. Name and subject index. Pp. 29–175 cover the years 1450 to 1600. Reviews: *Bulletin of the History of Medicine* 43 (1969), 485–86; *Medical History* 17 (1973), 433–34.

1399. Baas, Joh. Hermann. *Outlines of the History of Medicine and the Medical Profession.* Translated, and in conjunction with the author, revised and enlarged, by H. E. Handerson. Huntington, N.Y.: Robert E. Krieger, 1971. 2 vols. 1,173 pp. [First published in 1889 in one volume.]

Pp. 288–588 cover the fifteenth, sixteenth, and seventeenth centuries.

1400. Forbes, Thomas R. "Verbal Charms in British Folk Medicine." *Proceedings, American Philosophical Society* 115 (1971), (4), 293–316.

On English and Latin charms used in England and Scotland during the period 1300–1800. Documented.

1401. Wightman, William P. D. *The Emergence of Scientific Medicine.* Edinburgh: Oliver and Boyd, 1971. 109 pp.

Critical examination of *scientific* medicine from the time of the Hammurabi Code to the seventeenth century. Biographical notes. Reviewed in *Journal of the History of Medicine* 27 (1972), 101–3.

1402. Debus, Allen G., ed. *Science, Medicine and Society in the Renaissance. Essays to Honor Walter Pagel.* London: Heinemann, 1972. 2 vols.

Thirty-eight essays. Subjects include: Galen, Van Helmont, Bacon, Paracelsus, Newton, and Harvey. Reviewed in *TLS* 16 Aug 1974, p. 886, and *Isis* 66 (1975), 577–79.

1403. Putscher, Marielene. *Geschichte der medizinischen Abbildung. Von 1600 bis zur Gegenwart.* 2nd ed. Munich: Heinz Moos Verlag, 1972. 207 pp.; 304 illus., 17 pls.

Continuation of Herrlinger's work (item 1398). With a chronological list of sources, notes and index. Review: *Bulletin of the History of Medicine* 50 (1976), 632–34.

1404. Lindeboom, G. A. *Geschiedenis van de medische wetenschap in Nederland.* Bussum: Fibula van Dishoeck, 1973. 198 pp.

On the Netherlands' outstanding contribution to the advancement of medicine. Pp. 20–123 cover the sixteenth and seventeenth centuries. Pp. 178–92: Biographical notes and bibliography. Reviewed in *Medical History* 18 (1974), 392.

1405. Webster, Charles. *The Great Instauration: Science, Medicine and Reform, 1626–1660.* London: Duckworth, 1975. 630 pp.

Study of the scientific, medical, and social ideas of the English Puritans. Reviewed in: *English Historical Review* 91 (1976), 853–59; *TLS* 2 July 1976, pp. 810–12; and *Isis* 68 (1977), 485–87.

1406. Coulter, Harris L. *Divided Legacy: A History of the Schism in Medical Thought.* Vol. 1: *The Patterns Emerge, Hippocrates to Paracelsus.* Vol. 2: *Progress and Regress, J. B. Van Helmont to Claude Bernard.* Washington, D.C.: Wehawken Book Co., 1977. 2 vols.

Study of the conflict between the empirical and rationalist therapeutic philosophies which first emerged in the oldest works of the Hippocratic Corpus (500–400 B.C.). Vol. 1, pp. 339–507 and Vol. 2, pp. 1–274 cover the sixteenth and seventeenth centuries. Reviewed in *Isis* 69 (1978), 103–5 ("Thoughtful yet eclectic history of medical ideas, not of medical practice").

1407. García Ballester, Luis. *Medicina, ciencia y minorías marginadas: Los Moriscos.* Collección monografica, Universidad de Granada, 54. Granada: Universidad de Granada, 1977. 163 pp.

On the intellectual and social decline of Moslem medicine in sixteenth-century Spain. Includes an "analysis of the clash between Moorish medical practices—ritual incantations, oblations, and the use of astrology—and Catholic orthodoxy." Reviewed in *Isis* 70 (1979), 309–10.

1408. May, Jane O'Hara. *The Elizabethan Dyetary of Health.* Lawrence, Kans.: Coronado Press, 1977. 415 pp.

A study of Elizabethan ideas about the "preservation of health," based on sixteenth- and seventeenth-century texts. Contains a bibliography of primary and secondary sources.

1409. Webster, Charles, ed. *Health, Medicine and Mortality in the Sixteenth Century.* Cambridge: Cambridge Univ. Press, 1979. 394 pp.

Essays dedicated to the memory of Sanford Vincent Larkey (1898–1969), historian of Tudor medicine at Johns Hopkins University, Baltimore. Subjects include: Medical practitioners; Vernacular medical literature; Astrological medicine; Alchemical and Paracelsian medicine; The School of Padua. Pp. 371–80: Biographical notice and bibliography of Larkey (by Margaret Pelling). Reviews: *Medical History* 24 (1980), 471–73; *Journal of the History of Medicine* 36 (1981), 237–38.

1410–1411. Granjel, Luis S. *La medicina española renacentista.* Historia general de la medicina española, 2. Salamanca: Ed. Universidad de Salamanca, 1980. 289 pp.

———. *La medicina española del siglo XVII.* Historia general de la medicina española, 3. Salamanca: Ed. Universidad de Salamanca, 1978. 250 pp.

History of Spanish medicine during the Renaissance and seventeenth century. Study covers all aspects of medicine (including education, administration, and popular medicine). With bibliographies.

1412. Entry deleted.

1413. Bynum, W. F., and V. Nutton, eds. *Theories of Fever from Antiquity to the Enlightenment.* Medical History, Supplement no. 1. London: Wellcome Institute for the History of Medicine, 1981. 154 pp.

Seven essays, with index. Pp. 19–44: Fever pathology in the sixteenth century: tradition and innovation (Iain M. Lonie); pp. 45–70: Thomas Willis and the fevers literature of the seventeenth century (Don G. Bates); pp. 71–98: Sydenham versus Newton: the Edinburgh fever dispute of the 1690s between Andrew Brown and Archibald Pitcairne (Andrew Cunningham).

1414. Entry deleted.

1415. Münchow, Wolfgang. *Geschichte der Augenheilkunde.* Separatdruck aus "Der Augenarzt," Bd. 9 (2), ergänzte und überarbeitete Auflage. Stuttgart: Ferdinand Enke, 1984. 734 pp.

Pp. 189–280: History of ophthalmology in the sixteenth and seventeenth centuries.

1416. Schmitz, Rudolf, and Gundolf Keil, eds. *Humanismus und Medizin.* Mitteilungen der Kommission für Humanismusforschung, 11. Weinheim: Acta Humaniora, 1984. 198 pp.

"Collection of ten essays centred around the impact which a revived, new interest in classical texts and the humanist motto *ad fontes* had on medicine between the late Middle Ages and the Renaissance."

1417. Pagel, Walter. *Religion and Neoplatonism in Renaissance Medicine.* Marianne Winder, ed. London: Variorum Reprints, 1985. 346 pp.

Eleven papers published between 1935 and 1981.

1418. Wear, A., R. K. French, and I. M. Lonie, eds. *The Medical Renaissance of the Sixteenth Century.* Cambridge: Cambridge Univ. Press, 1985. 349 pp.

Twelve articles reviewing progress made by physicians other than Vesalius and Harvey. No bibliography, no index. Reviews: *Isis* 77 (1986), 374–75; *Medical History* 30 (1986), 359–60; *Sixteenth Century Journal* 17 (1986), 373–75.

1419. Cook, Harold J. *The Decline of the Old Medical Regime in Stuart London.* Ithaca, N.Y.: Cornell Univ. Press, 1986. 310 pp.

Examines how the medical practitioners of London reacted to the intellectual, political, and socioeconomic changes of the age variously called the "scientific revolution" or the "seventeenth century crisis." Pp. 287–301: Selected bibliography (primary and secondary sources).

1420. Carpenter, Kenneth J. *The History of Scurvy & Vitamin C.* Cambridge: Cambridge Univ. Press 1988. 288 pp. [First published 1986.]

Pp. 1–28: The explorers' sickness (1498–1700). Pp. 29–42: The writings of learned men (1540–1700). Pp. 255–78: References (in one alphabetical sequence).

1421. Nagy, Doreen Evenden. *Popular Medicine in Seventeenth-Century England.* Bowling Green, Ohio: Bowling Green State Univ. Press, 1988. 140 pp.

On the widespread system of health care which "existed alongside the small corps of a medical elite, traditionally credited with providing health care in England." Based on contemporary sources—diaries, casebooks, and medical literature. Bibliography (pp. 112–28).

1422. French, Roger, and Andrew Wear, eds. *The Medical Revolution of the Seventeenth Century.* Cambridge: Cambridge Univ. Press, 1989. 328 pp.

Eleven essays on different aspects of how medicine fared in an age of revolution. Forms a companion volume to 1418. Reviews: *Archives internationales d'histoire des sciences* 40 (1990), 153–59; *Isis* 81 (1990), 770–71; *Medical History* 34 (1990), 448–49.

ANATOMY. PHYSIOLOGY

BIBLIOGRAPHIES

1423. Haller, Albrecht von. *Bibliotheca anatomica, qua scripta ad anatomen et physiologiam facentia a rerum initiis recensentur.* Zurich: Orell, Gessner, Fuessli, et Socc., 1774–77. 2 vols. [Reprinted by Georg Olms, Hildesheim in 1969 with a foreword by Gunther Mann]

Classified annotated bibliography of anatomy and physiology from the beginnings to 1776. Arrangement chronological. Vol. 1 (pre-1700) is divided into seven chapters (Graeci; Arabes; Arabistae; Restauratores anatomes;

Schola italica; Animalium incisiones; Anatome humana) and subdivided (760 sections) by authors, groups of authors, disputations, and articles in journals. Annotations include bio-bibliographical information on authors, summary of contents of books and critical comments. Vol. 2 (post-1700) includes addenda, name index, and classified index of anonyma for the two vols. Some 15,000 titles in both vols.

1424. Choulant, Ludwig. *History and Bibliography of Anatomic Illustration*. Translated and annotated by Mortimer Frank. Further essays by Fielding H. Garrison, Mortimer Frank, Edward C. Streeter. With a new historical essay by Charles Singer and a bibliography of Mortimer Frank by J. Christian Bay. New York: Schuman, 1945. 435 pp. [Translation of the 1852 German ed. First published in Latin in 1842. Present ed. reprinted in 1962 by Hafner, New York]

A history and bibliography of representations of human anatomy by graphic means—from pre-Berengarius (1521) through Vesalius (1543) and Casserius (1627) to Albinus (1737). Pp. 1–21: Life of J. L. Choulant. Pp. 21A–21R: Beginnings of academic practical anatomy (C. Singer). Pp. 22–41: Historical introduction. Pp. 42–48: Anatomic illustration of antiquity and of the Middle Ages. Pp. 49–87: Manuscript anatomic illustration of the pre-Vesalian period. Pp. 88–260: Printed works up to 1700 (from Mondino de' Luzzi to Crisóstomo Martinez). Pp. 425–35: Name index. Reviews: *Janus II* 2 (1852), 162–67; *Isis* 4 (1922), 357–59; *Medical History* 7 (1963), 290.

1425. Durling, Richard J. "A Chronological Census of Renaissance Editions and Translations of Galen." *Journal of the Warburg and Courtauld Institutes* 24 (1961), 230–305.

Pp. 231–45: Historical sketch. Pp. 246–81: Short-title census of eds. and translations (1473–1600) with locations in major European libraries. Pp. 281–305: Indexes.

1426. Cushing, Harvey. *A Bio-bibliography of Andreas Vesalius*. 2nd ed. Hamden, Conn.: Archon Books, 1962. 264 pp. [First published in 1943]

Cushing's labour of love, prepared for the fourth centenary of the publication of the *Fabrica* (1543), was published posthumously by his colleague J. F. Fulton. This 2nd ed. by Madeleine Stanton has been considerably revised (in particular, the chapter on Vesaliana). The bibliography is in ten chapters. The first eight are each devoted to a work or group of works of Vesalius. (Ch. 2, pp. 10–43: The *Tabulae sex*; Ch. 6, pp. 73–153: The *Fabrica*). Ch. 9 contains an essay by Arturo Castiglioni on Fallopius and Vesalius. Ch. 10: Bibliography of Vesaliana (compiled by Mrs. John P. Peters. Pp. 218–64: Index of copies; Chronology of eds.; Addenda; Index of names. Reviews: *Isis*

35 (1944), 338–41; *The Library* (4), 26 (1945–46), 212–15; *Journal of the History of Medicine* 18 (1963), 300–1.

1427. O'Malley, C. D. "A Review of Vesalian Literature." *History of Science* 4 (1965), 1–14.

Surveys the publications intended to commemorate the fourth centenary of the birth and death of Vesalius (1514–64) and of the publication of the *Fabrica* (1543).

1428. Russell, K.F. *British Anatomy 1525–1800. A Bibliography of Works Published in Britain, America and on the Continent.* 2nd ed. Winchester, Hants.: St. Paul's Bibliographies, 1987. 245 pp. [First published 1963]

Some 900 items arranged alphabetically by author. Includes works of European authors translated into English or printed in Britain in their original language. Introduction (pp. xvii–xlix) contains a historical survey of anatomical literature in Britain. Reviews: *Isis* 56 (1965), 465; *The Library* (5), 21 (1966), 74–76.

1429. Keynes, Sir Geoffrey. *A Bibliography of the Writings of Dr. William Harvey 1578–1657.* Third ed. Revised by Gweneth Whitteridge and Christine English. Winchester, Hants.: St. Paul's Bibliographies, 1989. 136 pp.

First published in 1928 for the tercentenary of the first publication of the *De motu cordis.* 2nd revised ed. 1953. The present ed. takes into account recent scholarship (stimulated in part by the centenary celebrations of 1957 and 1978). New introduction by Whitteridge (pp. 1–20) including a note on MSS not published in Harvey's lifetime. The bibliography is in six sections (*De motu cordis*; *De circulatione sanguinis*; *De generatione animalium*; *Opera omnia*; MS works; Miscellanea) with bibliographical descriptions (revised by C. English). Each section has an introduction. Includes Keynes's preface to the 1953 ed. and his note "The reception of Harvey's doctrine during his lifetime, 1628–1657" (pp. 117–25). Pp. 129–32: Index of copies recorded.

MONOGRAPHS AND ARTICLES

1430. Foster, M. *Lectures on the History of Physiology during the Sixteenth, Seventeenth and Eighteenth Centuries.* New York: Dover, 1970. 310 pp. [First published by Cambridge Univ. Press in 1901]

Seven lectures (pp. 1–199) cover the history from Mundinus to Boerhaave. 1: Vesalius, his forerunners and followers. 2: Harvey and the circulation of the blood. The lacteals and lymphatics. 3: Borelli and the influence of the new physics. 4: Malpighi and the physiology of glands and tissues. 5: Van Helmont and the rise of chemical physiology. 6: Sylvius and his pupils; the physiology of digestion in the seventeenth century. 7: The English School of the seventeenth century; the physiology of respiration.

1431. Bayon, H. P. "William Harvey, Physician and Biologist: His Precursors, Opponents and Successors." *Annals of Science* 3 (1938), 59–118, 435–56; 4 (1939), 65–106, 329–89.

Explains the significance of Harvey's doctrines in relation to subsequent scientific progress and his role in "leading biology from mere observation and compilation into experimental and deductive study."

1432. Cole, F. J. *A History of Comparative Anatomy from Aristotle to the Eighteenth Century.* London: Macmillan and Co., 1944. 524 pp.

Pp. 49–311: Leonardo da Vinci to Ruysch. Pp. 312–442: Academies and societies; the Anatomy lesson. Pp. 443–63: The anatomical museum. Pp. 473–507: Biographical notes and bibliography. Reviewed in *Isis* 37 (1947), 112–14 and 38 (1948), 264–66.

1433. Fulton, John F. *Michael Servetus, Humanist and Martyr.* With a bibliography of his works and a census of known copies by Madeleine E. Stanton. New York: Herbert Reichner, 1953. 98 pp.

A study of the background and career of Servetus and the "lesser circulation." With an annotated bibliography of his works and a list of secondary sources. Review: *The Library* (5), 9 (1954), 277–78; *Isis* 45 (1954), 313–14.

1434. Rothschuh, K. E. *Geschichte der Physiologie.* Berlin: Springer-Verlag, 1953. 249 pp.

Pp. 27–68 cover the history of physiology from Leonardo da Vinci to Leeuwenhoek. Pp. 33, 36, 46 respectively contain chronological tables for anatomical discoveries, developments in physics and chemistry. Reviewed in *Isis* 45 (1954), 105–6.

1435. Bainton, Roland H. *Hunted Heretic: The Life and Death of Michael Servetus (1511–1553).* Boston: Beacon Press, 1953. 270 pp.

Biography with a chronology and bibliography. Review: *Journal of the History of Medicine* 9 (1954), 375–78.

1436. O'Malley, C. D. *Andreas Vesalius of Brussels 1514–1564.* Berkeley and Los Angeles: Univ. of California Press, 1964. 480 pp.

Biography with extracts from the *Fabrica*. Includes a chapter on pre-Vesalian anatomy. Reviews: *Isis* 56 (1965), 362–65; *Physis* 6 (1965), 454–69;

Title page of Andreas Vesalius's *De humani corporis fabrica*, Basel, 1543 (Bodleian Library, Oxford, Bl.16.Med.). ***Entries 1426, 1436.***

History of Science 4 (1965), 139–40; *TLS* 17 June 1965, p. 524; *Journal of the History of Medicine* 26 (1971), 87–92.

1437. Keynes, Geoffrey. *The Life of William Harvey.* Oxford: Clarendon Press, 1966. 483 pp.

Study of the life and work of Harvey. Appendix IV (pp. 447–55) contains a bibliographical survey of the reception of Harvey's doctrine during his lifetime, 1628–1657. Reviews: *Medical History* 11 (1967), 201–5; *Isis* 59 (1968), 99–101; *Journal of the History of Biology* 5 (1972), 189–204.

1438. Premuda, Loris. *Storia della fisiologia. Problemi e figure.* Udine: Del Bianco, 1966. 328 pp.

Pp. 77–130: History of physiology during the Renaissance and Baroque period, with bibliographies. Pp. 187–206: William Harvey and the birth of modern physiology.

1439. Pagel, Walter. *William Harvey's Biological Ideas: Selected Aspects and Historical Background.* Basle: Karger, 1967. 394 pp.

Analysis of Harvey's concepts of blood circulation and generation of animals. Reviewed in *Isis* 59 (1968), 101–2.

1440. Wolf-Heidegger, G., and Anna Maria Cetto. *Die anatomische Sektion in bildlicher Darstellung.* Basle: Karger, 1967. 612 pp.; 355 illus.

Outline history of anatomical dissection (pp. 1–119), with illustrations arranged in eight sections catalogued and annotated (pp. 121–392). Bibliographies (pp. 99–119, 383–92). Some two hundred illustrations relate to the period 1450–1700.

1441. Rothschuh, Karl E. *Physiologie: der Wandel ihrer Konzepte, Probleme und Methoden vom 16. bis 19. Jahrhundert.* Orbis academicus, Bd. II/15. Freiburg: Verlag Karl Alber, 1968. 407 pp.

Pp. 39–151: History of physiology in the sixteenth and seventeenth centuries. Pp. 336–66: Bibliography. Pp. 367–87: Biographical notes.

1442. Scarborough, John. "The Classical Background of the Vesalian Revolution." *Episteme* 2 (1968), 200–18.

On Vesalius and the Galenic tradition.

1443. Pagel, Walter. "William Harvey Revisited." *History of Science* 8 (1969), 1–31; 9 (1970), 1–41.

Aims at "removing Harvey from the Victorian pedestal of the modern view" and replacing him in the appropriate spiritual climate of the seventeenth century.

1444. Whitteridge, Gweneth. *Harvey and the Circulation of the Blood.* London: Macdonald, 1971. 269 pp.

Retraces the steps by which Harvey came to realise that the blood circulated throughout the whole body. Covers the period from his student days in Cambridge and Padua to the publication of his *De motu cordis* and the *De generatione animalium.*

1445. Temkin, Owsei. *Galenism: Rise and Decline of a Medical Philosophy.* Ithaca: Cornell Univ. Press, 1973. 240 pp.

Revised version of four lectures delivered at Cornell University in 1970. Pp. 134–92: Galenism since Vesalius. Pp. 193–227: Bibliography.

1446. Luyendijk Elshout, A. M. "Anthony Nuck (1650–1692), the 'Mercator' of the Body Fluids: A Review of his Anatomical and Experimental Studies." *Circa tiliam: studia historiae medicinae Gerrit Arie Lindeboom septuagenario oblata.* Leyden: E. J. Brill, 1974, pp. 150–64.

Nuck was a professor at Leiden and teacher of Boerhaave.

1447. Peumery, Jean-Jacques. "Les origines de la transfusion sanguine." *Clio medica* 9 (1974), 131–56, 215–50, 325–41.

On the first (seventeenth century) scientific experiments in blood transfusion. Bibliography.

1448. French, R. K. "Alexander Read [1586?–1641] and the Circulation of the Blood." *Bulletin of the History of Medicine* 50 (1976), 478–500.

Study of contemporary reaction to Harvey. Appendix contains Read's notes (in his copy of the *De motu cordis)* and a list of his library.

1449. Bylebyl, Jerome J., ed. *William Harvey and His Age: The Professional and Social Context of the Discovery of the Circulation.* Baltimore: Johns Hopkins Univ. Press, 1979. 154 pp.

Three essays on the various aspects of the complex interaction between Harvey's anatomical and physiological investigations and his role as a physician. Reviewed in *Isis* 71 (1980), 503–4.

1450. Frank, Robert G., Jr. *Harvey and the Oxford Physiologists. A Study of Scientific Ideas.* Berkeley: Univ. of California Press, 1980. 368 pp.

Study of seventeenth-century post-Harveian ideas concerning air, respiration, heat, blood, and heartbeat. Reviewed in *Isis* 73 (1982), 432–33.

1451. Brown, Theodore M. *The Mechanical Philosophy and the "Animal Oeconomy." A Study in the Development of English Physiology in the Seventeenth and Early Eighteenth Centuries.* New York: Arno Press, 1981. 384 pp.

A study of English physiological thought in the century after Harvey's discovery of the circulation of the blood. Ph.D. Thesis, Princeton University, 1968.

1452. Wear, Andrew. "Galen in the Renaissance." *Galen: Problems and Prospects.* A collection of papers submitted at the 1979 Cambridge conference, edited by Vivian Nutton. London: Wellcome Institute for the History of Medicine, 1981, pp. 229–62.

An introduction, concentrating on the continuation of the Galenic tradition: anatomy, physiology and the *rete mirabile*; medical method; medicine and astrology; Sanctorius Sanctorius (1561–1636).

1453. Keele, Kenneth D. *Leonardo da Vinci's Elements of the Science of Man.* New York: Academic Press, 1983. 385 pp.

An "attempt to construct a balanced picture of Leonardo's physical and physiological scientific research," based on extant MSS (including the Madrid Codices). Pp. 7–41: A scientific biography of Leonardo. Review: *History and Philosophy of the Life Sciences* 11 (1989), 339–43.

1454. Brazier, Mary A. B. *A History of Neurophysiology in the Seventeenth and Eighteenth Centuries: from Concept to Experiment.* New York: Raven Press, 1984. 230 pp.

Pp. 3–138 cover the seventeenth century. Reviewed in *Journal for the History of the Behavioural Sciences* 21 (1985), 192–93.

1455. Wilson, Luke. "William Harvey's *Prelectiones:* the Performance of the Body in the Renaissance Theater of Anatomy." *Representations* 17 (Winter 1987), 62–95.

On the *Prelectiones*—Harvey's lecture notes "intended not for publication, but to be read by their author while he conducted public dissections as Lumleian Lecturer at the Royal College of Physicians in London."

SURGERY

BIBLIOGRAPHIES

1456. Haller, Albrecht von. *Bibliotheca chirurgica, qua scripta ad artem chirurgicam facientia a rerun initiis recensentur.* Berne: Em. Haller, 1774–75. 2 vols. [Reprinted by Georg Olms, Hildesheim in 1971 with a foreword by G. Mann]

Classified annotated bibliography of the literature of surgery from the beginnings to the present. Arrangement chronological. Vol. 1 (up to 1710) is divided into six chapters (Graeci; Arabes; Arabistae; Instauratores; Schola italica; Chirurgia gallica) and subdivided (501 sections) by authors, groups of authors, disputations, and articles in journals. Annotations include bio-bibliographical information on authors, summary of contents of books, and critical comments. Vol. 2 includes addenda, name index, and classified index of anonyma. Some 12,500 titles in both vols.

1457. De Vigiliis von Creutzenfeld, Stephanus Hieronymus. *Bibliotheca chirurgica in qua res omnes ad chirurgiam pertinentes ordine alphabetico, ipsi vero scriptores, quotquot ad annum usque MDCCLXXIX innotuerunt, ad singulas materias ordine chronologico exhibentur . . .* Vienna: Johann Thoma, 1781. 2 vols.

Subject index to surgical literature—from Hippocrates to 1779—classified under some 200 headings (with cross references) from ABSCESSUS to VULNERA CAPITIS. Some 8,000 titles. The bibliographical details are sometimes very brief.

1458. Pazzini, Adalberto. *Bio-bibliografia di storia della chirurgia.* Rome: Edizioni Cosmopolita, 1948. 525 pp.

Pp. 65–156: Surgery in the Renaissance and seventeenth century. (Two chapters each with introductory note and bio-bibliography. Notes on 133 persons from Pietro da Montagna to William Chesleden.) Pp. 371–450: Bibliography.

1459. Poletti, G. B. *De re dentaria apud veteres, sive Repertorium bibliographicum, in quo libri omnes de re dentaria ab arte typographica inventa usque ad annum MCM typis . . . enumerantur.* 2nd ed., revised. Milan: Gorlich, 1951. 214 pp.

List of books on dental subjects, partly annotated. Includes some six hundred pre-1700 titles. Review: *Rivista di storia della scienza* 42 (1951), 279.

1460. Strömgren, Hedvig Lidforss. *Index of Dental and Adjacent Topics in Medical and Surgical Works before 1800.* Library Research Monographs, Vol. 4, University Library, Copenhagen. Copenhagen: Munksgaard, 1955. 255 pp.

Bibliography of some 2,000 works, with brief notes, arranged alphabetically by author. Pp. 217–55: Subject index.

MONOGRAPHS AND ARTICLES

1461. Gurlt, Ernst. *Geschichte der Chirurgie und ihre Ausübung. Volkschirurgie—Alterthum—Mittelalter—Renaissance.* Berlin: Hirschwald, 1898. 3 vols.

Vol. 2 (pp. 265–926) and Vol. 3 (pp. 1–455) cover the history of surgery in Europe during the years 1500–1625. Treatment is topo-biographical, with illustrations of surgical instruments. Vol. 3, pp. 457–809: Summary.

1462. Gnudi, Martha Teach, and Jerome Pierce Webster. *The Life and Times of Gaspare Tagliacozzi, Surgeon of Bologna 1545–1599. With a Documented Study of the Scientific and Cultural Life of Bologna in the Sixteenth Century.* New York: Herbert Reichner, 1950. 538 pp.

Professor of Anatomy in Bologna, Tagliacozzi was the first to describe and practice what is today known as plastic surgery. His *De curtorum chirurgia per insitionem* (1597) was a landmark in the history of surgery.

1463. Huard, Pierre, and Mirko Drazen Grmek. *La chirurgie moderne. Ses débuts en Occident: XVIe, XVIIe, XVIIIe siècles.* Paris: Dacosta, 1968. 253 pp.

History of surgery in the West from Ambroise Paré (1510?–1590) to John Hunter (1718–83). Pp. 7–22: Scientific basis of surgery. Pp. 23–44: Evolution of the major problems of surgery. Pp. 47–126: Surgical specialities. Pp. 137–226: Plates. Pp. 227–42: Bio-bibliographical notes. Pp. 243–53: Bibliography. Reviewed in *Bulletin of the History of Medicine* 44 (1970), 94–95.

1464. Dulieu, Louis. *La chirurgie à Montpellier de ses origines au début du XIXe siècle.* Avignon: Les Presses Universelles, 1975. 345 pp.

A history of surgery in Montpellier, copiously illustrated. Pp. 49–202: From the Renaissance to 1792. Pp. 229–315: Biographical notices on surgeons and barber-surgeons from the twelfth century to the Revolution. Pp. 319–32: Bibliography.

1465. Wangensteen, Owen H., and Sarah D. Wangensteen. *The Rise of Surgery. From Empiric Craft to Scientific Discipline.* Minneapolis: Univ. of Minnesota Press, 1978. 785 pp.

History of general surgery in 26 chapters, arranged systematically. Pp. 581–720: Notes and bibliographies. Reviewed in *Isis* 72 (1981), 129–30.

1466. Hoffmann-Axthelm, Walter. *Die Geschichte der Zahnheilkunde.* 2nd ed., revised. Berlin: Quintessenz Verlags, 1985. 496 pp. [English translation of the 1973 German ed. was published in 1981]

Pp. 155–219 contain in one chapter the history of dentistry in the sixteenth and seventeenth centuries. Documented.

1467. Dumaître, Paule. *Ambroise Paré [c. 1510–1590] chirurgien de quatre rois de France.* Paris: Librairie Académique Perrin, 1986. 348 pp.

Well-documented biography, based on manuscript and printed sources, of the father of French surgery. With a chronology and a bibliography of his works and his biographies. Reviewed in *History and Philosophy of the Life Sciences* 11 (1989), 138–39.

1468. De Moulin, Daniel. *A History of Surgery with Emphasis on the Netherlands.* Dordrecht: Martinus Nijhoff, 1988. 408 pp.

A chronological survey from antiquity to the present. Pp. 64–147: The Renaissance; The golden age. Pp. 187–259: Practical surgery in the seventeenth and eighteenth centuries. Pp. 384–401: Bibliography.

OBSTETRICS. GYNAECOLOGY. EMBRYOLOGY

BIBLIOGRAPHIES

1469. Power, D'Arcy. "The Birth of Mankind or the Woman's Book. A Bibliographical Study." *The Library* (4), 8 (1927), 1–37.

On the earliest known English work on midwifery—*The byrth of mankynde translated out of Laten into Englysshe* (1540)—and its later eds. (up to 1654). The German original *Der swangern frauwen und hebammen Roszgarten* by Eucharius Rösslin (1513), was reprinted many times. Other eds: Latin, 1532, 1536, 1537; Dutch, 1516; French, 1536; Czech, 1519.

1470. Hellman, Alfred M. *A Collection of Early Obstetrical Books. An Historical Essay with Bibliographical Descriptions of Thirty-seven Items, including Twenty-five Editions of Roesslin's* Rosengarten. Additions by C. Doris Hellman. New Haven, Conn., 1952–79. 2 pts. 79 + 11 pp.

Contains descriptions of forty-six books published between 1513 and 1635, with introductory note and list of references.

1471. Sapori, Giuliana. *Il fondo di medicina antica della Biblioteca Ginecologica Emilio Alfieri.* Con un ricordo di Emilio Alfieri [1874–1949] del Prof. Luigi Belloni. Milan: Università degli Studi, 1975. 421 pp.

Catalogue of historic medical books (2,110 titles of which 1,319 are pre-1700) collected by a gynaecologist. The nucleus of the collection (now at the University of Milan) relates to the study of gynaecology.

1472. Hageln, Ove, ed. *The Byrth of Mankynde otherwyse named The Womans Booke. Embryology, Obstetrics, Gynaecology through Four Centuries. An Illustrated and Annotated Catalogue of Rare Books in the Library of the Swedish Society of Medicine.* Stockholm: Svenska Läkaresällskapet, 1990. 176 pp.

Catalogue of ninety-two books of which twenty-six were published before 1700. Review: *Isis* 84 (1993), 197–98.

MONOGRAPHS AND ARTICLES

1473. Fassbender, Heinrich. *Geschichte der Geburtshülfe.* Jena: G. Fischer, 1906. 1,028 pp.

Pp. 105–243 contain a study of the major works on midwifery composed during the years 1460 to 1730.

1474. Weindler, Fritz. *Geschichte der gynäkologisch-anatomischen Abbildungen.* Dresden: Zahn und Jaensch, 1908. 186 pp.; 122 illus.

A history of illustrations in works on gynaecology. No bibliography, but sources of illustrations are given. Pp. 59–166 cover the period from Johannes de Ketham (1491) to Johannes Swammerdam (1637–80). Review: *Janus* 13 (1908), 399–400.

1475. La Torre, Felice. *L'utero attraverso i secoli. Da Erofilo ai nostri giorni. Storia,—Iconografia—Struttura—Fisiologia. Con speciale accenno alla funzione gestatrice.* Città di Castello: Unione Arti Grafiche, 1917. 831 pp., 560 fig.; 22 pls.

Illustrated history of our knowledge of the womb. Pp. 163–82: Mondino de' Luzzi (1316) to Alessandro Achillino (1463–1512). Pp. 183–452: Berengario Carpi (1523) to Antonio Nuch (1697). The work has been described by Sarton as entertaining but distasteful, and the author discursive and garrulous, *Isis* 5 (1923), 279.

1476. Adelmann, Howard B., ed. *The Embryological Treatises of Hieronymus Fabricius of Aquapendente.* Ithaca, N.Y.: Cornell Univ. Press, 1942. 883 pp.

Facsimile ed. of (i) *De formatione ovi et pulli* (1621) and (ii) *De formato foetu* (1604). With introduction, translation, and commentary. Pp. 6–134: The life of Fabricius; Brief sketch of the history of embryology before Fabricius; Analysis of the embryological treatises; Bibliographical note.

1477. Ricci, James V. *The Development of Gynaecological Surgery and Instruments. A Comprehensive Review of the Evolution of Surgery and Surgical Instruments for the Treatment of Female Diseases from the Hippocratic Age to the Antiseptic Period.* Philadelphia, Pa.: The Blakiston Co., 1949. 594 pp.

A survey of the historical background of gynaecological surgery and instruments by a clinician. Pp. 75–168: Renaissance and seventeenth century. Copiously annotated and illustrated. Reviewed in *Bulletin of the History of Medicine* 27 (1953), 588–89.

1478. Ricci, James V. *The Genealogy of Gynaecology. History of the Development of Gynaecology throughout the Ages, 2000 B.C.–1800 A.D., with Excerpts from the Many Authors Who Have Contributed to the Various Phases of the Subject.* Second ed., enlarged and revised. Philadelphia: The Blakiston Co., 1950. 494 pp. [First published in 1943]

The 1st ed. was criticized by O. Temkin for its inaccuracies in *Bulletin of the History of Medicine* 16 (1944), 422–24. This ed. has been completely revised. Pp. 237–337 cover the period 1453–1700. With bibliography, notes and name index.

1479. O'Neill, Ynez Violé. "Giovanni Michele Savonarola [1385–c. 1468]; an Atypical Renaissance Practitioner." *Clio Medica* 10 (1975), 77–93.

One of Savonarola's most important medical works, *Ad mulieres ferrarienses de regimine pregnantium*, a treatise on obstetrics and gynaecology (published only in 1952), was written not in Latin, as would have been expected of a university professor, but in a lively local dialect and offered to the women of Ferrara.

1480. Eccles, Audrey. *Obstetrics and Gynaecology in Tudor and Stuart England.* London: Croom Helm, 1982. 145 pp.

Examines the "rise of new ideas in obstetrics and the fall of old ones in the literature meant for practitioners" during the period 1540–1740. Reviewed in *Bulletin of the History of Medicine* 58 (1984), 426–27.

MATERIA MEDICA

1481. Schelenz, Hermann. *Geschichte der Pharmazie.* Berlin: Julius Springer, 1904. 934 pp. [Reprinted by Georg Olms, Hildesheim in 1965]

This work was considered a classic in its time. Pp. 387–540 cover the history of pharmacology from Paracelsus to 1700. Pp. 829–934: Name and subject index. Review: *Ambix* 11 (1963), 159.

1482. Roddis, Louis H. "Garcia da Orta: The First European writer on Tropical Medicine and a Pioneer in Pharmacognosy." *Annals of Medical History*, NS, 1 (1929), 198–207.

Notes on his life and work.

1483. Scouloudi, I. "Sir Theodor Turquet de Mayerne, Royal Physician and Writer, 1573–1655." *Proceedings, Huguenot Society of London* 16 (1940), (3), 301–37.

Turquet de Mayerne, physician to Henry IV of France, settled in England in 1611. He played an important part in the production of the first English *Pharmacopoeia* (1618).

1484. Benedicenti, Alberico. *Malati, medici e farmacisti. Storia dei rimedi traverso i secoli e delle teorie che ne spiegano azione sull'organismo.* 2nd ed. Milan: Hoepli, 1947–51. 2 vols. 1,457 pp. [First published 1924–25]

A documented and copiously illustrated history of medication. The author was professor of pharmacology at the University of Genoa. Pp. 421–975 cover the period 1450 to 1700. Pp. 1435–55: Indexes (names and subjects). No bibliography. Review: *Rivista di storia della scienza* 43 (1952), 279.

1485. Urdgang, George. "The Development of Pharmacopoeias. A Review with Special Reference to the *Pharmacopoeia Internationalis* [1951]." *Bulletin of the World Health Organisation* 4 (1951), 577–603.

On the origins, nature, and development of pharmacopoeias. Several pages are devoted to the antidotaria and pharmacopoeias of the fifteenth to seventeenth centuries.

1486. Guerra, Francisco. *Nicolás Bautista Monardes, su vida y su obra (ca. 1493–1588).* Mexico, D.F.: Compañia Fundidora de Fierro y Acero de Monterrey, S.A., 1961. 226 pp.; illus.

Study of the life and work of the Spanish physician through whose writings the South American *materia medica* began to be known in Europe. He is considered to be one of the founders of pharmacognosy (science dealing with the sources, physical characteristics, uses, and doses of drugs) and experimental pharmacology. His works were translated during the sixteenth and early seventeenth centuries into Latin, English, Italian, French, and Dutch. Pp. 83–183: Bibliographical survey of his works.

1487. Florkin, Marcel, ed. *Materia Medica in the XVIth Century. Proceedings of a Symposium of the International Academy of the History of Medicine Held at the University of Basel, 7th September 1964.* Analecta Medico-Historica 1. Oxford: Pergamon Press, 1966. 80 pp.

Six papers: "Dioscorides and Renaissance materia medica" (J. Stannard); "Remarques sur l'utilisation des animaux dans la matière médicale" (J. Théodorides and M. D. Grmek); "Drugs from the Indies and the political economy of the century" (P. Guerra); "Le gaïac à Strasbourg" (E. Wickersheimer); "Materia medica in India" (D. V. Subba Reddy). Bibliographies.

1488. Schneider, Wolfgang. *Geschichte der pharmazeutischen Chemie.* Weinheim: Verlag Chemie, 1972. 376 pp.

Traces (pp. 29–133) the progress from pharmaceutical alchemy to pharmaceutical chemistry, from the Middle Ages to 1700. Includes lists of pharmaceutical preparations taken from the author's *Lexikon zur Arzneimittelgeschichte*, Vol. 3 (Frankfurt-am-Main, 1968). With notes, subject and author indexes.

1489. Dulieu, Louis. *La pharmacie à Montpellier, de ses origines à nos jours.* Avignon: Les Presses Universelles, 1973. 343 pp.; illus.

Pp. 31–111: History of pharmacy in Montpellier from the Renaissance to 1700. Pp. 228–90: Biographies of apothecaries. Pp. 315–30: Bibliography.

1490. Sanchez Teller, M. C., F. Guerra, and J. L. Valverde. *La doctrina farmacéutica del Renacimento en la obra de Francisco Hernández, c. 1515–1587.* Estudios del Departamento de Historia de la Farmacia y Legislacíon Farmacéutica, Universidad de Granada, 6. Granada, 1979. 138 pp.

Physician to Philip II of Spain, Hernández spent seven years in Mexico studying the flora and fauna. His studies constitute a vital contribution to the materia medica of the Renaissance. Pp. 41–133: Table of medicinal plants of Mexico which have been described by Hernández. See also items 1328 and 1494.

1491. Bosch, Klaus. *Zur Vorgeschichte chemiatrischer Pharmakopöepräparate im 16/17 Jahrhundert.* Veröffentlichungen aus dem pharmaziegeschichtlichen Seminar der Technischen Universität Braunschweig, Bd. 21. Brunswick, 1980. 224 pp.

An analysis of the pharmacological writings of seventeen major (and several minor) authors from Paracelsus to Adrian von Mynsicht and of their treatment of eighty-eight selected drugs. Pp. 209–21: Bibliography.

1492. Kajdański, Edward. "Receptarum sinensium liber of Michael Boym." *Janus* 73 (1986–90), 105–24.

Notes on Boym (1612–59), Polish Jesuit, author of *Flora sinensis* (Vienna, 1656), and his work on Chinese medicine, *Medicus sinicus.* A part of the latter was published as *Specimen medicinae sinicae* (Frankfurt-am-Main, 1682) and another part as *Clavis medicae* (Nuremberg, 1686).

1493. Leibrock-Plehn, Larissa. *Hexenkräuter oder Arznei. Die Abtreibungsmittel im 16. und 17. Jahrhundert.* Heidelberger Schriften zur Pharmazie- und Naturwissenschaftsgeschichte, Bd. 6. Stuttgart: Wissenschaftliche Verlagsgesellschaft, 1992. 238 pp.

A study of the use of herbs for abortion during the sixteenth and seventeenth centuries. With a glossary of plant names (Latin and German) and a list of primary and secondary sources.

1494. López Piñero, José M., José Luis Fresquet Febrer, María Luz López Terrada, and José Pardo Tomás. *Medicinas, drogas y alimentos vegetales del nuevo mondo. Textos y imágenes españolas que* *los introdujeron en Europa.* Madrid: Ministerio de Sanidad y Consumo, 1992. 388 pp.

History of the introduction into Europe of the *materia medica* of the New World. In three sections: The first descriptions of American plants; The first scientific studies—of Nicolás Monardes and of Francisco Hernández; The diffusion of the knowledge of American *materia medica* in Europe. Each section contains: an account of the sources and their authors; descriptions of the plants, drugs, etc.; anthology of texts and drawings; bibliography. No index. See also items 1328 and 1490.

1495. Thulesius, Olav. *Nicholas Culpeper, English Physician and Astrologer.* New York: St. Martin's Press, 1992. 210 pp.

Modern biography of Culpeper (1616–54), author of *Midwifery* (1651) and *The Herbal* (1652).

PSYCHOLOGY. PSYCHIATRY. NEUROLOGY

BIBLIOGRAPHIES

1496. Laehr, Heinrich. *Die Literatur der Psychiatrie, Neurologie und Psychologie von 1459–1799.* Berlin: Georg Reimer, 1900. 3 vols.

Chronologically arranged bibliography, with occasional notes, Vol. 1 (751 pp.) covers pre-1700 literature. Vol. 3 (271 pp.): Name and subject index.

1497. Schüling, Hermann. *Bibliographisches Handbuch zur Geschichte der Psychologie. Das 17. Jahrhundert.* Berichte und Arbeiten aus die Universitätsbibliothek Giessen, 5. Giessen, 1964. 292 pp.

Bibliography of the considerable literature on different aspects of human behavior and experience. (Psychology as an independent scientific discipline did not exist in the seventeenth century.) Alphabetically arranged by author, with indexes.

1498. Brozek, Josef. "Contemporary West European Historiography of Psychology." *History of Science* 13 (1975), 29–60.

A bibliographical survey (some 150 references).

1499. Watson, Robert I., Sr. *The History of Psychology and the Behavioural Sciences: A Bibliographical Guide.* New York: Springer, 1978. 241 pp.

Some 800 titles, mostly annotated, classified in five main divisions: General resources; Historical accounts; Methods of historical resources; Historiographic fields; Historiographic theories. No index. Of marginal interest for the study of pre-1700 psychology. Review: *Journal of the History of the Behavioural Sciences* 15 (1970), 190–92.

1500. Viney, Wayne, Michael Wertheimer, and Marilyn Lou Wertheimer. *History of Psychology: A Guide to Information Sources.* Psychology Information Guide Series, Vol. 1. Detroit, Mich.: Gale Research, 1979. 502 pp.

Some 3,000 entries, mostly annotated, arranged in five sections (50 subsections): General sources; Histories of psychology; Systems and schools of psychology; Histories and major works on selected subjects; Histories of related fields. English language sources only. Each major section has a brief explanatory note. Pp. 421–502: Indexes (name, title and subject). Review: *Journal of the History of the Behavioural Sciences* 17 (1981), 431–32.

MONOGRAPHS AND ARTICLES

1501. Edgar, Irving I. "Shakespeare's Medical Knowledge with Particular Reference to His Delineation of Madness. A Preliminary Survey of Critical Opinion." *Annals of Medical History,* NS, 6 (1934), 150–68.

Contains a bibliography of 53 books and 126 articles.

1502. Zilboorg, Gregory. *A History of Medical Psychology.* In collaboration with George W. Henry. New York: W. W. Norton and Co., 1941. 606 pp.

Pp. 144–280 cover the period 1450 to 1700—from the *Malleus maleficarum* to Stahl's *Theoria medica vera.* Pp 144–74: The blows of the witches' hammer. Pp. 174–244: The first psychiatric revolution (Vives, Paracelsus, Agrippa, Weyer, Bodin). Pp. 245–80: The age of reconstruction. Review: *Isis* 34 (1942), 189–90.

1503. Hunter, Richard Alfred, and Ida Macalpine. *Three Hundred Years of Psychiatry, 1535–1860. A History Presented in Selected English Texts.* Hartsdale, N.Y.: Carlisle Publishing, Inc., 1982. 1,107 pp. [First published 1963 by Oxford Univ. Press]

Intended to serve as a source book and a historical outline of clinical psychiatry. Although confined to English sources, American texts and foreign texts (in translation) which played their part in shaping psychiatric thought are included. Most texts are preceded by introductions. Pp. 1–274 cover the sixteenth and seventeenth centuries, containing some one hundred texts from Bartholomaeus Anglicus (1535) to Herwig (1699).

1504. McHenry, Lawrence C. *Garrison's History of Neurology*. Revised and enlarged with a bibliography of classical, original, and standard works in neurology. Springfield, Ill.: Charles C. Thomas, 1969. 552 pp. [Revision of an article first published in 1925]

Broad survey of the background of neurology from antiquity to 1900. Pp. 25–89: Middle Ages, Renaissance, and seventeenth century.

1505. Kail, Aubrey C. *The Medical Mind of Shakespeare*. Balgowlah, N.S.W.: Williams and Wilkins, 1986. 320 pp.

An analysis of Shakespeare's plays shows that he had a profound knowledge of contemporary physiology and psychology. "He employed medical terms in a manner which would have been beyond the powers of any ordinary playwright or physician." Pp. 299–300: Bibliography. Illustrated.

EPIDEMIC DISEASES

BIBLIOGRAPHIES

1506. Girtanner, Christoph. *Abhandlung über die venerische Krankheit*. Göttingen: J. C. Dieterich, 1788–89. 3 vols.

Vols. 2 and 3 of this treatise on venereal disease consist of an annotated bibliography (933 pp.). Vol. 2 covers the literature of the period 1493 to 1700 (704 items). Vol. 2, pp. xiii–xl: Chronological index of authors, 1495–1789. At the end of Vol. 3 there is an index (unpaginated) of names for Vols. 2 and 3. Vol. 1, pp. 3–58: History of venereal disease.

J. K. Proksch's monumental bibliography, *Die Litteratur über die venerischen Krankheiten* (Bonn: Peter Hanstein, 1889–1900), 5 vols. (including supplement and author index) covering the literature from 1494 to 1899 and containing some 40,000 titles, is not annotated, but is classified in 190 sections (including history).

1507. Haeser, Heinrich. *Lehrbuch der Geschichte der Medicin und der epidemischen Krankheiten.* 3rd ed. Jena: Dufft; Gustav Fischer, 1875–82. 3 vols. [Reprinted by Olms Verlag, Hildesheim, 1963]

Vols. 1–2: Chronological history of medicine through the early nineteenth century. Vol. 3: History of epidemic diseases. Vol. 2, pp. 1–469 and Vol. 3, pp. 213–447 cover the Renaissance and seventeenth century. Each volume has an analytical list of contents. Bibliographical notes in text. Index of names.

MONOGRAPHS AND ARTICLES

1508. Sudhoff, Karl. *Aus der Frühgeschichte der Syphilis. Handschriften und Inkunabelstudien epidemiologische Untersuchung und kritische Gänge.* Studien zur Geschichte der Medizin, Heft 9. Leipzig: Barth, 1912. 175 pp.

The sudden appearance (and identification) of syphilis in Europe at the end of the fifteenth century resulted in a flood of writings—from Edicts, prognostications and prayers to poems and prescriptions. These fourteen studies by Sudhoff are based on documents of the period 1495 to 1505 (49 MSS cited). He has also edited ten tractates (printed in Germany and Italy during the years 1495–98) which show the earliest attempts to treat the disease as a separate and special medical topic: *The Earliest Printed Literature of Syphilis Being Ten Tractates from the Years 1495–1498.* In complete facsimile with an introduction and other accessory material, ed. by Karl Sudhoff and adapted by Charles Singer (Florence: R. Lier and Co., 1925), xlviii + 352 pp.

1509. Jeanselme, Edouard. *Histoire de la syphilis: son origine, son expansion. Progrès réalisés dans l'étude de cette maladie depuis la fin du XVe siècle jusqu'à l'époque contemporaine. Traité de la syphilis,* Vol. 1. Paris: Doin, 1931. 432 pp.

Well-documented study by a historian with clinical experience. Pp. 34–298 cover the years 1492–1700. No bibliography, no index. Review: *Isis* 19 (1933), 249–52.

1510. Goodman, Hermann. *Notable Contributors to the Knowledge of Syphilis.* New York: Froben Press, 1944. 144 pp.

Historical survey of the study and treatment of syphilis, and also of the discussion of the origins of the disease (pre- or post-Columbian). Pp. 11–61 cover the years 1492 to 1700. No bibliography.

1511. Creighton, C. *A History of Epidemics in Britain.* 2nd ed. With additional material by D. E. C. Eversley, E. Ashworth Underwood,

A Malafranczos morbo gallo꜡ preseruatio ac Cura a Bartholo= meo Stëbĕr Uiennenfi artium ꝛ medicine doctore nuper edita.

Woodcut illustration of urine examination and treatment of syphilitic couple. Title page of Bartholomaeus Steber's *A malafranczos morbo Gallorum, preservatio ac cura* (Vienna, 1497–98). From Karl Sudhoff and Charles Singer, *The Earliest Printed Literature on Syphilis*, Florence, 1925, p. 263 (Bodleian Library, Oxford, 25834 d.14). **Entries 1508–13.**

Lynda Ovenall. London: Frank Cass & Co., 1965. 2 vols. [First published 1891–94]

Includes three introductory essays: Epidemiology as social history; Charles Creighton, the man and his work; A select bibliography of epidemiological literature since 1894.

Vol. 1: From A.D. 664 to the Great Plague (1666). Pp. 237–692: The sweating sickness, 1485–1551; Plague in the Tudor period; Gaol fevers, influenzas, and other fevers in the Tudor period; The French pox; Small pox and measles; Plague, fever, and influenza from the accession of James I to the Restoration; Sicknesses of early voyages and colonies; The Great Plague of London and the last of plague in England.

Vol. 2: From the extinction of the plague to the present time. Pp. 692–706: Index.

Essay review by F. H. Garrison in *Bulletin, New York Academy of Medicine* 4 (1928), 469–76.

1512. Guerra, Francisco. "The Dispute over Syphilis: Europe versus America." *Clio Medica* 13 (1978), 39–61.

A fresh look at the problem in the light of modern science. Pp. 57–61: Bibliography. See also Corinne Shear Wood, "Syphilis in Anthropological Perspective," *Social Science and Medicine* 12 (1978), 47–55.

1513. Quetel, Claude. *History of Syphilis*. Cambridge: Polity Press, 1990. 342 pp. [Translated from the 1986 French edition]

A synthesis of the historical material on the subject. Ch. 1–4 (pp. 9–105): A terrifying affliction (1493–1519); A much-disputed origin; The great pox (Sixteenth century); From pestilence to disease (The seventeenth and eighteenth centuries). Pp. 280–342: Notes, bibliographical notes chronology and index. Reviews *TLS* 5 Oct. 1990, p. 1051; *Choice* 28 (1991), 803.

PARACELSIAN MEDICINE

BIBLIOGRAPHIES

1514. Sudhoff, Karl. *Bibliotheca Paracelsica. Besprechung der unter Hohenheims Namen 1527–1893 erschienenen Druckschriften. Versuch einer Kritik der Echtheit der Paracelsischen Schriften, 1. Theil.*

Berlin: Georg Reimer, 1894. 722 pp. [Reprinted by Akademie Verlag, Graz in 1958]

Bibliography of the works of Paracelsus. Pp. 3–697: Books published up to 1700. Pp. 698–703: List of Sudhoff's writings on Paracelsus. Pp. 703–10: Index of publishers and printers. Pp. 711–22: Index of names. Review: *Bulletin of the History of Medicine* 33 (1959), 480–82.

1515. Sudhoff, Karl. *Nachweise zur Paracelsus-Literatur.* Acta Paracelsica 1–5, Supplement. Munich, 1932. 68 pp.

Bibliography of secondary literature on Paracelsus of the period 1527–1930.

1516. Weimann, Karl Heinz. *Paracelsus-Bibliographie 1932–1960. Mit einem Verzeichnis neu entdeckter Paracelsus-Handschriften (1900–1960).* Im Auftrage des Paracelsus-Kommission. Kosmosophie, 2. Wiesbaden: F. Steiner, 1963. 100 pp.

List of works by and about Paracelsus. Updates Sudhoff's bibliography of 1932 (item 1515). Reviews: *Ambix* 11 (1963), 52; *Medical History* 7 (1963), 394–95; *Sudhoffs Archiv* 47 (1963), 505–6.

MONOGRAPHS AND ARTICLES

1517. Leicester, Henry M. *Development of Biochemical Concepts from Ancient to Modern Times.* Cambridge, Mass.: Harvard Univ. Press, 1974. 286 pp.

Pp. 81–110: Paracelsus and the beginnings of iatrochemistry. The transitional seventeenth century. Reviewed in *Sudhoffs Archiv* 62 (1978), 207.

1518. Pagel, Walter. *Paracelsus. An Introduction to Philosophical Medicine in the Era of the Renaissance.* 2nd revised ed. Basel: S. Karger, 1982. 399 pp. [First published 1958]

An account of sixteenth-century philosophical medicine with Paracelsus as the center. Includes chapters on the life of Paracelsus and his sources. Reviews: *Isis* 50 (1959), 274–76; *Medical History* 27 (1985), 320; *Ambix* 31 (1984), 51.

1519. Schneider, Wolfgang. *Mein Untergang mit Paracelsus und Paracelsisten. Beiträge zur Paracelsusforschung, besonders auf arzneimittelgeschichtlichem Gebiet.* Frankfurt-am-Main: Govi-Verlag, 1982. 199 pp.

Collection of twenty-two papers published between 1952 and 1982. With indexes (medicines; names of persons; works of Paracelsus).

1520. Trevor-Roper, Hugh. "The Paracelsian Movement." *Renaissance Essays*. By Hugh Trevor-Roper. London: Secker & Warburg, 1985, pp. 149–99.

On the confrontation between Paracelsus and the traditional "Galenists" of the medical schools.

1521. Braun, Lucien. *Paracelsus. Alchemist—Chemiker, Erneuerer der Heilkunde. Eine Bildbiographie.* Zurich: Schweizer Verlaghaus, 1988. 158 pp. [First published in French in 1988]

An outline of the life and work of Paracelsus. With biographical notes on persons mentioned in the text, a chronological table, and copious illustrations.

1522. Jacobi, Jolande, ed. *Paracelsus: Selected Writings.* 2nd ed., reprinted Princeton, N.J.: Princeton Univ. Press, 1988. 290 pp. Translated from the German (Zurich, 1942)]

Pp. xxv–lxxii: Paracelsus, his life and work. Pp. 235–44: Table of contents of the Sudhoff-Matthiessen ed. of the complete works of Paracelsus. Pp. 245–66: Glossary of unfamiliar Paracelsian terms. Pp. 267–79: Bibliography of eds. of Paracelsus, and of biographical and critical works. The text is accompanied by 148 illustrations, mostly contemporary.

1523. Debus, Allen G. *The French Paracelsians: the Chemical Challenge to Medical and Scientific Tradition in Early Modern France.* Cambridge: Cambridge Univ. Press, 1992. 247 pp.

Study of chemical philosophy in early modern France. Review: *Medical History* 36 (1992), 355–56.

MEDICAL EDUCATION

1524. Puschmann, Theodor. *A History of Medical Education.* Introduction by Erwin H. Ackerknecht. New York: Hafner, 1966. 650 pp. [Facsimile of the 1891 ed. Translated from the German ed. of 1889]

"Remains the only serious treatise in the field up to this day . . . . Extremely rich in otherwise inaccessible detail" (Introduction). Pp. 285–439 deal with medical teaching in the sixteenth, seventeenth and eighteenth centuries.

1525. Wickersheimer, C. A. Ernst. *La médecine et les médecins en France à l'époque de la Renaissance.* Paris: Maloine, 1906. 693 pp.

On the teaching and practice of medicine in Renaissance France.

1526. Frank, Mortimer. "Medical Instruction in the Seventeenth Century." *Journal of the American Medical Association* 64 (1915), 1373–80.

Topics covered: Anatomy (dissections), physiology, surgery, practical obstetrics; museums, medical schools.

1527. Burckhardt, Albrecht. *Geschichte der medizinischen Fakultät zu Basel, 1460–1900.* Basel: Friedrick Reinhardt, 1917. 495 pp.

History of the oldest medical faculty this side of the Alps. Pp. 1–220 cover the years from the foundation to 1730. Includes biobibliographical studies of the leading personalities (including Paracelsus, Caspar Bauhin, Felix Platter, Theodor Zwinger II, J.J. Harder). On Vesalius's visit to Basel, see pp. 43–44.

1528. Castiglioni, Arturo. *The Renaissance of Medicine in Italy.* The Hideyo Noguchi lectures. Publications of the Institute of the History of Medicine. The Johns Hopkins University, Ser. 3, Vol. 1. Baltimore, Md.: Johns Hopkins Press, 1934. 91 pp.

1. The dawn of the Renaissance in the life, art and science of Italy. The thought of Leonardo. 2. The flowering of medical studies at the Italian universities from Berengario to Cesalpino. 3. The legacy of scientific renaissance and the main currents of thought from Fracastoro to Galileo. Index of proper names. No bibliography.

1529. Delaunay, Paul. *La vie médicale aux XVIe, XVIIe et XVIIIe siècles.* Paris: Editions Hippocrate, 1935. 556 pp.

The social and professional history of medicine. Topics: students; examinations; private and professional life of the doctor; professional associations; teaching; religion and medicine; the doctor and politics; the doctor in society; intellectual life; the practice of medicine. Each chapter has a bibliography.

1530. Copeman, W. S. C. *Doctors and Disease in Tudor Times.* London: Dawson's of Pall Mall, 1960. 186 pp.

Lectures delivered before the Royal College of Physicians of London (1958–59). Surveys the Tudor medical scene: the evolution of the profession; status and practice of Tudor medicine; medical education; scientific basis of Tudor medicine; art of diagnosis; diseases; treatment; public health and hygiene.

1531. Hahn, André, Paule Dumaître, and Janine Samion-Contet. *Histoire de la médecine et du livre médicale à la lumière des collections de la Bibliothèque de la Faculté de Médecine de Paris.* Paris: Editions Olivier Perrin, 1962. 433 pp.; 263 illus.

Pp. 11–22: The Library of the Paris Faculty of Medicine from its origins (1395) to 1732. Pp. 33–79: Medicine in the 15th century; The discovery and spread of printing; Incunables; Medical incunables in the Library. Pp. 81–197: The sixteenth century. Pp. 199–260: The seventeenth century. Pp. 377–94: Bibliography. Review: *Medical History* 7 (1963), 285–86.

1532. Dulieu, Louis. *La vie médicale estudiantine à Montpellier pendant la Renaissance.* Conférence donnée au Palais de la Découverte, D.124. Paris, 1968. 43 pp.

The life of medical students in Montpellier, based on contemporary records.

1533. O'Malley, C. D., ed. *The History of Medical Education.* Berkeley: University of California Press, 1970. 548 pp.

See item 236.

1534. Debus, A. G., ed. *Medicine in Seventeenth-Century England: A Symposium Held at UCLA in Honor of C. D. O'Malley.* Berkeley: University of California Press, 1974. 485 pp.

Topics covered include: Galenism; Paracelsian medicine; Embryology; Endocrinology; Pediatrics; Medical statistics; Quackery; Medicine and anatomy at Oxford and Cambridge; Medicine and the Royal Society. With a bibliography of O'Malley's works. Reviews: *Journal of the History of Medicine* 30 (1975), 170–72; *TLS* 14 Mar 1975, p. 283.

1535. Dulieu, Louis. *La médecine à Montpellier. Le moyen-âge. La Renaissance.* Avignon: Les Presses Universelles, 1975–79. 2 vols.

On the medical school, its organization, its teaching, its teachers, and the hospitals. With illustrations, biographical notes, and bibliography.

1536. Maddison, Francis, Margaret Pelling, and Charles Webster, eds. *Essays on the Life and Work of Thomas Linacre c. 1460–1524.* Linacre studies. Oxford: Clarendon Press, 1977. 416 pp.

Subjects covered: Linacre's years in Italy 1487–1499; College of Physicians of London; Galenism; Founding of the Linacre lectures; Medical education in England; Medical humanism. Pp. 290–536: Bibliographical survey of Linacre's works (by Giles Barber). Pp. 337–53: Published references to Thomas Linacre (by Margaret Pelling). Reviews: *Isis* 69 (1978), 295; *British Journal for the History of Science* 11 (1978), 290.

1537. Nutton, Vivian. "John Caius and the Linacre Tradition." *Medical History* 23 (1979), 373–91.

An investigation of the humanist tradition in medicine (i.e., the return to the classical sources of humoral therapy, and, in particular, to Hippocrates and Galen) in England and a closer examination of John Caius in the context of sixteenth-century European medicine.

1538. French, Roger. "Medical Teaching in Aberdeen from the Foundation of the University [1495] to the Middle of the Seventeenth Century." *History of Universities* 3 (1983), 127–57.

On the oldest medical chair in Britain. No annotations.

1539. Siraisi, Nancy C. *Avicenna in Renaissance Italy: The Canon and Medical Teaching in Italian Universities after 1500.* Princeton: Princeton Univ. Press, 1987. 410 pp.

Study of the role of the *Canon* in the university teaching of philosophy of medicine and physiological theory. Pp. 361–76: Latin eds. and commentaries of the *Canon* published after 1500. Pp. 377–95: Selected bibliography. Reviewed in *Isis* 80 (1989), 520–21.

1540. McVaugh, Michael, and Nancy G. Siraisi, eds. "Renaissance Medical Learning: Evolution of a Tradition." *Osiris* (2), 6 (1990), 1–244.

Nine essays (with an introduction) on the theme of medical knowledge in Western Europe between the twelfth and sixteenth centuries. The last three (pp. 161–234) relate to: Giovanni Argenterio and sixteenth-century medical innovation (Siraisi); Girolamo Mercuriale's *De modo studendi* (R. J. Durling); The reception of Fracastoro's theory of contagion (V. Nutton). Pp. 236–44: Index.

MEDICAL BIOGRAPHY

BIBLIOGRAPHIES AND BIOGRAPHICAL DICTIONARIES

1541. Munk, William. *The Roll of the Royal College of Physicians of London; Comprising Biographical Sketches of All the Eminent Physicians, Whose Names Are Recorded in the Annals from the Foundation of the College in 1518 to Its Removal in 1825, from Warwick Lane to Pall Mall East.* 2nd ed., revised and enlarged *Vol. 1: 1518 to 1700.* London: published by the College, 1878. 520 pp.

Notes—of varying length, from five lines to several pages—on some seven hundred physicians.

1542. New York Academy of Medicine. *Catalog of Biographies in the Library of the New York Academy of Medicine.* Boston, Mass.: G. K. Hall, 1960. 165 pp.

Some three thousand catalogue cards (single biographies of physicians and scientists) arranged alphabetically by biographee.

1543. Talbot, C. H., and E. A. Hammond. *The Medical Practitioners in Medieval England: A Biographical Register.* London: Wellcome Historical Medical Library, 1965. 503 pp.

Includes particulars of some three hundred practitioners who flourished during the period 1450 to 1518. Pp. 451–503: Index of personal names, places, and subjects.

1544. Wellcome Institute for the History of Medicine. *Subject Catalogue of the History of Medicine and Related Sciences.* [Vols. 14–18] *Biographical Sections 1–5.* Munich: Kraus International Publications, 1980. 5 vols.

Biographical works arranged alphabetically by biographee (Aagard–Zybelin). Entries for each person are arranged in reverse chronological order. Dates of birth and death given.

1545. Lindeboom, G. A. *Dutch Medical Biography. A Biographical Dictionary of Dutch Physicians and Surgeons, 1475–1975.* Amsterdam: Rodopi, 1984. 2,243 pp.

Includes some five hundred notices of persons who lived during the years 1475–1700.

BIOGRAPHIES

1546. Bucknill, Sir John Charles. *The Medical Knowledge of Shakespeare.* London: Longman, 1860. 292 pp.

Study by a physician of the "extent of Shakespeare's knowledge of medicine and the degree to which he was influenced by medical trains of thought."

1547. Palmer, W. M. "Cambridgeshire Doctors in the Olden Time." *Proceedings, Cambridgeshire Antiquarian Society* 15 (1911), 200–79.

Notes on medical practice and practitioners (including drugs, fees, and medical libraries) of the sixteenth and seventeenth centuries. Includes three lists of books—belonging to John Thomas (d. 1545), Robert Pickering (d. 1552), and Thomas Lorkin (d. 1591).

1548. Sulble, H. *Quelques charlatans célèbres au XVIIe siècle.* Toulouse: E. H. Guitard, 1922. 145 pp.

Study based on official documents and texts. Includes: Review of charlatanism up to the seventeenth century. Characteristics of charlatanism in the seventeenth century. Bio-bibliographical notices on some sixty charlatans. Secret remedies and *charlatanesque* publications. Pp. 132–42: Bibliography. Reviewed in *Bulletin, Société française d'histoire de médecine* 17 (1923), 131–33.

1549. Le Fanu, W. R. "Huguenot Refugee Doctors in England." *Proceedings of the Huguenot Society, London* 19 (1956), (4), 113–27.

Biographical notes. Includes some thirty refugees who arrived between the years 1560 and 1700.

1550. Simpson, Robert Ritchie. *Shakespeare and Medicine.* Edinburgh: E. & S. Livingstone, 1959. 267 pp.

"The objects of the study were to record the medical references in the plays and poems, to assess Shakespeare's knowledge of the medicine of his day, to consider the uses to which he put it, and to discuss some of the possible sources of his medical knowledge." In 18 chapters, with subject index.

1551. Dewhurst, Kenneth. *John Locke (1632–1704), Physician and Philosopher. A Medical Biography with an Edition of the Medical*

Notes in His Journals. London: Wellcome Historical Medical Library, 1963. 331 pp.

Annotated ed. of the medical notes in Locke's journals (1675–98) with commentary and index. The journals are of special interest as he knew most of the famous physicians and scientists of his day. Reviews: *Bulletin of the History of Medicine* 39 (1965), 90; *English Historical Review* 80 (1965), 839.

1552. Lindeboom, G. A. *Herman Boerhaave: the Man and His Work*. London: Methuen, 1968. 452 pp.

Papers read at a symposium commemorating the tercentenary of Boerhaave's birth. In two parts. Part 1: Biographical (childhood, student, physician, teacher). Part 2: Personality; philosophical and theoretical medical views; clinician, physician, botanist, chemist; his international audience; Bibliography.

1553. Reti, Ladislao, and William C. Gibson. *Some Aspects of Seventeenth-Century Medicine & Science*. Papers read at a Clark Library Seminar, October 12, 1968. Los Angeles: William Andrews Clark Memorial Library, University of California, 1969. 46 pp.

Van Helmont's studies of the chemistry of digestion, those of Boyle on respiration, and Christopher Wren's contribution to medical science.

1554. Wykes, Alan. *Doctor Cardano, Physician Extraordinary*. London: Frederick Muller, 1969. 187 pp.

Popular biography of Cardano, a "sketch of the man moving within the epoch of the Renaissance, Reformation and Counter Reformation." Reveals some aspects of his nature suppressed by Morley (item 483) and Waters in the interests of "Victorian pudency." Unfortunately errs in keeping Tartaglia alive after 1557.

1555. Burmeister, Karl Heinz. *Achilles Pirmin Gasser, 1505–1577. Arzt und Naturforscher, Historiker und Humanist. 1: Biographie. 2: Bibliographie. 3. Briefwechsel*. Wiesbaden: Guido Pressler, 1970. 3 vols.

Detailed study of the polymath from Lindau (Lake Constance), editor of the *Narratio prima* (2nd ed., 1541) of Rheticus, and the *De magnete* (1558) of Peter Peregrinus. Among his correspondents were Gesner and Vesalius. Vol. 1, pp. 121–29, contain a description of his collection of books—2,884 volumes (907 medical) acquired by Ulrich Fugger. Reviews: *Ambix* 18 (1971), 219–20; *Isis* 62 (1971), 543–44.

1556. Antonioli, Roland. *Rabelais et la médecine*. Etudes Rabelaisiennes, 12. Travaux d'Humanisme et de la Renaissance, 143. Geneva: Droz, 1976. 394 pp.

On medicine, pharmacy, and anti-Arabism in the early sixteenth century and the role of medicine in the life and work of Rabelais.

1557. Lindeboom, G. A. *Descartes and Medicine.* Amsterdam: Rodopi, 1978. 134 pp.

"Historical introduction to the study of Descartes' attitude to medicine and his relations with contemporary physicians in the country where he sought refuge and lived during the years of the full development of his faculties." With a bibliography.

1558. Pagel, Walter. *Joan Baptista Van Helmont: Reformer of Science and Medicine.* Cambridge: Cambridge Univ. Press, 1982. 219 pp.

Synthesis of Van Helmont's scientific and medical discoveries and ideas, with his cosmology and religious philosophy. Reviewed in *British Journal of the History of Science* 17 (1984), 106–7.

1559. Fierz, Markus. *Girolamo Cardano 1501–1576: Physician, Natural Philosopher, Mathematician, Astrologer, and Interpreter of Dreams.* Boston: Birkhauser, 1983. 202 pp. Translated from the 1977 German ed., with additions.

An "attempt to acquaint the modern reader with the philosophy and scientific investigations of the universal scholar Girolamo Cardano." Focuses interest on Cardano the physician, natural philosopher, and physicist. Reviewed in *Annals of Science* 41 (1984), 90.

1560. Pagel, Walter. *From Paracelsus to Van Helmont. Studies in Renaissance Medicine and Science.* Edited by Marianne Winder. London: Variorum Reprints, 1986. 350 pp.

Fifteen papers (published between 1931 and 1981), mostly relating to Van Helmont.

1561. Hoeniger, F. David. *Medicine and Shakespeare in the English Renaissance.* Newark: Univ. of Delaware Press, 1992. 404 pp.

Intended for students of the literature, history, and thought of Elizabethan and Jacobean England. Pp. 17–67: Medicine and medical practitioners in the age of Shakespeare. Pp. 69–127: Major medical philosophies and systems. Pp. 129–78: Physiology and psychology: the body and how it functions. Pp. 179–272: Pathology, diagnosis, and therapy. Pp. 273–338: Three Shakespearean plays examined in the light of literary and medical traditions. Pp. 347–86: Notes and references. Pp. 387–404: Indexes.

The Scientific Revolution

BIBLIOGRAPHIES AND SOURCE BOOKS

1562. Hall, Marie Boas, ed. *Nature and Nature's Law. Documents of the Scientific Revolution*. London: Macmillan, 1970. 381 pp.

Thirty-nine selected texts of the period 1500–1800, arranged under: a new age and an old tradition; the astronomical revolution; the new philosophy; experimental innovation (physiology; microscopy; pneumatics); scientific societies; the Newtonian triumph (optics; mechanics; the Newtonian system); the organization of matter (the mechanical philosophy; chemical theories). With an introductory essay on the Scientific Revolution. Reviewed in *TLS* 12 Feb. 1971, p. 190.

1563. Rossi, Paolo, ed. *La rivoluzione scientifica da Copernico a Newton*. Turin: Loescher, 1973. 417 pp.

A source book in twelve chapters covering all aspects of the Scientific Revolution. With introductory notes and bibliographies.

1564. Krafft, Fritz. "Renaissance der Naturwissenschaften: Naturwissenschaften der Renaissance. Ein Überblick über die Nachkriegs-literatur." *Humanismusforschung seit 1945. Ein Bericht aus interdisziplinärer Sicht*. Mitteilung 2, Deutsche Forschungsgemeinschaft, Kommission für Humanismusforschung. Boppard: Harald Boldt, 1975, pp. 111–83.

Survey of post-1945 literature. Pp. 161–83: Bibliography.

1565. Rei, Dario. *La rivoluzione scientifica. Scienza e società in Europa tra il XV e il XVII secolo*. Turin: Società Editrice Internazionale, 1975. 158 pp.

Source book. Pp. 12–125: Forty-one documents arranged in six chapters, with introductory notes. Pp. 128–50: The historiography of the Scientific Revolution (with extracts).

1566. Hall, Marie Boas. "Il rinascimento scientifico." *Il rinascimento: interpretazioni e problemi. A Eugenio Garin nel suo settantesimo compleanno*. Bari: Laterza, 1979, pp. 323–52. [English translation published by Methuen, London, 1982]

Survey of recent studies in Renaissance science.

1567. Hunter, Michael. "[Restoration Science and Its Content] Bibliographical Essay." *Science and Society in Restoration England* (Cambridge: Cambridge Univ. Press, 1981), pp. 198–219.

Critical introduction to existing scholarship. Some 250 books and articles cited. Arrangement follows that of the chapters in the book. See item 342.

1568. Westfall, Richard S. "Teaching the History of Science: Resources and Strategies. The Scientific Revolution." *History of Science Society Newsletter* 15 (3) (1986), 14a–14f.

Guide to the study of the Scientific Revolution—a bibliographical survey—by a leading scholar. Some ninety titles.

MONOGRAPHS AND ARTICLES

1569. Whitehead, A. N. *Science and the Modern World*. Cambridge: University Press, 1927. 265 pp. [First published 1926. Several reprints]

"Study of some aspects of Western culture during the past three centuries, in so far as it has been influenced by the development of science." Pp. 1–24: The origins of modern science; pp. 49–70: The century of genius. Revised version of the Lowell lectures delivered at Boston. Whitehead (Fellow of Trinity College, Cambridge; professor of applied mathematics, Imperial College, London; professor of philosophy, Harvard University) was co-author with Bertrand Russell of *Principia Mathematica* (1910). Reviewed in *Nature* 117 (1926), 847–50.

1570. Strong, Edward W. *Procedures and Metaphysics: A Study in the Philosophy of Mathematical-Physical Science in the Sixteenth and Seventeenth Centuries*. Berkeley: Univ. of California Press, 1936. 301 pp.

Argues that "the great revival in the mathematical sciences which culminated in the work of Galileo was not primarily inspired by Renaissance neoplatonism with its metaphysical idealization of mathematics but was due to the development of methodological procedures within the restricted subject-matter of the sciences themselves." Reviewed in *Isis* 29 (1938), 110–13.

1571. Maier, Anneliese. *Die Mechanisierung des Weltbilds im 17. Jahrhundert.* Forschungen zur Geschichte der Philosophie und der Pädagogik, Heft 18. Leipzig: F. Meiner, 1538. 54 pp. [Republished in *Zwei Untersuchungen zur Nachscholastischen Philosophie,* 2nd ed., Rome, 1968, pp. 1–67]

"Takes issue with the interpretation of the scientific revolution as a linear historical process initiated by the revival of atomism and Galileo's innovations in mechanics, which, accompanied by the increased use of experimentation, gradually produced a new body of scientific thought dependent only on its own laws and assumptions " (Steven D. Sargent). See also item 932.

1572. Koyré, Alexandre. *Etudes galiléennes. 1. A l'aube de la science classique. 2. La loi de la chute des corps. 3. Galilée et la loi d'inertie.* Histoire de la pensée. Paris: Hermann, 1939. 3 vols.

Outlines the structural patterns of the old and new world-views and studies the changes effected by the revolution of the seventeenth century. Reviewed in *Isis* 33 (1942), 654–56.

1573. Wolf, A. *A History of Science, Technology, and Philosophy in the Sixteenth and Seventeenth Centuries.* 2nd ed. prepared by Douglas McKie. London: George Allen and Unwin, 1950. 692 pp. [First published 1935]

A collection of short essays on the history of various scientific disciplines in 26 chapters arranged mainly by subject. The author was professor at University College, London (retired 1941) and the book was intended to meet the needs of students pursuing courses [M.Sc.] in the history, methods, and principles of science at the University. Reviews: *Science* 83 (1936), 262–64 ("profusely illustrated mine of useful and delightful information"—Sigerist); *Isis* 24 (1935), 164–67 ("badly arranged and very deficient from the purely historical point of view"—Sarton). In the 2nd ed., prepared by the author's successor, a number of errors were corrected and the bibliographies extended. In his review—*Isis* 44 (1953), 167—I. B. Cohen noted that, in spite of the shortcomings pointed out by Sarton, the work was still indispensable for studying the period, "nothing even vaguely resembling it being in existence." Also reviewed in *TLS* 19 Oct. 1951, p. 664.

1574. Mason, S. F. "The Scientific Revolution and the Protestant Reformation." *Annals of Science* 9 (1953), 64–87, 154–75.

Studies of Calvin and Servetus in relation to the new astronomy and the theory of the circulation of the blood; and of Lutheranism in relation to iatrochemistry and German nature philosophy.

1575. Hall, A. Rupert. *The Scientific Revolution, 1500–1800. The Formation of Modern Scientific Attitude.* London: Longmans, 1954. 390 pp.

"Character study, rather than a biographical outline, of the scientific revolution." Pp 1–302: Science in 1500; new currents in the sixteenth century; attack on tradition—mechanics and astronomy; experiment in biology; the principles of science in the early seventeenth century; the organization of scientific inquiry; technical factors in the Scientific Revolution; the *principate* of Newton; descriptive biology and systematics. Pp. 375–79: Bibliographical notes. Reviewed in: *Isis* 46 (1955), 304–5; *Annals of Science* 11 (1955), 99–101; *Centaurus* 4 (1955), 171–74. Revised and corrected ed., with enlarged bibliography, published in 1962 (394 pp.). See also item 1626.

1576. Butterfield, Herbert. *The Origins of Modern Science, 1300–1800.* 2nd ed. London: Bell and Wyman, 1957. 242 pp. [First published in 1949]

The Scientific Revolution as seen by a "general historian." Revised version of a series of lectures "delivered for the History of Science Committee in Cambridge in 1948" by the author, who was Regius Professor of Modern History in the University. Reviewed in *TLS* 25 Nov. 1949, p. 761.

1577. Daumas, Maurice. "Les sciences physiques aux XVIe et XVIIe siècles." *Histoire de la science.* Ed. Maurice Daumas. Encyclopédie de la Pléiade. Paris: Gallimard, 1957, pp. 837–82.

Leonardo's influence; Sixteenth-century chemistry; Optics; Acoustics; Pneumatics; Unification of chemical theories.

1578. Hooykaas, R. "The Rise of Modern Science: When and Why?" *British Journal for the History of Science* 20 (1957), 453–73.

States that the rise of modern science had two main causes: (1) the new natural history and the methodological epistemological changes connected with it; (2) "the transition from an organistic to a mechanistic view of the world, a change closely connected with experimental philosophy and the contribution made to it by engineers, physicians, alchemists, cartographers, pilots and instrumentalists."

1579. Koyré, Alexandre. *From the Closed World to the Infinite Universe.* Baltimore: Johns Hopkins Univ. Press, 1957. 313 pp.

Discussion on the "works of a few great thinkers [from Nicholas of Cusa and Marcellus Palingenius to Newton and Leibniz] who, in deep understanding of its primary importance, have given their full attention to the fundamental problem of the structure of the world." Reviewed in *Isis* 49 (1958), 363–66.

1580. Kuhn, Thomas S. *The Copernican Revolution. Planetary Astronomy in the Development of Western Thought.* Cambridge, Mass.: Harvard Univ. Press, 1957. 295 pp.

Examines the conceptual changes in cosmology, physics, philosophy, and religion. Reviewed in *Isis* 49 (1958), 366–67 and *Scripta Mathematica* 24 (1959), 330–31.

1581. Lenoble, Robert. "Origines de la pensée scientifique moderne." *Histoire de la science.* Maurice Daumas, ed. Encyclopédie de la Pléiade. Paris: Gallimard, 1957, pp. 369–534.

The medieval heritage; the Doctors Faust; Francis Bacon; from Leonardo to Copernicus and Kepler; Galileo; the "explosion" of mechanism; man and science in the seventeenth century. Bibliography.

1582. Espinasse, Margaret. "The Decline and Fall of Restoration Science." *Past and Present* 14 (1958), 71–90.

"In the history of science, the years immediately after 1688 mark the beginning of a new era."

1583. Madden, Edward H., ed. *Theories of Scientific Method: The Renaissance through the Nineteenth Century.* Seattle: Univ. of Washington Press, 1960. 346 pp.

Six (out of eleven) studies deal with: Renaissance astronomy and natural science; Bacon; Descartes; Hobbes; Newton. Reviewed in *TLS* 26 Aug. 1960, p. 551.

1584. Dijksterhuis, E. J. *The Mechanization of the World Picture.* Oxford: Clarendon Press, 1961. 539 pp. [Translated from the 1950 Dutch ed.]

"The mechanization of the world picture during the transition from ancient to classical science meant the introduction of a description of nature with the aid of mathematical concepts of classical mechanics; it marks the beginning of the mathematization of science, which continues at an ever-increasing pace in the twentieth century." Pp. 223–501: The prelude to the growth of classical science; the evolution of classical science; physics, chemistry, and philosophy of nature in the seventeenth century. Epilogue. Reviewed in: *Isis* 42 (1951), 66–67; *Science* 134 (1961), 1684; *TLS* 24 Nov. 1961, p. 847.

1585. Jones, Richard Foster. *Ancients and Moderns: A Study of the Rise of the Scientific Movement in Seventeenth-Century England.* 2nd ed., with an index, new preface, and minor revisions. St. Louis:

Washington University Studies, 1961. 354 pp. [First published in 1936]

Study of the "thought movement in the seventeenth century to which modern science in England traces its source." In three parts: The Renaissance (scientific attitude of the Elizabethans; the decay of nature; the Bacon of the seventeenth century; the Gilbertian tradition, 1600–1640). The Puritan era (the advancement of learning and piety; the revolt from Aristotle and the ancients; projects, inventions, and the progress of science). The Restoration (the defence of the experimental philosophy; the "Bacon-faced generation"). Pp. 273–354: notes and index. No bibliography. Reviewed in *Isis* 26 (1936), 171–72.

1586. Koyré, Alexandre. *La révolution astronomique: Copernicus, Kepler, Borelli.* Paris: Hermann, 1961. 525 pp.

Study of: the heliocentrism of Copernicus; celestial dynamics of Kepler; the unification of celestial and terrestrial physics by Borelli. Reviewed in *Isis* 53 (1962), 517–19. English translation (1973) reviewed in: *Isis* 66 (1975), 116; *Annals of Science* 32 (1975); 301–2; *Sixteenth Century Journal* 5 (1974), 131; *Archives internationales d'histoire des sciences* 25 (1975), 162–64.

1587. Randall, J. E. *The School of Padua and the Emergence of Modern Science.* Padua: Antenore, 1961. 141 pp.

On the influence of Italian Aristotelianism on Galileo. Criticized by Giorgio de Santillana in *Isis* 54 (1963), 300–2. Reviewed in *American Historical Review* 67 (1961–62), 483–84.

1588. Rhys, Henley Howell, ed. *Seventeenth Century Science and the Arts.* Essays by Stephen Toulmin, Douglas Bush, James Ackermann, and Claude V. Palisca. Princeton, N.J.: Princeton Univ. Press, 1961. 137 pp.

The William J. Cooper Foundation lectures, Swarthmore College, 1960. Four essays on how the scientific revolution of the seventeenth century affected other spheres of human thought and creation, particularly literature, music, and the arts. Reviewed in *Isis* 54 (1963), 412 and *TLS* 24 Aug 1962, p. 643.

1589. Boas, Marie. *The Scientific Renaissance 1450–1630.* The Rise of Modern Science, 2. London: Collins, 1962. 380 pp.

Argues that the period 1450–1630 constitutes a definite stage in the history of science—"an era of profound change" marking a break with the past. Analyses the effect that the recovery of Greek texts, the voyages of discovery, and the invention of printing had on European man's view of the world. Surveys (in eleven chapters) the scientific knowledge acquired during the period, beginning with the eds. of Ptolemy's *Geography*, Peuerbach's *Theoricae novae planetarum* and Regiomontanus's *Epitome* and ending with Harvey's *De motu cordis* and Galileo's *Dialogo* and *Discorsi*. Reviewed in *Isis* 56 (1965), 240–42.

1590. Grant, Edward. "Hypotheses in Late Medieval and Early Modern Science." *Daedalus* 91 (1962), 599–612.

"Although certain similarities may be discerned between the methodologies [of medieval and seventeenth-century science], they were yet separated by a fundamental difference which had significant consequences on the scientific outlook of the two periods." Comments by Benjamin Nelson on pp. 612–16.

1591. Grant, Edward. "Late Medieval Thought, Copernicus, and the Scientific Revolution." *Journal of the History of Ideas* 23 (1962), 197–220.

"Copernicus is really the initiator of a very basic attitude which came to be held in some form or other by most of the great figures of the Scientific Revolution—namely, that fundamental principles in the form of hypotheses or assumptions about the universe must be physically true, and incapable of being otherwise."

1592. Feuer, Lewis S. *The Scientific Intellectual, The Psychological and Sociological Origins of Modern Science*. New York: Basic Books, 1963. 441 pp.

Seeks to trace the "evolution of the scientific intellectual as a human type." Pp. 1–182 relate to the Scientific Revolution. Reviewed in *Isis* 56 (1965), 369–70.

1593. Hall, A. Rupert. *From Galileo to Newton, 1630–1720*. The Rise of Modern Science, 3. London: Collins, 1963. 380 pp.

Traces the changes that took place in the spirit and ideas of science between the publications of Galileo's *Dialogo* (1632) and of the final ed. of Newton's *Principia* (1726). With notes and bibliography. Reviewed in *Isis* 56 (1965), 367–78 and *TLS* 12 July 1963, p. 506.

1594. Kearney, H. F. "Puritanism, Capitalism and the Scientific Revolution." *Past and Present* 28 (1964), 81–101.

On the sociological interpretation of the Scientific Revolution in Christopher Hill's *Intellectual Origins of the English Revolution* (abbreviated version of item 320). The latter's reply is in *Past and Present* 29 (1964), 88–97.

1595. Kearney, Hugh F., ed. *Origins of the Scientific Revolution.* London: Longmans, 1964. 159 pp.

In three parts: (1) Extracts from works of modern historians on: the Renaissance; the role of experiment; the mathematical revolution; the Middle Ages; social and religious considerations. (2) Select documents on the same topics. (3) Editor's note on wider issues. Reviewed in *TLS* 4 June 1964, p. 474.

1596. Bernal, J. D. *Science in History.* 3rd ed. London: Watts, 1965. 1,039 pp.

Panoramic study, by the well-known British Marxist physicist, "relating advances in science to political and economic developments in society." Based on a series of lectures given at Ruskin College, Oxford in 1948. Pp. 251–357: The Scientific Revolution. Critical reviews in: *Isis* 48 (1954), 471–73; *Centaurus* 4 (1956), 285–93; *British Journal for the History of Science* 3 (1966), 188; *Annals of Science* 22 (1966), 142–44.

1597. Harré, R., ed. *Early Seventeenth Century Scientists.* Oxford: Pergamon Press, 1965. 188 pp.

Topics covered: ideas of scientific method (Bacon, Descartes); the rudiments of biochemistry (Van Helmont), physics (Gilbert, Galileo, and Kepler), and physiology (Harvey). Reviewed in *TLS* 16 June 1966, p. 540.

1598. West, J. F. *The Great Intellectual Revolution.* London: Murray, 1965. 132 pp.

"Tries to show how the growth of science [in the seventeenth century], the political history of the times, the changes in literary expression, and the philosophies put forward, were all interrelated and interacting, and both produced and responded to a climate of opinion which was changing so radically during the course of the century as to amount to an intellectual revolution, perhaps the most important in human history, and one which is still far from complete." Reviewed in *TLS* 19 Aug. 1965, p. 721.

1599. Zambelli, Paola. "Rinnovamento umanistico, progresso tecnologico e teorie filosofiche alle origini della rivoluzione scientifica." *Studi storici* 6 (1965), 507–46.

Critical analysis of some recent studies of the Scientific Revolution.

1600. Koyré, Alexandre. *Etudes d'histoire de la pensée scientifique.* Paris: Presses Universitaires de France, 1966. 372 pp.

Collection of twenty articles written between 1930 and 1961 on the origins of modern science (with special reference to: Leonardo da Vinci, Tartaglia, Benedetti, Galileo, Gassendi, Cavalieri, Pascal). Reviewed in: *Isis*

60 (1969), 111–12; *TLS* 31 Aug. 1973. p. 994; *Archives internationales d'histoire des sciences* 25 (1975), 136–40.

1601. Popkin, R. H. "Skepticism, Theology and the Scientific Revolution in the Seventeenth Century." *Problems in the Philosophy of Science*, edited by Imre Lakatos and Alan Musgrave. Amsterdam: North Holland Publishing Co., 1968, pp. 1–39.

Paper read at the International Colloquium in the Philosophy of Science, London, July 11–17, 1965 (with discussion). The author's thesis is that "we have suffered, and still suffer, from a distorted account of the relations of religion and science in the seventeenth century."

1602. Briggs, Robin. *The Scientific Revolution of the Seventeenth Century*. Seminar Studies in History. London: Longman, 1969. 121 pp.

"This series provides studies with supporting documents of important topics in history." Places the scientific revolution in historical perspective, providing an analysis and a critical narrative of the main developments. With extracts from primary sources and a bibliography.

1603. Crombie, A. C. "Historians and the Scientific Revolution." *Physica* 11 (1969), 167–80.

On seventeenth- and eighteenth-century historians and the modern historiography of science.

1604. Gusdorf, Georges. *La révolution galiléenne*. Les sciences humaines et la pensée occidentale, 3. Paris: Payot, 1969. 2 vols.

General introduction to the movement of ideas in the seventeenth century. No index. Reviewed in *British Journal for the Philosophy of Science* 22 (1971), 76–77 and *TLS* 5 June 1969, p. 604.

1605. Bullough, Vern L., ed. *The Scientific Revolution*. European Problem Studies. New York: Holt, Rinehart and Winston, 1970. 129 pp.

Eighteen studies arranged under: background; the Renaissance and the development of science; economic, social, and technological factors; continuity of scientific development; a possible reconciliation of viewpoints. Includes suggestions for further reading. Reviewed in *Isis* 62 (1971), 245.

1606. Kearney, Hugh. *Science and Change, 1500–1700*. London: Weidenfeld and Nicolson, 1971. 253 pp.

Argues that "the key to interpreting the origins and course of the Scientific Revolution is to be found in three distinctive traditions or paradigms—

the organic, magical and mechanistic." Pp. 241–48: bibliographical essay. Reviewed in *TLS* 14 Jan. 1972, p. 37.

1607. Rossi, Paolo. *Aspetti della rivoluzione scientifica.* Collana di Filosofia, 13. Naples: Morano, 1971. 424 pp.

Nine articles written on various aspects of the Scientific Revolution (Decline of astrology; Bacon and the Bible; "Profile" of Galileo Galilei, Thomas Burnet). Reviewed in *Isis* 63 (1972), 444–45.

1608. Westfall, Richard S. *The Construction of Modern Science: Mechanisms and Mechanics.* New York: John Wiley, 1971. 171 pp.

Intended for undergraduates. Explores the founding of modern science under the combined influence of the Platonic-Pythagorean tradition and the mechanical philosophy. Suggestions for further reading (pp. 160–65).

1609. Russo, François. "Les études newtoniennes d'Alexandre Koyré." *Archives de philosophie* 37 (1974), 107–32.

Analysis of Koyré's *Etudes newtoniennes* (item 520), *La révolution astronomique* (item 1586) and *Etudes galiléennes* (item 1572).

1610. Délorme, S., ed. *Avant, avec, après Copernic: la représentation de l'Univers et ses conséquences épistémologiques.* XXXIe Semaine de Synthèse, 1–7 juin 1973. (Centre Internationale de Synthèse, Paris). Paris: Blanchard, 1975. 439 pp.

Copernicus's world, his part in the Scientific Revolution, and his influence on the conception of man's place in the universe. Forty-five papers of which twenty-six relate to the sixteenth and seventeenth centuries. Reviewed in *Archives internationales d'histoire des sciences* 26 (1976), 321–22.

1611. Oberman, Heiko A. "Reformation and Revolution: Copernicus's Discovery in an Era of Change." *The Cultural Context of Medieval Learning.* John Emery Murdoch and Edith Dudley Sylla, eds. Boston Studies in the Philosophy of Science, 26. Dordrecht: Reidel, 1975, pp. 397–435.

Proposes "a new view of the first encounter between the Protestant Reformation and Copernicanism" and indicates "a common background to these two sixteenth-century reform movements in nominalism and in the medieval campaign *contra vanam curiositatem.*"

1612. Righini Bonelli, N. L., and W. R. Shea, eds. *Reason, Experience and Mysticism in the Scientific Revolution.* New York: Science History Publications, 1975. 320 pp.

Seventeen essays (resulting from a symposium) on seventeenth-century science—chemical philosophy, alchemy, mechanics, astronomy, iatromechanics, anatomical microscopy, mathematics—several devoted to Galileo and Newton. Reviews: *Nature* 258 (1975), 30–31; *Revue d'histoire des sciences* 29 (1976), 175–76; *Physis* 17 (1975), 149–51; *Isis* 68 (1977), 156–58; *Technology and Culture* 17 (1976), 793–95.

1613. Cohen, I. Bernard. "The Eighteenth-Century Origins of the Concept of Scientific Revolution." *Journal for the History of Ideas* 37 (1976), 257–88.

The author was unable to find any references to revolutions in the sciences in the literature prior to 1700. See item 1631.

1614. Guerlac, Henry. *Essays and Papers in the History of Modern Science.* Baltimore: Johns Hopkins Univ. Press, 1977. 540 pp.

A collection of articles written over a period of twenty-five years. Pp. 69–242 relate to Newtonian science (nine papers). Reviewed in *TLS* 23 June 1978, p. 698.

1615. Jacob, J. R. *Robert Boyle and the English Revolution: A Study in Social and Intellectual Change.* New York: Franklin, 1977. 240 pp.

A study of Boyle's thought "in the context of the world in which it evolved." Reviewed in *Isis* 70 (1979), 314–15.

1616. Copenhaver, B. P. "Essay Review of: Robert S. Westman and J. E. McGuire, *Hermeticism and the Scientific Revolution*: Papers Read at a Clark Library Seminar, March 9, 1974 (Los Angeles, California: William Andrew Clark Memorial Library, University of California at Los Angeles, 1977)." *Annals of Science* 75 (1978), 527–31.

See items 651 and 1618.

1617. Debus, A. G. *Man and Nature in the Renaissance.* Cambridge: Cambridge Univ. Press, 1978. 159 pp.

An introduction to science and medicine of the period 1450–1650 with suggestions for further reading. Reviewed in *Isis* 70 (1979), 588–92.

1618. Schmitt, Charles B. "Reappraisals in Renaissance Science." *History of Science* 16 (1978), 200–14.

Essay review of Westman and McGuire, *Hermeticism and the Scientific Revolution* (1977), item 651.

1619. Cohen, I. Bernard. *The Newtonian Revolution. With Illustrations of the Transformation of Scientific Ideas.* Cambridge: Cambridge Univ. Press, 1980. 404 pp.

Studies of the history and concept of revolution in the sciences and of some of the main features of the *Principia*. Based on a series of lectures given in 1966 at Belfast. With notes, bibliography and indexes (pp. 290–404). Reviews: *Annals of Science* 38 (1981), 609–10; *History of Science* 20 (1982), 140–44.

1620. Copenhaver, B. P. "Jewish Theologies of Space in the Scientific Revolution: Henry More, Joseph Raphson, Isaac Newton and Their Predecessors." *Annals of Science* 37 (1980), 489–548.

"Students of early modern science might profit from an encounter with the Christianized and geometricized Cabala of More and Raphson."

1621. Merchant, Carolyn. *The Death of Nature: Women, Ecology and the Scientific Revolution.* San Francisco: Harper and Row, 1980. 348 pp.

Argues that the present world view which reconceptualizes reality as a machine rather than a living organism has "sanctioned the domination of both nature and women." Stresses the need to reevaluate the contributions of the founding "fathers" of modern science—Francis Bacon, Harvey, Descartes, Hobbes, and Newton—and to reappraise the fate of alternative philosophies shaped by the organic world view. Reviews: *Ambix* 28 (1981), 57–58; *Isis* 72 (1981), 287–88; *Journal of Modern History* 54 (1982), 66–70. Essay review by Walter Pagel, "Mother Nature and the Scientific Revolution," *History of Science* 19 (1981), 148–53.

1622. Randles, W. G. L. *De la terre plate au globe terrestre. Une mutation épistémologique rapide (1480–1520).* Cahiers des Annales, 38. Paris: Armand Colin, 1980. 120 pp.

On the "general acceptance of a spherical earth by the world of scholars during the age of discovery." Reviews: *Isis* 73 (1982), 136–37; *Revue d'histoire des sciences* 35 (1982), 167–68; *Revue de synthèse* 102 (1981), 191–93.

1623. Grant, Edward. *Much Ado about Nothing. Theories of Space and Vacuum from the Middle Ages to the Scientific Revolution.* Cambridge: Cambridge Univ. Press, 1981. 456 pp.

Pp. 148–81: Extracosmic, infinite void space in sixteenth- and seventeenth-century scholastic thought. Pp. 182–255: Infinite space in nonscholastic thought during the sixteenth and seventeenth centuries. Pp. 352–418: Notes. Pp. 419–37: Bibliography. Pp. 439–56: Index. Reviews: *Science* 214 (1981), 785–86; *Journal for the History of Astronomy* 13 (1982), 217–18; *American*

Historical Review 87 (1982), 1044; *American Scientist* 70 (1982), 99–100; *Isis* 75 (1984), 383.

1624. Carugo, Adriano, and Alistair C. Crombie. "The Jesuits and Galileo's Ideas of Science and of Nature." *Annali dell'Istituto e Museo di Storia della Scienza* 8 (2), (1983), 1–68.

Studies the "relation of the ideas developed by Galileo of science and of nature to the scholastic revival of Aristotelianism and Thomism, promoted by the Council of Trent and articulated in the late sixteenth and early seventeenth centuries by the Jesuits." Documented.

1625. Deleted.

1626. Hall, A. Rupert. *The Revolution in Science 1500–1700.* London: Longman, 1983. 373 pp.

A re-thinking and re-writing of the author's *The Scientific Revolution* (1954)—item 1575—based on the "solid, scholarly work" of the last thirty years. With notes but no bibliography. Reviews: *Choice* 21 (1983), 593; *Nature* 306 (1983), 121; *British Journal for the History of Science* 17 (1984), 233; *History and Philosophy of the Life Sciences* 10 (1988), 198–99.

1626A. Harman, P. M. *The Scientific Revolution.* London: Methuen, 1983. 35 pp.

A brief, up-to-date account of the subject for students preparing for Advanced Level examinations. With suggestions for further reading.

1627. Van Berkel, Klaas. *Isaac Beeckman (1588–1637) en de mechanisering van het wereldbeeld. Isaac Beeckman (1588–1637) and the Mechanization of the World Picture (with a summary in English).* Nieuwe Nederlandse bijdragen tot de geschiedenis der geneeskunde en der natuurwetenschappen, No. 9. Amsterdam: Rodopi, 1983. 348 pp. [Thesis, University of Utrecht, 1983]

Biography of Beeckman, an analysis of his *Journal*, and a discussion of the impact of his religious opinions and social background on his mechanistic world view. Reviews: *Technology and Culture* 25 (1984), 334–35; *Isis* 76 (1985), 273–74.

1628. Feingold, Mordechai. *The Mathematicians' Apprenticeship; Science, Universities and Society in England, 1560–1640.* Cambridge: Cambridge Univ. Press, 1984. 248 pp.

Refutes the view that Oxford and Cambridge played no part in the Scientific Revolution of the seventeenth century. Reviews: *British Journal for the*

History of Science 18 (1985), 212–17; *Journal for the History of Astronomy* 16 (1985), 56–59; *Annals of Science* 43 (1986), 297.

1629. Meinel, Christoph. *In Physicis Futuram Saeculum Respicio: Joachim Jungius und die Naturwissenschaftliche Revolution des 17. Jahrhunderts.* Veröffentlichung der Joachim Jungius-Gesellschaft der Wissenschaften Hamburg, 52. Göttingen: Vandenhoeck und Ruprecht, 1984. 44 pp.

Attempts to place Jungius's programme of work for the years 1622–38 in the context of the Scientific Revolution. On Joachim Jungius (Junge), see Hans Kangro's article in *DSB*, Vol. 7, pp. 193–96.

1630. Sivin, Nathan. "Why the Scientific Revolution Did Not Take Place in China—Or Didn't It?" *Transformation and Tradition in the Sciences. Essays in Honor of I. Bernard Cohen.* Everett Mendelsohn, ed. Cambridge: Cambridge Univ. Press, 1984, pp. 531–54. [Originally published in *Chinese Science* 5 (1982), 45–66]

Examines the "Scientific Revolution problem" (formulated by Joseph Needham—"Why did modern science, the mathematization of hypotheses about Nature, with all its implications for advanced technology, take its meteoric rise *only* in the West at the time of Galileo?") and asks what assumptions about the European tradition of science encourage us to take this problem more seriously than its intrinsic merits justify.

1631. Cohen, I. Bernard. *Revolution in Science.* Cambridge, Mass.: Belknap Press, 1985. 711 pp.

Examines critically some of the major revolutions in science from the sixteenth to the twentieth century. Pp. 1–47: Science and revolution; pp. 49–101: Historical perspectives on "revolution" and "revolution in science"; pp. 105–94: Scientific revolutionaries of the seventeenth century; pp. 389–404: The historians speak. Reviews: *New York Times Book Review* 21 Apr. 1985, p. 29; *Nature* 315 (1985), 433–34; *Annals of Science* 43 (1986), 564–66; *British Journal for the History of Science* 19 (1986), 340–42; *Journal of the History of Behavioural Sciences* 23 (1987) 244–47.

1632. Westfall, Richard S. "Science and Patronage: Galileo and the Telescope." *Isis* 76 (1985), 11–30.

On patronage as a seventeenth-century category, a profitable subject of study.

1633. Neher, André. *Jewish Thought and the Scientific Revolution of the Sixteenth Century: David Gans (1541–1613) and His Times.* The

Littman Library of Jewish Civilization. Oxford: Oxford Univ. Press, 1986. 285 pp. [Translated from the 1974 French ed.]

"Comprehensive study of the work of David Gans and his intellectual environment both in its Jewish aspect and with regard to the great spiritual revolutions of the sixteenth century (the discovery of America, the round-the-world voyages, the Renaissance, the Reformation, the astronomical discoveries from Copernicus to Galileo)." Reviewed in *TLS* 5 Jun. 1987, 603, and *Isis* 81 (1990), 105–7.

1634. Porter, Roy. "The Scientific Revolution: A Spoke in the Wheel?" *Revolution in History.* Edited by Roy Porter and Mikuláš Teich. Cambridge: Cambridge Univ. Press, 1986, pp. 290–316.

Urges that the standard notion of *The Scientific Revolution* ("initially the brain-child and shibboleth of a specific cluster of scholars emerging during the 1940s") be treated with reservation. "Its tendency to privilege scientific thought . . . rightly captures science's importance, yet conjures up a mystified, ahistorical myth of its role in making the modern world" (p. 310).

1635. Redondi, Pietro, ed. "Science: the Renaissance of a History. Proceedings of the International Conference Alexandre Koyré, Paris, Collège de France, 10–14 June 1986." *History and Technology* [Special issue] 4 (1–4), (1987), 1–582.

Thirty-four papers by former students and collaborators of Koyré, presented at a conference commemorating the twentieth anniversary of his death. Koyré is generally considered the founder of the history of science as a historical discipline. It was he who established the "concept of the Scientific Revolution of the seventeenth century as the turning point in the history of science between antiquity and the present age."

1636. Jacob, Margaret C. *The Cultural Meaning of the Scientific Revolution.* New York: Alfred A. Knopf, 1988. 275 pp.

Seeks to explain the historical process by which scientific knowledge became an integral part of Western culture during the seventeenth and eighteenth centuries. Pp. 10–135: The first prophets of the new sciences; The social meaning of Cartesianism; Science in the crucible of the English Revolution; The Newtonian enlightenment. Pp. 255–61: Bibliographical essay.

1637. Stayer, Marcia Sweet, ed. *Newton's Dream.* McGill: Queen's Quarterly, 1988. 135 pp.

Series of lectures delivered in Kingston, Ontario, commemorating the Tercentenary of the *Principia.* Of these, two lectures are relevant to our period: "Newton and the Scientfic Revolution" (Westfall, pp. 4–18) and "Isaac Newton, Explorer of the Real World" (A. P. French, pp. 50–77).

1638. Yoder, Joella G. *Unrolling Time. Christian Huygens and the Mathematization of Nature.* Cambridge: Cambridge Univ. Press, 1988. 238 pp.

Study of Huygens and the creation of the theory of evolutes. Pp. 180–233: Notes and bibliography. Review: *Annals of Science* 47 (1990), 102–3.

1639. Wendel, Günther, ed. *Naturwissenschaftliche Revolution im 17. Jahrhundert.* Hrsg. im Auftrage des Beirates für Wissenschaftsgeschichte beim Ministerium für Hoch-und Fachschulwesen der DDR. Beiträge zur Wissenschaftsgeschichte, 7. Berlin: VEB Deutscher Verlag der Wissenschaften, 1989. 276 pp.

Nineteen papers on different aspects of the Scientific Revolution read at a seminar held in Siebenlehn (23/26 June 1986). Topics include: concept and content of the Scientific Revolution; Descartes; astronomy and astrology; mathematics; matter, aether and force; chemical technology; shipbuilding and shipping; cartography; medicine.

1640. Grmek, Mirko D. *La première révolution biologique. Réflexions sur la physiologie et la médecine du XVIIe siècle.* Paris: Payot, 1990. 358 pp.

Study based partly on articles in journals and papers read at conferences. In three sections: quantitative experiment; mechanical explanation; inductive method. Pp. 317–47: Bibliography. Pp. 348–55: Name index. Review: *British Journal for the History of Science* 24 (1991), 255–56.

1641. Lindberg, David C., and Robert S. Westman, eds. *Reappraisals of the Scientific Revolution.* Cambridge: Cambridge Univ. Press, 1990. 557 pp.

Collection of essays which evolved out of a symposium held during the History of Science Society meeting in Los Angeles in December 1981. (1) "Conceptions of the Scientific Revolution, from Bacon to Butterfield: A Preliminary Sketch" (David C. Lindberg). (2) "Conceptions of Science in the Scientific Revolution" (Ernan McMullin). (3) "Metaphysics and the New Science" (Gary Hatfield). (4) "Proof, Poetics, and Patronage: Copernicus's Preface to *De revolutionibus*" (Robert S. Westman). (5) "A Reappraisal of the Role of the Universities in the Scientific Revolution" (John Gascoigne). (6) "Natural Magic, Hermeticism, and Occultism in Early Modern Science" (Brian P. Copenhaver). (7) "Natural History and the Emblematic World View" (William B. Ashworth). (8) "From the Secrets of Nature to Public Knowledge" (William Eamon). (9) "Chemistry in the Scientific Revolution: Problems of Language and Communication" (Jan V. Golinski). (10) "The New Philosophy and Medicine in Seventeenth-century England" (Harold J. Cook). (11) "Science and Heterodoxy: An Early Modern Problem Reconsidered" (Michael Hunter). (12) "Infinitesimals and Transcendent Relations: the Mathematics of Motion in the Late Seventeenth Century" (Michael S. Mahoney). (13) "The Case of Mechanics: One Revolution or Many?" (Alan Gabbey). Pp. 528–51: Index.

1641A. Harman, P. M., and Alan E. Shapiro, eds. *The Investigation of Difficult Things: Essays on Newton and the History of the Exact Sciences in Honour of D. T. Whiteside.* Cambridge: Cambridge Univ. Press, 1992. 531 pp.

Collection of twenty essays (in five sections) by Whiteside's colleagues, of which 1–14 are relevant to the years before 1700. Includes a select bibliography of his writings.

1642. Porter, Roy, and Mikuláš Teich, eds. *The Scientific Revolution in National Context.* Cambridge: Cambridge Univ. Press, 1992. 305 pp.

Ten essays by leading scholars bringing out the importance of specifically national contexts (Italy, France, Germany, Netherlands, Poland, Spain and Portugal, England, Bohemia, Sweden, Scotland) for understanding the Scientific Revolution. With an introduction (pp. 1–10) by the editors. Pp. 288–305: Index.

1643. Huff, Toby E. *The Rise of Early Modern Science: Islam, China, and the West.* Cambridge: Cambridge Univ. Press, 1993. 409 pp.

Study by a sociologist of the "long-standing question of why modern science arose only in the West and not in the civilizations of Islam and China, despite the fact that medieval Islam and China were more scientifically advanced." Pp. 365–85: Selected bibliography; pp. 387–407: Index.

1644. Cohen, H. Floris. *The Scientific Revolution. A Historiographic Inquiry.* Chicago: Univ. of Chicago Press, 1994. 662 pp.

A historiographical survey and an analysis of the concept of the Scientific Revolution. Includes a discussion of the non-emergence of early modern science outside Western Europe and of the reasons why the Scientific Revolution eluded China. Pp. 527–662: Notes, bibliography, and index. Review: *Nature* 372 (1994), 51–52.

1645. Field, J. V., and Frank A. J. L. James, eds. *Renaissance and Revolution: Humanists, Scholars, Craftsmen and Natural Philosophers in Early Modern Europe.* Cambridge: Cambridge Univ. Press, 1994. 309 pp.

"Collection of fifteen essays which open up new perspectives on some of the problems presently seen to be associated with the Scientific Revolution." With "Afterword: Retrospection on the Scientific Revolution" by A. R. Hall. Review: *TLS* 6 Jan. 1995, p. 10.

1646. Hunter, Michael, ed. *Robert Boyle Reconsidered.* Cambridge: Cambridge Univ Press, 1994. 231 pp.

Collection of twelve essays. With bibliography of writings on Boyle published since 1940 and index. Review: *TLS* 6 Jan. 1995, p. 10.

Author Index

à Beughem, C. 125
ABHB: annual bibliography of the history of the printed book 113
Ackermann, J. S. 1588
Ackermann, S. 321
Actas da IV Reunião internacional de História da Náutica 1123
Adams, F. D. 1019
Adams, H. M. 106
Adams, T. R. 1063
Adamson, J. W. 226
Adelmann, H. B. 1476
Agassi, J. 999
Aitchison, L. 994
Aiton, A. S. 58
Aiton, E. J. 811, 868, 885
Albion, R. G. 1057
Albuquerque, L. de 1124
Alden, J. 42
Alexander, M. V. C. 280
Alexander, P. 449
Allan, M. 1315
Allen, D. C. 695
Allen, D. E. 1267
Allen, P. 255, 256
Allison, A. F. 114
Almagià, R. 165, 1073, 1089
Alvarez Lopez, E. 1294
American Historical Association 9
American Library Association 1, 90
Amman, P. J. 968
Amsler, J. 1110
Anderson, A. H. 178

Anderson, F. H. 413
Anderson, R. C. 1158
Andrade-Lima, D. de 1326
Andreski, S. 660
Anglo, S. 648
Antonioli, R. 1556
Arber, A. 1344
Arber, E. 87
Arents, G. 1254
Ariew, R. 903
Ariotti, P. E. 431
Arnoldi, G. 217
Artelt, W. 1354
Artz, F. B. 288
Ashcraft, R. 181
Ashworth, W. B. 1641
Aston, D. 319
Aston, T. H. 276
Atkinson, G. 1049
Axtell, J. L. 264

Babini, J. 495
Babuscio, J. 79
Bachmann, A. R. 1370
Bagrow, L. 1067B, 1104
Baillie, G. H. 1196
Bainton, R. H. 511, 1435
Balay, R. 1
Baldwin, S. A. 1345
Ball, W. W. R. 244, 738
Ballard, T. 150
Balmer, H. 1024
Bangert, W. V. 324
Barbieri, F. 729
Barbosa, A. 338
Bargallo, M. 1023
Barlow, W. P. 38
Barnadas, J. M. 1191

Barnard, E. A. B. 167
Baron, M. 770
Bartels, J. 1020
Bassett, D. A. 1015
Bateman, C. 148
Bates, D. G. 1412
Battersby, W. J. 231
Bay, J. C. 124
Bayon, H. P. 1431
Bechler, Z. 559
Bedel, C. 502
Bedini, S. A. 1224
Beer, A. 878, 879
Beer, P. 878, 879
Bell, S. P. 31
Ben David, J. 323
Benedicenti, A. 1484
Benetti, F. Z. 1380
Benjamin, A. E. 1003
Bennett, H. S. 127, 132, 134
Bennett, J. A. 880, 881, 960, 1243
Benoit, P. 1186
Berkel, K. van 1627
Bernal, J. D. 1596
Bernard, F. 149
Bernard-Maitre, H. 719, 854
Bernard Quaritch Ltd. 213
Berry, W. T. 129
Besson, A. 145
Besterman, T. 5, 6, 125
Biagioli, M. 820
Bialas, V. 1026, 1029, 1036
Bibliografia italiana di storia della scienza 27A

Bibliografia italiana di storia delle matematiche 729

Bibliografia nautica portuguesa 1050

Bibliographia Cartesiana 396

Bibliographic index 7

Bibliographie d'Aristote 389

Bibliographie de l'histoire des universités françaises 219

Bibliographie internationale de l'histoire des universités 214

Bibliographie internationale de 1'humanisme 105

Bibliography of books . . . on or connected with historical studies on scientific instruments 1199

Bibliography of the history of technology 18, 1135

Bibliography of the philosophy of technology 1139

Biblioteca Casanatense 615

Bienkowska, B. 872

Bierens de Haan, D. 709, 733, 734

Bietenholz, P. G. 407

Bigmore, E. C. 86

Bignami Odier, J. 191

Bill, E. G. W. 277

Billings, P. 1058

Bird, D. T. 1382

Bishop, W. J. 168

Biswas, A. K. 1027

Black, R. 302

Blackwell, C. T. 404, 462

Blackwell, R. J. 439

Blake, J. B. 1355

Blay, M. 955

Blume, B. 969

Blumenberg, H. 453

Blunt, W. 1298

Boas, M. (*see also* Hall, M. Boas) 415, 506, 993, 1589

Bodleian Library 202

Boehm, L. 368

Boerhaave, H. 152

Bologa, V. L. 60

Bolton, H. C. 977

Bömer, A. 169

Boncompagni, B. 816A

Bortolotti, E. 755

Bos, E. J. M. 278, 549, 781

Bosanquet, E. F. 684, 685, 692

Bosch, K. 1491

Bottazzi, F. 575

Bourdier, F. 1261

Bourlière, F. 1256

Boutroux, 922

Bouvet, M. 1295

Bowen, J. 237

Boxer, C. R. 1080, 1100, 1311

Boyce, G. C. 14

Boyd, William 238

Boyer, C. B. 752, 759, 929, 930

Bradbury, S. 423, 1222

Bradner, L. 511

Braun, L. 1521

Braunmühl, A. von 741

Brazier, M. A. B. 1454

Brebner, J. B. 1079

Bretschneider, B. 1275

Bridson, G. D. R. 1270, 1270A

Briggs, F. 1602

British Library 39, 40

British Museum 39, 1099

British Museum Library 39

British Museum (Natural History) 1307

British Society for the History of Science 35

Britten, J. 1280

Broc, N. 1118

Brook, W. H. 34, 982

Brockliss, L. W. B. 291, 294

Brody, J. 98

Brooke, J. H. 353

Brooks, J. E. 1254

Brooks, R. A. 119

Brown, Harcourt 357, 544

Brown, J. 1234

Brown, L. A. 1088

Brown, T. M. 1451

Browne, E. J. 554

Brozek, J. 1498

Brucker, J. 408

Brugmans, H. L. 489

Brundell, B. 454

Brunet, P. 66, 1021

Bruni Celli, B. 1368

Brush, S. G. 1013

Bucknill, J. C. 1546

Buisson, F. B. 211

Bulletin cartésien 398

Bulletin signalétique 27

Bullough, V. L. 1605

Bunch, B. 84

Burckhardt, A. 1527

Burckhardt, F. 1201, 1202

Burke, J. G. 424, 561

Burmeister, K. H. 835, 1103, 1555

Burtt, E. A. 410

Bush, D. 1588

Butterfield, A. D. 1016

Butterfield, H. 1576

Büttner, M. 332

Butts, R. E. 437

Bylebyl, J. 1449

Bynum, W. F. 554, 1413

Cabeen, D. C. 98

Caillet, A. L. 606

Cajori, F. 748

Callmer, C. 191

Callot, E. 1300

Cambridge Bibliography of English Literature 111

Cambridge History of Renaissance Philosophy 408, 458

Camden, C. 693, 694

Cameron, H. C. 361

Campbell, L. B. 1157

Campbell, T. 1065

Canguilhem, G. 435

Cantor, G. N. 596, 958, 1003
Cantor, M. 740
Capocaccia, A. A. 1149
Capp, B. 689
Cappelletti, V. 484
Carl Zeiss-Stiftung 1244
Carmody, F. J. 686
Carozzi, A. V. 1012
Carpenter, K. J. 1420
Carpenter, N. C. 966
Carroll, P. T. 56
Carter, J. 107
Carter, R. B. 444
Carugo, A. 1624
Caspar, M. 836, 858
Castiglioni, A. 1426, 1528
Castillejo, D. 555
Caullery, M. 502
Cavazza, M. 895
Caverni, R. 484, 1204
CDI Ten Year Cumulation 28
Centore, F. F. 935
Cetto, A. M. 1440
Channell, D. F. 1144
Chaplin, A. 251
Chapman, S. 1020
Charbonnier, P. J. 923
Charnock, J. 1154
Chiarugi, A. 1302
Chilton, D. 1216
Choulant, L. 1424
Chrisman, M. U. 141
Christianson, G. E. 565
Christianson, J. R. 913
Christie, J. R. R. 1003
Cinq cents ans de bibliographic hippocratique 1367
Cipolla, C. M. 536, 1150, 1174
Clagett, M. 506, 799, 946
Clair, C. 133
Clark, A. J. 46
Clark, C. 486
Clark, G. 362
Clarke, A. G. 704
Clarke, D. M. 680
Clavelin, M. 943
Clay, R. S. 1208

Clercq, S. de 1127
Clulee, N. H. 649, 674, 774
Coats, A. M. 1318
Cochetti, M. 121, 597
Cochrane, E. 545
Cockle, M. J. D. 1130
Codd, L. E. 1330
Coffin, C. M. 491
Cohen, H. F. 278, 974, 1644
Cohen, I. B. 350, 352, 550, 866, 937, 952, 1613, 1619, 1631
Cole, F. J. 1432
Cole, G. W. 40A
Coleman, D. C. 1170
Collingwood, R. G. 412
Collison, R. 48
Colloque international de Tours 434
Compère, M. M. 293
Comprehensive Dissertation Index 28
Consiglio d'Europa 340
Contant, J. P. 989
Cook, C. 75, 77, 83
Cook, H. J. 1641
Coolidge, J. L. 753
Copeman, W. S. C. 1530
Copenhaver, B. P. 652, 1616, 1620, 1641
Coral, L. 115
Corsi, P. 21
Cortesão, A. 1082, 1089, 1092
Cosenza, M. E. 99
Cossali, P. 730
Costabel, P. 479
Costello, W. T. 257
Cotter, C. H. 1108
Coulter, H. L. 1406
Court, T. H. 1208
Coutinho, G. 1090
Cowan, J. M. 1291
Coyne, G. V. 568, 583, 896
Craig, H. 1137
Cranz, F. E. 97, 401
Crapulli, G. 425
Creasy, J. S. 1265

Creighton, C. 1511
Cressy, D. 271
Crestois, P. 1303
Creutzenfeld, S. H. de V. von 1457
Critical Bibliography of the History of Science 25
Crombie, A. C. 299, 423, 497, 498, 506, 787, 1603, 1624
Crone, G. R. 1109
Crone, H. G. T. 1066
Crosby, A. W. 1321
Crosland, M. P. 540
Crossley, D. 1195
Crowc, M. J. 612
Cultura, scienze e tecniche nella Venezia del cinquecento 578
Current Bibliography in the History of Technology 1134
Current Bibliography in the Philosophy of Technology 1140
Curry, P. 666, 704
Curtis, M. H. 258
Cushing, H. 1426
Cuvier, G. 1271
Cyclopedia of Education 224
Czartoryski, P. 193, 887

Dainville, F. de 287, 290, 502
Dannenfelit, E. H. 1317
Daremberg, C. 1387
Darmstaedter, L. 57
Datta, B. 61
Dauben, J. E. 729
Daumas, M. 360, 502, 1147, 1172, 1228, 1577
David, F. N. 762
David, J. Ben 323
Davies, G. 12
Day, A. 2
De Marco, F. 191
De Morais e Sousa, L. 1077
De Morgan, A. 707
De Moulin, D. 1468

De Piero, A. 1071
De Santillana, G. 506, 859
De Schepper, M. 122
De Silva, D. 1065A
De Ursis, S. 508
De Vecchi, B. 160
De Vigilis von
    Creutzenfeld, S. H.
    1457
De Virville, A. D. 1304
DeVorkin, D. H. 841
De Waard, C. 1232
Deacon, M. 1106, 1112
Deacon, R. 634
Deahl, J. 221
Dear, P. 455
Debrock, G. 587
Debus, A. G. 265, 632,
    677, 1000, 1004, 1402,
    1523, 1534, 1617
Defossez, L. 1213
Dehergne, J. 472
Dekker, E. 904, 912
Del Lungo, C. 484
Delambre, J. 843
Delaunay, P. 1284, 1312,
    1529
Delorme, J. 78
Delorme, S. 1610
Desmond, R. 1353
Deubner, F. 721
Deubner, H. 721
Devresse, R. 175
Dewhurst, D. W. 833
Dewhurst, K. 1551
Di Pietro, G. 182
Di Trocchio, B. 484
Diaz-Plaja, G. 467
Dickins, B. 173
Dickreiter, M. 972
Diffie, B. W. 1114
Dijksterhuis, E. J. 499,
    506, 771, 930, 1584
Dissertation Abstracts
    International 29
Dobbs, B. J. T. 647
Dobrzycki, J. 869
Doggett, R. 1239
Donahue, W. H. 892
Donatelli, L. 575
Donnelly, M. C. 1218

Doorman, G. 1162
Dostrovsky, S. 971
Drabkin, I. E. 934
Drake, S. 418, 859, 934,
    936, 947, 970
Dreyer, J. L. E. 845, 846
Dufferin, K. E. 352
Duffy, C. 1181
Dugas, R. 928
Duhem, P. 863
Dulieu, L. 1316, 1464,
    1489, 1532, 1535
Dumaître, P. 1467
Dumas, F. R. 633
Dunn, R. 705A
Dunn, R. M. 79
Durand, D. B. 626
Durbin, P. T. 551
Durkheim, E. 289
Durling, R. J. 1374, 1384,
    1425, 1540
Duval, M. 1333
Duveen, D. I. 980

Eade, J. C. 700
Eales, N. B. 1375
Eames, W. 88
Eamon, W. 1184,1641
Easlea, B. 656
Eccles, A. 1480
Edelman, N. 98
Edgar, I. I. 1501
Edgerton, S. Y. 825
Edwards, A. W. F. 816
Edwards, C. H. 797
Edwards, P. 405
Ehrman, A. 128
Eisenstein, E. 139, 142
Elia, A. d' 569
Elia, P. d' 1101
Elia, P. M. d' 508
Ellenberger, F. 1040
Elmer, P. 204
Emboden, W. A. 1347
Emden, A. B. 214
Equipe Descartes 398
Erman, W. 212
Espinasse, M. 1582
Estienne, H. 123
Evans, J. 621
Evans, R. J. W. 367

Farrell, A. P. 229
Farrington, B. 414
Fasbender, H. 1473
Fastlicht, S. 1328
Fattori, M. 450
Favaro, A. 153, 484
Febvre, L. 137
Feingold, M. 278, 823,
    1628
Feisenberger, H. A. 189
Feldhaus, F. M. 1145
Ferguson, E. S. 18, 1135
Ferguson, J. 978
Ferguson, W. K. 511
Ferrone, V. 347
Feuer, L. S. 1592
Field, J. V. 701, 704, 909,
    1645
Fierz, M. 1559
Figala, K. 650
Fincham, J. 1155
Findlen, P. 302A
Finocchiaro, M. A. 451,
    580
Firpo, M. 347
Fisch, M. H. 359
Fischer, H. 130, 524
Fisher, N. W. 34
Fladt, K. 763
Fletcher, J. M. 218, 220,
    221
Florkin, M. 1487
Folkerts, M. 782, 788, 914
Fontoura da Costa, A.
    1050, 1084
Forbes, E. G. 186, 830,
    882
Forbes, R. J. 512, 1164
Forbes, T. R. 1400
Foster, M. 1430
Fournier, P. 1289
Franci, R. 805, 812
Francis, W. W. 1376
Frank, Mortimer 1526
Frank, R. G. 269, 548,
    1450
Freeman, E. J. 1359
Freeman, R. B. 1264
Freeman-Grenville,
    G. S. P. 73
French, A. P. 1637

French, P. J. 641
French, R. 282, 1422, 1538
French, R. K. 1418, 1448
Fresquet Febrer, J. L. 1494
Freudenthal, G. 1006
Frumkin, M. 1163, 1166
Fuchs, G. F. C. 976
Fulton, J. F. 1426, 1433
Funkenstein, A. 348

Gabbey, A. 596, 1641
Gabriel, A. B. 215
Gabrieli, G. 356
Gallois, L. 1069
Galluzzi, P. 369, 577, 949, 1193
Gamba, E. 817
Gamble, R. C. 354A
Garcia Ballester, L. 17, 1407
Gardner, F. L. 604, 605, 683
Garin, E. 699, 1566
Garrison, F. H. 1369, 1393
Gascoigne, J. 823, 1641
Gascoigne, R. M. 82
Gaskell, P. 197
Gautier, L. 1388
George, W. 1329, 1349
Gerhardt, C. I. 790
Gerl, A. 914
Gerlo, A. 103, 112, 122
Geymonat, L. 418
Ghiretti, F. 575
Giard, L. 456, 462
Gibson, R. W. 392
Gieryn, T. F. 350
Gilbert, N. W. 417
Gille, B. 1152, 1167, 1172
Gillispie, C. C. 54, 470
Gingerich, O. 870, 893, 900, 910
Giovannozzi, G. 484
Giraud, L. 1278
Girtanner, C. 1506
Giusti, E. 809
Gjertsen, D. 576
Glaisher, J. W. L. 746
Gliozzi, M. 65
Gnudi, M. T. 1462

Godbolt, S. 1359
Godwin, J. 654
Goldschmidt, E. P. 163
Goldsmith, V. F. 114
Goldstein, T. 1030
Goldstine, H. H. 789
Golinski, J V. 1003, 1005, 1641
Goodison, N. 1231
Goodman, A. 592
Goodman, D. 600
Goodman, D. C. 584
Goodman, H. 1510
Goodman, M. M. 44
Gould, H. 531
Grafe, I. 1192
Grafton, A. 592, 823, 875
Grande, S. 1074
Granjel, L. S. 1410
Grant, E. 426, 438, 897, 917A, 1590, 1591, 1623
Granzow, U. 1125
Grässe, J. G. T. 602
Grassi, G. 842
Graves, F. P. 228
Gray, G. J. 468
Greaves, R. L. 263
Greene, E. L. 1336
Grendler, M. 198
Grendler, P. F. 138, 302
Grewe, K. 1011
Grierson, P. 194
Grmek, M. D. 1463, 1487, 1640
Grosse, H. 742
Grössing, H. 808
Growoll, A. 88
Grun, B. 74
Guédes, M. 1261
Guenée, S. 219
Guerlac, F. 385, 1614
Guerra, B. 1486, 1512
Guerra, P. 1487
Guerrini, A. 1269
Guerrini, M. 482
Guilmartin, J. B. 1180
Guiness Book of Records 85
Gunawardene, D. C. 1320
Gunn, M. 1330
Gunther, R. T. 254, 1205, 1209, 1281, 1283

Gunther, R. W. D. 1210
Günther, S. 295, 1068
Gurlt, E. 1461
Gusdorf, G. 1604
Guye, S. 1226
Guyot, L. 1089
Gwynn, R. D. 344

Hacking, I. 785
Haeser, H. 1507
Hageln, O. 1472
Hahn, A. 1531
Halbronn, J. 704
Hald, A. 824
Hale, J. R. 53, 1175
Halkett, S. 117
Hall, A. R. 318, 506, 581, 585, 593, 601, 801, 876, 926, 930, 1575, 1626
Hall, G. 98
Hall, M. Boas 377, 382, 385, 439, 518, 585, 1562, 1566 (*see also* Boas, M.)
Hallam, E. 463
Haller, A. von 1248, 1360, 1423, 1456
Halleux, R. 1009
Hallyn, F. 594
Hambly, M. 1245
Hammond, E. A. 1543
Hamy, E. T. 285, 1274
Hanen, M P. 658
Hannaway, O. 998
Hansen, A. 62
Hansen, B. 653
Hardy, J. E. 1015
Harig, G. 556
Harman, P. M. 1626A, 1641A
Harré, R. 1597
Harrison, E. 905
Harrison, J. 176, 195
Hart, C. 1179, 1187
Hartley, H. 376
Harvey, A. P. 1015
Harvey, J. M. 2
Hassinger, E. 216
Hatfield, G. 1641
Hauber, E. D. 602
Hawks, E. 1286

Hay, C. 818
Hay, D. 243, 592
Heawood, B. 1075
Heesakkers, C. L. 122
Heilbron, J. L. 890, 950, 956
Heinekamp, A. 402, 815
Heischkel, E. 1354
Hellemans, A. 84
Heller, M. 568, 583
Hellman, A. M. 1470
Hellman, C. D. 852, 858, 862
Hellman, G. 1007, 1017, 1206
Henderson, J. 712
Heniger, J. 1346, 1351
Heninger, S. K. 432, 469, 886, 1025
Henrey, B. 1262
Henry, J. 404, 462, 585, 596, 669
Herivel, J. 519
Herrlinger, R. 1398
Hess, H. J. 199
Hesse, M. B. 931
Hessel, A. 170
Hessen, B. 533
Higham, R. 1138
Hill, C. 315, 319, 320, 329
Hill, H. O. 1217
Hillard, D. 45
Hirschberg, J. 1389
Hocker, S. H. 1268
Hodge, M. J. S. 596
Hodgen, M. T. 514
Hodgkiss, A. G. 1064
Hoefer, F. 844, 985
Hoeniger, F. D. 1561
Hoeven, J. van der 63
Hofmann, J. E. 758, 765, 768, 775, 776, 777, 783, 790
Hogart, R. C. 616
Holder, L. 1229
Hollander, B. 1392
Hollister-Short, G. J. 1188
Holmes, G. 337
Holmyard, E. J. 1146
Home, R. W. 920

Hooke, J. 151
Hooykaas, R. 325, 507, 1578
Hoppen, K. T. 364, 384
Horn, E. 212
Horn, J. M. 33
Horst, A. J. van der 1350
Horst, G. C. 602
Hoskin, M. 911
Hoskin, M. A. 896
Houghton, W. E. 492, 493
Houston, R. A. 242
Houzeau, J. C. 687, 829, 833
Howard, R. C. 1331
Howard, R. 1001, 1337
Howland, A. C. 625
Howse, D. 1058, 1059, 1240
Howson, G. 274
Huard, P. 1105, 1463
Huddy, J. 1119
Huff, T. E. 1643
Hufman, W. H. 675
Hughes, B. B. 183
Hulme, E. W. 1156
Hunger, F. W. T. 1285
Hunt, F. V. 948
Hunter, M. 272, 342, 386, 387, 388, 588, 704, 1567, 1641
Hunter, R. A. 1503
Hutchinson, E. 661, 704
Hutin, S. 638
Hutton, C. 731
Hutton, S. 404, 462
Hyamson, A. M. 51

Impey, O. 570
Irsay, S. d' 286
Isis Current Bibliography 26
Isis Guide to the History of Science 56
Ito, Y. 1041
IUHPS—Scientific Instruments Commission 1199

Jackson, B. D. 1250

Jackson, L. L. 744
Jacob, J. R. 334, 341, 383
Jacob, M. C. 334, 335, 341, 1636
Jacobi, J. 1522
Jacobs, P. M. 32
Jacques, D. 1350
Jacquot, J. 1305
Jaki, S. L. 863, 888
James, F. A. J. L. 1645
Jaramillo-Arango, J. 1297
Jarman, T. L. 259
Jayawardene, S. A. 20, 116, 728, 779
Jayne, S. 96, 171
Jeanselme, E. 1509
Jensen, M. 480, 1178
Jilek, F. 1151
Jilkova, J. 1151
Johnson, F. R. 171, 253, 466, 750, 832, 849, 855
Johnson, J. K. 481
Jolly, C. 207
Jones, R. F. 1585
Jordan, G. 1143
Joy, L. S. 457
Julia, D. 293

Kail, A. C. 1505
Kajdanski, E. 1492
Kangro, H. 997
Kargon, R. H. 421
Karpinski, L. C. 58, 717, 718, 1131
Karrow, R. W. 1067B
Karsten, M. C. 1300A, 1310
Kaunzner, W. 914
Kearney, H. 266, 1606
Kearney, H. F. 320, 1594, 1595
Keele, K. D. 1453
Keeler, M. F. 12
Keevil, J. J. 1396
Keil, G. 1416
Kelly, J. T. 915
Kenney, C. E. 1085
Ker, N. R. 276
Kernchen, D. 3
Kernchen, H. J. 3

Kessler, E. 408, 456, 458, 601B
Keynes, G. 145, 1263, 1429, 1437
Kibre, P. 162
Kies, C. H. 617
Kiessling, N. 208
King, E. J. 238
King, H. C. 1214
Kittelson, J. M. 221
Klaaren, E. M. 336
Klebs, A. C. 92
Klein, J. 769
Klemm, F. 1171, 1176
Kline, M. 778
Klossowski de Rola, S. 676
Knappich, W. 705
Knobel, E. B. 828
Knobloch, E. 785
Knott, C. G. 745
Kocher, P. H. 310
Koeman, C. 1055, 1066
Koestler, A. 860
Kohl, B. G. 120
Kokomoor, F. W. 714, 749
Kolb, R. 190
Körber, H. G. 1220
Koreny, F. 1343
Kors, A. C. 646
Koyré, A. 502, 520, 866, 927, 1572, 1579, 1586, 1600
Krafft, F. 552, 873, 1564
Kranzberg, M. 1148
Kraus, G. 1273
Kravath, F. F. 1037
Kraye, J. 408
Kren, C. 478, 619
Kristeller, P. O. 97, 102, 104, 391, 419
Krivatsy, P. 1385
Krogt, P. van der 1127
Kruger, M. S. 1594
Kubrin, D. 659
Kuehl, W. B. 30
Kuhn, T. S. 1580
Kuhner, D. 1008
Kunitzsch, B. 901

La Charité, R. C. 119
La Roncière, C. de 1083
La Torre, F. 1475
Labowsky, L. 131, 196
Lach, D. F. 517, 595
Lacoarret, M. 720
Lacroix, H. de 1133
Ladendorf, H. 477
Laehr, H. 1496
Laeven, A. H. 595A
Laing, J. 117
Laird, W. R. 961
Laissus, Y. 1261, 1319
Lakatos, I. 1601
Lalande, J. 827
Lancaster, A. 687, 833
Landis, D. C. 42
Landsberg, H. E. 1013
Landwehr, J. 1067A
Lane, F. C. 1176A
Langer, W. L. 72
Larkey, S. V. 253, 466, 849
Laslett, P. 176
Laudan, L. 422
Laurie, S. S. 225
Lawes, J. 55
Lawler, J. 157
Lawrence, G. H. M. 1258
Lawson, J. 262
Le Fanu, W. R. 1549
Lea, H. C. 625
Lebrun, B. 292
Leedham-Green, E. S. 205
Lefranc, A. 284
Legée, G. 1261
Legré, L. 1277
Leibrock-Plehn, L. 1493
Leicester, H. M. 990, 1517
Leifth-Ross, P. 1339
Lennep, J. van 668
Lenoble, R. 411, 502, 1022, 1581
Levack, B. P. 672
Levine, M. 11
Lewis, C. 951
Lewis, G. 276
Leyh, G. 169
Liddell, J. R. 164
Lindberg, D. C. 349, 937, 945, 958, 962, 1641

Lindeboom, G. A. 1366, 1404, 1545, 1552, 1557
Linden, S. J. 984
Linet, J. 45
List, M. 838, 839
Livingstone, D. N. 1129
Locke, A. W. 965
Lohne, J. A. 798
Lohr, C. H. 403, 456, 458
Long, P. O. 1189
Lonie, I. M. 1418
López, R. S. 511
López Piñero, J. M. 17, 547, 1126, 1494
López, Terrada, M. L. 1494
Lowry, M. 140
Lunsingh Scheurleer, T. H. 303
Luyendijk Elshout, A. M. 1446

Macalpine, I. 1503
Maccagni, C. 525
Mach, E. 921
Mackay, A. 592
Mackensen, L. von 1235
Mackey, R. 1139
Macomber, H. P. 856
Macphail, I. 609
Madden, E. H. 1583
Maddison, F. R. 1197, 1227, 1536
Maeyama, Y. 877
Maffioli, C. S. 589
Mahoney, M. S. 780, 823, 1641
Maier, A. 932, 1571
Maks, C. S. 1159
Malclès, L. N. 2, 126
Malloch, A. 1376
Maloney, G. 1367
Mandrou, R. 338
Manfrè, G. 4
Manley, D. 85
Manno, A. 578
Manuel, F. 528
Manuel, F. E. 330, 513
Marañón, G. 467
Marcan, P. C. 474
Marcorini, E. 586

Marcus, J. 1334
Marinoni, A. 482, 579, 807
Martins, H. J. 137, 144
Martin, R. 681
Martinet, M. 479
Mason, S. F. 1574
Mathias, P. 326
Maule, A. F. 1326
Maurice, K. 1236
May, J. O'Hara 1408
May, K. O. 724
May, W. E. 1229
Mayer, A. 69
Mayer, O. 1236
Maylender, M. 356
McColley, G. 850
McGregor, A. 570
McGuire, J. E. 523, 651, 957, 1616, 1618
McHenry, L. C. 1504
McKie, D. 376, 1573
McLachlan, H. 252
McLean, A. 136
McMullin, E. 439, 526, 908, 1641
McNally, P. F. 143
McNeil I. 1153
McVaugh, M. 1540
Meadows, A. J. 864
Mees, J. 1072
Meinel, C. 459, 670, 1629
Meissner, F. L. 1361
Meli, D. B. 601A
Menato, M. 597
Merchant, C. 1621
Merkel, I. 677
Merton, R. K. 308, 350, 352
Metraux, A. 1089
Mett, R. 914
Metzger, H. 986, 988
Meyer, K. 873
Meyer, R. W. 416
Meyer-Steineg, T. 1397
Miall, L. C. 1279
Michel, H. 1223, 1226
Michel, P. H. 874
Middleton, W. E. K., 365, 1212, 1219, 1221
Midelfort, H. C. E. 608

Mieli, A. 484, 495
Mikami, Y. 64
Milanesi, M. 1121
Milkau, F. 169
Mitcham, C. 1139
Mitchell, A. C. 1018
Mollat, M. 1096
Molloy, P. M. 1014
Monroe, P. 224
Monseigny, A. M. 1319
Montebelli, V. 817
Monter, E. W. 636, 642
Monti, M. T. 1383
Moore, J. 706
Morello, N. 577
Morgan, P. 179, 200
Morison, S. E. 1111
Morley, H. 483
Morton, A. G. 1332
Morton, L. T. 1359, 1369
Mottelay, P. F. 918
Mugnai Carrara, D. 210A
Muir, P. H. 107
Mullay, M. 2
Mullens, W. H. 1251
Müller, K. 402
Müller-Jahncke, W. D. 702
Multhauf, L. S. 1190
Multhauf, R. P. 983, 996
Munby, A. N. L. 115
Münchow, W. 1415
Munk, W. 1541
Murdoch, Tessa 345
Musgrave, A. 1601
Myers, J. N. L. 172

Nagy, D. E. 1421
Nauert, C. G. 420
Naux, C. 764
Navire et l'économie maritime 1096
Needham, J. 230, 510, 861
Nef, J. U. 1173
Negri, L. 577
Neher, A. 1633
Neu, J. 23, 26, 981
Neugebauer, O. 898
New Cambridge Bibliography of English Literature 111

Nicholl, C. 657
Nickson, M. A. E. 209
Nicolson, M. H. 490, 503, 509, 522
Nobis, H. M. 851
Nordenskiöld, A. B. 1046
Nordenskiold, E. 1287
North, J. D. 590
Notestein, W. 620
Numbers, R. L. 349
Nutton, V. 1413, 1452, 1537, 1540

Oakshott, W. 180
Oates, J. C. T. 172
Oberman, H. A. 1611
O'Brien, P. J. 1322
Ochs, K. H. 571
O'Day, R. 275
Olby, R. C. 596
Oldroyd, D. R. 430, 1033
Olivieri, B. 445
Olschki, L. 485
O'Malley, C. D. 236, 530, 1364, 1427, 1436, 1533
Omont, H. 154
O'Neill, Y. V. 1479
Orchard, T. N. 847
Orme, N. 270, 279
Ornstein, M. 355
Ortroy, F. van 1048
Osler, M. J. 346, 658
Osler, W. 208, 1376

Pacey, A. 1141
Pachella, R. 610
Paetow, L. J. 14
Pagano, S. M. 566
Pagel, J. L. 1390
Pagel, W. 1395, 1417, 1439, 1443 1518, 1558, 1560
Paget-Thomlinson, E. W. 1217
Palisca, C. V. 967, 1588
Palm, L. C. 278, 589, 1335
Palmer, R. 301
Palmer, W. M. 1547
Palter, R. 532, 562
Panowsky, E. 511

Pantzer, K. 109
Papp, D. 495
Paredi, A. 201
Pares, J. 1246
Parias, L. H. 292, 1110
Parker, D. 697
Parker, I. 250
Parkinson, C. L. 80
Parkinson, E. 1377
Parks, G. B. 1094
Parr, J. 696
Parry, J. H. 1107
Parry, R. B. 1067
Parshall, K. H. 819
Parsons, W. B. 1161
Partington, J. R. 995
Pas, P. W. van der 1324
Pascoe, L. C. 76
Pastor, L. 305
Patai, R. 682A
Paulsen, F. 296, 297
Paxton, J. 81
Pazzini, A. 1362, 1458
Pearson, E. S. 794
Pearson, K. 794
Peddie, R. A. 8
Pedersen, O. 896
Petersen, T. M. 1326
Pedretti, C. 482, 1192
Pelling, M. 1356, 1536
Pelseneer, J. 59, 314
Pennington, D. 339
Penrese, B. 1091
Pepe, L. 729
Pepper, R. 246
Percopo, E. 690
Perez-Ramos, A. 460
Perkins, C. R. 1067
Peset Reig, M. 17
Peters, E. 646
Peters, H. J. M. W. 1066
Peters, Mrs. John P. 1426
Petit, G. 1309
Petzholdt, J. 5
Peumery, J. J. 1447
Philip, I. 202
Phillips, L. B. 50
Picatoste y Rodriguez, F. 464
Picolet, G. 906
Picutti, E. 820A

Pighetti, C. 468
Pinto, J. 1241
Pirro, A. 964
Pitt, J. C. 437
Platen, M. von 191
Poletti, G. B. 1459
Pollard, A. W. 109
Pollard, G. 128
Poni, C. 369
Poole, H. E. 129
Popkin, R. H. 441, 1601
Porter, R. 554, 1010, 1634, 1642
Poschmann, B. 871
Posthumus Meyjes, G. H. M. 219
Poulle, E. 1237
Poupard, P. 451, 580
Powell, K. 75
Power, D'A. 1469
Powicke, F. M. 214
Premuda, L. 1438
Price, D. J. de S. 506
Prieto, C. 1031
Prior, M. E. 623
Pritchard, A. 614
Pritzel, G. A. 1249
Proksch, J. K. 1506
Pumfrey, S. 599
Pursell, C. W. 1148
Purver, M. 378
Puschmann, T. 1524
Putscher, M. 1403

Qaisar, A. J. 1182
Quéniart, J. 292
Quetel, C. 1513

Rabb, T. K. 317
Rahn, K. 1326
Raimondi, E. 368
Raistrick, A. 309
Rambaldi, E. I. 821
Randall, J. H. 391, 1587
Randier, J. 1238
Randles, W. G. L. 1622
Rashdall, H. 214
Rattansi, P. M. 380, 523
Rauffner, J. A, 867
Rauschert, S. 1320
Rava, B. 356

Raven, C. E. 1296, 1299
Raven, C. F. 311
Ravetz, J. R. 596
Read, C. 10
Read, J. 992
Redgrave, J. R. 109
Redondi, B. 343, 351, 1635
Redwood, J. 546
Reeds, K. M. 1327, 1352
Rei, D. 1565
Reich, K. 914
Reti, L. 184, 1553
Rey Pastor, J. 747
Reynolds, T. S. 1185
Rhodes, D. E. 187, 1340
Rhys, H. H. 1588
Riccardi, P. 708, 710
Ricci, J. V. 1477, 1478
Ricci, V. 1035
Richards, J. B. 1002
Richardson, E. C. 90
Rider, P. R. 109
Rider, R. E. 727
Riedl, J. O. 390
Righini, G. 889
Righini Bonelli, M. L. 182, 370, 1612
Riley, L. W. 394
Riondato, E. 572
Rioux, G. 233
Ristow, W. W. 1060
Ritter, F. 739
Ritz, W. 1288
Rivinus, A. 1370
Rizzo, T. 1008
Robbins, R. H. 607, 612
Roberts, J. 682
Robinson, A. H. 1128
Rochot, B. 502
Roddis, L. E. 1482
Roero, C. S. 822
Roger, J. 535, 1032
Rogers, G.A.J. 422
Rohde, A. 1207
Rohde, E. S. 1252, 1290
Roller, D. H. D. 44
Rome, A. 715
Ronchi, V. 937, 944
Roos, C. 1355

Rose, P. L. 188, 772, 786, 791, 792, 938
Rosen, E. 835, 837, 851, 902
Rosenberg, B. M. 871
Rossi, P. 429, 635, 1563, 1607
Rossi, P. L. 599
Rostenberg, L. 177, 210
Rothschuh, K. E. 1434, 1441
Rouleau, F. 510
Rowbottom, M. E. 316
Rowley, G. 487
Rowse, A. L. 542, 639, 643
Ruderman, D. B. 678
Ruestow, E. G. 941
Ruge, S. 1044
Rusk, R. R. 235
Russell, K. F. 1428
Russell, P. E. 1122
Russo, A. 575, 760, 1136
Russo, F. 19, 312, 1609
Rytz, W. 1288

Sabra, A. I. 933
Sachs, J. von 1277
Sachse, W. L. 13
Saint-Lager, J. B. 1272
Sallander, X. 1371
Samion-Contet, J. 1531
Sanabria, S. L. 1183
Sanchez Canton, F. J. 166
Sanchez Tellez, M. C. 1490
Sanders, P. M. 810
Sanderson, M. 1059
Sanz, C. 1053
Sapori, G. 1471
Sargent, S. D. 932
Sarton, G. 16, 26, 93, 487, 501, 505, 511
Sasaki, C. 813
Savage, F. G. 1282
Savelli, R. 1378
Savoie, R. 1367
Sayle, C. 158
Scala, G. W. 381
Scaliger, J. J. 192

Scarborough, J. 1348, 1442
Scarre, G. 673
Schaffer, S. 588, 704, 919, 959
Schatzbergs, W. 481
Schelenz, A. 1481
Scherz, G. 1028
Scheurer, P. B. 587
Schevill, F. 487
Schiebinger, L. 591
Schilder, G. 1127
Schlicke, P. 2
Schlosser, L. 573
Schmeidler, F. 851, 914
Schmid, A. 1255
Schmid, K. A. 222
Schmitt, C. B. 239, 240, 241, 300, 401, 408, 427, 442, 446, 458, 476, 773, 1618
Schmitz, E. H. 953
Schmitz, R. 298, 552, 1416
Schneer, C. J. 954
Schneider, G. 1194
Schneider, I. 806
Schneider, W. 1488, 1519
Schofield, C. J. 894
Schreiber, W. L. 1253
Schuler, R. M. 613
Schüling, H. 101, 395, 428, 1497
Schulte, E. 1062
Schuster, J. A. 596
Schwab, M. 389
Scott, J. F. 757
Scouloudi, I. 1483
Scoville, W. C. 1165
Scriba, C. J. 766, 790
Sebba, G. 396
Secret, F. 630
Sédillot, L. A. 283
Selin, H. 482A
Sergescu, P. 496, 756
Serrai, A. 121, 597
Shapin, S. 350, 581, 959
Shapiro, A. E. 576, 823, 949, 1641A
Shapiro, B. 354, 548
Shapiro, B. J. 268, 322, 447

Shea, W. R. 143, 534, 1612
Sheynin, O. B. 784, 793
Shirley, J. W. 537, 543, 557, 563
Shirley, R. W. 1061
Simon, G. 698
Simon, I. 823
Simon, J. 261
Simpkins, D. M. 723
Simpson, R. R. 1550
Singer, C. 1146, 1508
Singer, D. W. 853
Singleton, C. S. 527
Sinisgalli, R. 795
Siraisi, N. G. 1539, 1540
Sivin, N. 1630
Skelton, R. A. 1097
Skinner, Q. 408
Slaughter, M. M. 560
Slawinski, M. 599
Sloan, P. R. 436
Smeur, A. J. E. M. 722
Smiles, S. 304
Smit, P. 1260, 1323, 1365
Smith, A. G. R. 327
Smith, C. S. 424, 1008, 1177
Smith, D. E. 711, 716, 816A
Smith, B. 1391
Smith, P. 488
Snelders, H.A.M. 278, 564, 1335
Soppelsa, M. 433
Sosef, M. S. M. 1351
Spargo, J. W. 91
Sparn, W. 456, 458
Spaulding, T. M. 1131
Stadter, P. A. 185
Stahl, W. H. 1051
Stannard, J. 1487
Stanton, M. 1426
Stapleton, D. H. 1142
Stark, E. 216
Starnes, De W. T. 95
Stayer, M. S. 1637
Stearn, W. T. 1259, 1308, 1314
Stearns, R. P. 1301
Steck, M. 726
Steinberg, S. H. 81

Steinschneider, M. 89
Stelling-Michaud, S. 217
Stephanides, M. 70
Stephenson, B. 907
Stevens, H. N. 1047
Stevenson, D. 679
Stevenson, J. 83
Sticker, B. 873
Still, G. B. 1394
Stillman, J. M. 987
Stillwell, M. B. 108, 1257
Stimson, A. 1247
Stimson, D. 373, 374, 848
Stone, L. 260
Stones, G. B. 409
Stornaiolo, C. 156
Strand, K. Aa 878
Streseman, E. 1325
Strickland, S. W. 352
Stringer, G. G. 471
Strömgren, H. L. 1460
Stroup, A. 371
Struik, D. J. 558, 1341
Sudhoff, K. 1397, 1508, 1514, 1515, 1516, 1548
Suster, G. 671
Svendsen, K. 504
Swann, H. K. 1251
Swerdlow, N. M. 875, 883, 898
Swetz, F. J. 816A
Syfret, R. H. 375
Sylla, D. 1611
Symons, J. 1358

Tabarroni, G. 484
Talbot, C. H. 1543
Tamny, M. 596, 957
Tanner, R. C. H. 802, 803
Tatham, A. F. 1064
Taton, R. 515, 857, 911
Taylor, A. 38, 94
Taylor, E. G. R. 1078, 1081, 1093, 1095
Taylor, F. S. 1211
Teall, J. L. 628
Teich, M. 1634, 1642
Teigen, P. M. 143
Temkin, O. 1445
Théodorides, J. 1261, 1309

Thiel, B. 705
Thimm, W. 871
Thomas, K. 339, 640, 1338
Thomas-Stanford, C. 713
Thompson, J. W. 123, 487
Thompson, L. S. 203
Thompson, S. J. 1042
Thompson, S. P. 1076
Thoren, V. E. 913
Thorndike, L. 622
Thornton, J. L. 135
Thrower, N. J. W. 1113
Thulesius, O. 1495
Tillyard, E. M. W. 494
Timpanaro, S. 484
Tobey, J. L. 399
Tocanne, B. 440
Todhunter, I. 732
Tomás, J. P. 1494
Tonelli, G. 49, 397
Tooley, R. V. 1115
Toomey, A. F. 6
Toti Rigatelli, L. 805
Totok, W. 3, 400
Toulmin, S. 1588
Transue, P. J. 221
Treutlein, P. 735, 736
Trevor-Roper, H. 575, 664, 1520
Tricot-Royer, J. 159
Troncarelli, B. 667
Tully, R. I. J. 135
Turner, A. 1242
Turner, G. L'E. 423, 1198, 1200, 1244
Tyacke, N. 339
Tyacke, S. 1116, 1119, 1120

Ueberweg, F. 400
Ullman, B. L. 185
Ultee, M. 209
Underwood, E. A. 855
Unger, F. 737
Urdgang, G. 991, 1485
Usher, A. P. 1168

Valverde, J. L. 1490
Van Egmond, W. 725
Van Helden, A. 370, 899, 1230, 1232

Van Kley, E. J. 517, 595
Varet, G. 393
Veendorp, H. 1293
Veltman, K. 1117
Vénard, M. 292
Vera, B. 67
Verginelli, V. 618
Vervliet, H. D. L. 112, 122
Vetter, Q. 68
Vickers, B. 582, 665
Victor, J. M. 796
Villoslada, R. G. 500
Viney, W. 1500
Visser, R. P. W. 278, 549
Vorsterman van Oijen, G. A. 1203
Vorstius, D. 161

Waerden, B. L. van der 814
Waite, R. A. 481
Walderman, W. 870
Walford, A. J. 2
Walker, D. P. 627, 973, 975
Wall, C. 361
Wallace, W. A. 443, 567, 963
Wallace, W. J. 1034
Wallis, H. 1128
Wallis, P. 475
Wallis, R. 475
Wangenstein, O. H. 1465
Wangenstein, S. D. 1465
Ward, D. C. 1012
Warner, D. J. 840
Waters, D. W. 1098
Watson, A. G. 682
Watson, F. 223, 245, 246, 247, 248, 249
Watson, G. 111
Watson, R. I. 1499
Wauwermans, H. E. 1070
Wear, A. 596, 1418, 1422, 1452
Webster, C. 267, 273, 331, 333, 363, 366, 379, 663, 1357, 1405, 1409, 1536
Webster, J. P. 1462
Weil, A. 780

Weimanns K. H. 1516
Weindler, F. 1474
Weindling, P. 21
Weiss, B. 24
Weiss, R. 232
Weitzel, R. 3
Wellisch, H. 473
Wells, E. B. 118
Wendel, G. 1639
Wertheimer, M. 1500
Wertheimer, M. L. 1500
West, J. F. 1598
Westcott, W. W. 604,
    605, 683
Westfall, R. S. 313, 351,
    553, 658, 939, 1568,
    1608, 1632, 1637
Westman, R. S. 651, 884,
    891, 910, 1616, 1618,
    1641
Weyant, R. G. 658
Wheatley, H. B. 157
White, A. D. 306
White, J. J. 1270
Whitehead, A. N. 1569
Whiteside, D. T. 543, 576,
    598, 761, 767, 916
Whitrow, M. 22
Whittaker, E. T. 925
Whitteridge, G. 1444
Wickersheimer, C. A. E.
    1525
Wickersheimer, E. 1487

Wiener, P. P. 406
Wightman, W. P. D. 100,
    328, 516, 1401
Wilkinson, R. S. 174, 629
Willey, B. 307
Williams, E. N. 52
Williams, L. 1169
Williams, M. I. 37
Williams, N. 71
Williams, T. I. 1146
Willmoth, F. 1039, 806
Wilson, C. 911
Wilson, C. A. 865   *See
Wilson, D. 538   below*
Wilson, D. K. 281
Wilson, L. 1455
Winder, M. 688
Wing, D. G. 110
Winius, G. D. 1114
Winsor, J. 1043
Wisan, W. 539
Witten, L. C. 610
Wolf, A. 1573
Wolf-Heidegger, G. 1440
Wollgast, S. 461
Wong, Ming 1105
Wood, D. N. 1015
Woods, C. S. 1512
Wood, P. B. 387, 959
Woodward, W. H. 227
Woolf, H. 385
Wormald, F. 172
Wright, C. E. 172

Wright, C. T. H. 41
Wright, R. 1292
Wroth, L. C. 1086
Wroughton, J. 77
Wunderlich, H. 1233
Wussing, H. 914
Wyatt, T. 1169
Wykes, A. 1554
Wyman, C. W. H. 86

Yale University Library
    609, 610
Yates, F. A. 234, 358, 631,
    637, 645, 655
Yeldhan, F. A. 751
Yeomans, D. K. 917
Yoder, J. G. 1638

Zambelli, P. 703, 1599
Zaunick, R. 1370
Zeller, M. C. 754
Zetterberg, J. P. 804
Zeuthen, H. G. 743
Ziggelaar, A. 940
Zilboorg, G. 624, 1502
Zilsel, E. 924
Zinner, A. 831, 834, 851,
    914, 1215
Zischka, G. A. 47
Zögner, L. 1062
Zorzi, M. 206
Zoubov, V. P. 529
Zycinski, J. 568, 583

Yates

637   Theatre of the World
645   Rosicrucia E.

Subject Index

Suffixes *b* and *c* are used to distinguish bibliographies and catalogues from other items.

Cross-references, whether from a main heading or a subheading, have been placed as *see also* references at the end of the entry after all the subheadings (with personal names arranged in a separate sequence). In general, they are *not* made: (a) from specific headings to general ones (e.g., from algebra to mathematics); (b) between etymologically related headings (e.g., educationists and education); (c) from the history of a subject to the subject; or, (d) to individuals from their countries of origin (or adoption).

"Science" and "Renaissance science" have not been indexed in detail and cross-references to and from "Scientific Revolution" have been limited to specific entries.

Aberdeen University 100, 282, 1538
academies 19, 69, 166, 321, 369, 1072,
　　1135, 1432
　　dissenting 250, 252, 274
　　French 219, 355, 357, 358, 371, 385,
　　　　1319
　　German 355, 367, 368
　　Italian 302A, 355, 356, 365, 368–370,
　　　　484, 578, 618
　　non-conformist 250, 252, 274
　　scientific 166, 355, 369, 371
　　see also scientific societies
aeronautics 1179, 1187
Africa 73, 482A*b*, 1075, 1348
　　Southern 1273, 1280, 1300A, 1310,
　　　　1320, 1330, 1351
Agricola, Georg 990, 1021
agriculture 8*b*, 13*b*, 119, 127, 132, 467,
　　603*b*, 1265
Agrippa von Nettesheim, Cornelius 420,
　　1502
alchemy 486, 552, 629, 644, 652, 657, 668,
　　676

bibliographies 5, 8, 602, 603, 611,
　　614, 616, 619, 977, 984
catalogues 174, 609, 610, 618, 978,
　　979, 980
dictionaries 49, 406
Europe 340, 619, 667, 670
history 665, 682A, 992, 1004, 1409,
　　1612
Newton 555, 619, 647, 650, 658,
　　1641A
see also chemistry; illustrations; oc-
　　cult science
algebra 716, 717*b*, 727*b*, 739, 768, 769,
　　778
history 731, 736, 743, 760, 812, 814,
　　819; Italy 730, 786, 805
see also arithmetic; Bombelli; Car-
　　dano
almanacs 689, 692, 695, 915
bibliographies 5, 684, 685, 689, 693,
　　716, 833
see also prognostication(s)
America 40A*b*, 42*b*, 58, 174, 482A*b*, 915,
　　1023, 1086, 1321

discovery 1037, 1075, 1079, 1089, 1111
plants from 1292, 1300, 1348
see also Brazil; materia medica; Mexico; New World, plant collecting
anatomy 160, 480, 577, 1423–1455, 1526, 1612
   bibliographies 8, 1423, 1424, 1428, 1432, 1440
   dissection 1440, 1455, 1526
   history 1387, 1424, 1428, 1434
   illustrations 575, 1270, 1392, 1424, 1440
   Renaissance 340, 505, 545, 945, 1452
   teaching 285, 1283, 1432, 1436, 1455, 1526, 1534
   see also Galen; Harvey; Vesalius
anthropology 285, 514
anti-Atistotelian movements 253, 1395
apothecaries *see* materia medica
Arabic literature 89, 686, 945, 1360*b*, 1423*b*, 1456*b*, 1556
Archimedes 154, 799, 946
   see also Greek texts
architecture 8*b*, 53, 69, 166, 572, 643, 709*b*, 771, 1189
   Wren 471, 531, 880
   see also Naval history; Leonardo da Vinci
Aristotelianism 14b, 241, 445, 572, 584, 798, 862, 1624
   Galileo and 445, 1567, 1624
   reactions to 253, 897, 1395
   Renaissance 419, 434, 446, 456, 458, 462
   universities 241, 291, 433, 434
Aristotle 389*b*, 394*b*, 401*b*, 403*b*, 431, 438, 938, 961, 1585
   see also Greek texts; philosophy
arithmetic 127, 132, 134, 578, 735, 737
   bibliographies 8, 211, 707, 711, 716, 722, 725
   history 716, 737, 742, 744, 746, 778, 816A
   libri d'abaco 725, 730, 786, 805, 812, 816A, 819
   practical 716, 742, 779, 812

teaching 211, 721, 725, 735, 737, 742, 744, 751, 805
see also algebra; educational books; mathematicians
artisans 304, 344, 424, 506
Ashmole, Elias 189, 570
Asia 61, 73, 482A*b*, 1046, 1075
   Europe and 517, 595, 1065A*b*, 1067A*b*, 1182
   see also China; flora; India; Japan
astrologers 187, 573, 652, 690, 695, 697, 700, 804
   see also Cardano; Dee; Kepler
astrology 406, 552, 639, 644, 649, 652, 683–705, 890, 895, 1090
   bibliographies 5, 8, 603, 605, 606, 611, 617, 683, 686–688, 696
   catalogues 166, 187, 842
   decline of beliefs 599, 640, 1607
   England 486, 659, 692, 694–697, 700, 704, 705A, 804
   Germany 688, 691, 698, 701
   history 510, 605, 683, 691, 702, 704, 705
   medical 687, 689, 695, 702, 1407, 1409, 1452
   Renaissance 340, 599, 667, 695, 696, 699, 704
   Scientific Revolution and 665, 1607
   see also almanacs; Kepler; occult science; prognostication(s)
astronomers 502, 835*b*–839*b*, 880, 890, 891, 906
   Jesuit 715, 851, 854
   see also Brahe; Bruno; Copernicus; Galilei; Kepler; Newton; Rheticus; Wren
astronomical instruments 588, 841*b*, 906, 1197*b*, 1198*b*, 1209*b*, 1215, 1220, 1237, 1243
   astrolabes 1090, 1209, 1247
   collections 198, 1205, 1207, 1215, 1225, 1235
   telescopes 490, 503, 903, 1162; history 911, 1214, 1230, 1232
   see also instruments; observatories
astronomy 119, 174, 340, 510, 525, 716, 771, 808, 827–917, 1641A

bibliographies 827, 829, 832–834, 852; general 5, 8, 108, 119, 709, 716; of history 833, 835–839, 841; stars 87, 828, 840
books 127, 132, 134, 211, 249, 855
catalogues 830, 831, 842, 1379
eclipses 687, 693
England 127, 132, 134, 832, 915, 1205, 1240
English literature and 486, 700, 847, 850, 864
nautical 1090, 1108
practical 845, 882, 915
see also comets; Copernicanism; history of astronomy; libraries (pre-1700); planetary theories; Scientific Revolution; Kepler
atlases 1055, 1056, 1065
atomism 343, 406, 409, 421, 457, 459, 596
Avicenna 1539
axiomatic method 428

Bacon, Francis 226, 320, 392*b*, 400*b*, 429, 460*b*, 484, 492, 935, 1402, 1621
Great Instauration 414, 1405
methodology 430, 935, 1583, 1597
occult science 635, 665
philosophy 413, 414, 450
Scientific Revolution and 1581, 1597, 1607, 1621
Baldi, Bernardino 817, 934
ballistics 798, 923, 926, 1233
see also mechanics
Bär, Nicolai Reymers (Ursus) 902
Barba, Alvaro Alonso 1191
Barozzi (Barocius), Francesco 792, 799
Barrow, Isaac 723, 753, 813, 823
Basle University 1352, 1527
Beeckman, Isaac 1627
Belgium 59, 1070
see also Netherlands
Belon, Pierre 1276, 1284, 1333
Benedetti, Giovanni Battista 578, 934, 1600
Bernoulli family 800, 816, 822
Bessarion, Cardinal 131, 154, 196, 206
bibliographies 8, 39–42, 125, 126; pre-1700 94

see also bibliographies of bibliographies; bibliographies of biographies; *and under individual names and subjects*
bibliographies of bibliographies 1–7, 121, 126
medicine 1355, 1359
bibliographies of biographies 50, 51, 724, 977, 1542
humanism 99, 103, 112
science 22, 23, 25, 26, 1365, 1542
see also bio-bibliographies
bibliography 125, 126
bibliophiles, 145, 163, 168, 209, 1376
Billingsley, Sir Henry 723, 774, 821
bio-bibliographies
collective: cartographers 1067B; humanists 103, 112, 122; Jesuits 472; philosophers 390, 403; physicians 1527, 1548; scientists 54, 390, 470, 496, 1251; surgeons 1458, 1463; theologians 390
individuals: Frisius 1048; Gasser 1555; Gesner 473; Leibniz 402; Orta 1482; Paracelsus 1514, 1522; Paré 1467; Ries 721; Vesalius 1426
see also bibliographies of biographies
biographies 53, 54, 407, 689, 1054*c*–1056*c*, 1116, 1135*b*, 1161; 1172; 1489
see also bibliographies of biographies; biobibliographies; educationists; humanists; mathematicians; medicine; physicians; Renaissance; surgeons; *and under individual names*
biology 69, 502, 525, 575, 1283, 1287, 1300, 1395, 1640
bibliographies 1260, 1261, 1300, 1364, 1365
see also botany
Birch, Thomas 381
Biringuccio, Vannoccio 572, 990, 1176, 1177

blood 596, 1430, 1433, 1439, 1444, 1447–1449, 1574
Bodin, Jean (1530–96) 1502
Bodleian Library, Oxford 146, 202, 208
Boerhaave, Hermann 152, 988, 1314, 1360, 1446, 1552
Bologna 369, 589, 1272, 1273, 1462
    mathematics 755, 777, 779, 803, 819
Bombelli, Rafael 777, 779, 803, 819
book-keeping, double entry 771
book trade 101, 107, 123, 128, 138, 140, 141, 144, 157
    bibliographies 88, 113, 128, 135, 145
    catalogues 87, 88, 128, 213, 603, 611; sale 115, 148–150, 210
    England 88, 127, 132, 134, 177, 210
    Italy 138, 140, 340, 572
    scientific 38, 127, 132, 134, 135, 145, 210
    see also medicine; printing
Borelli, Giovanni Alfonso 1430, 1586
botanical expeditions 1289, 1296, 1304, 1308, 1318, 1330, 1333
    see also plant collecting
botanical gardens 502A, 1259, 1273, 1281, 1292, 1300, 1302
    Edinburgh 1291
    Leiden 1285, 1293
    Marburg 298
    Montpellier 1278
    Padua 198, 1302, 1340
    Paris 1295
    see also education in France; Jardin des Plantes
botanists 1276, 1277, 1280, 1281, 1285, 1323, 1336
    Jesuit 1275, 1289, 1492
botany 371, 1001, 1259, 1303, 1308, 1341
    bibliographies 121, 1248–1250, 1257, 1262–1265, 1269
    early books 121, 1257, 1286, 1352
    education 298, 1303, 1314, 1323, 1352
    history 69, 1260, 1261, 1304, 1332
    humanism and 552, 1327
    illustrations 1266, 1270, 1343

see also flora; France; herbals; herbaria; Leiden University; plant collecting; Dioscorides
Bovelles (Bouelles, Bovillus), Charles de 796, 810
Boyle, Robert 415, 492, 669, 959, 1553, 1615
    chemistry 990, 993, 995, 1003, 1034
    life and work 316, 449, 518, 999
Brahe, Tycho 845, 859, 860, 894, 902, 910, 911, 913
Brazil 1289, 1326
Brosse, Guy de la see La Brosse, Guy de
Bruno, Giordano 400, 631, 651, 853, 874, 908
Bürgi, Jost 1213, 1235
Burton, Robert 208, 574

cabbala 38, 521, 606, 611, 630, 652, 667, 678, 1620
    see also occult science
Caius, John 194, 1537
calculus 752, 761, 778, 781, 783, 797, 800, 806, 809
    infinitesimal 583, 770, 801, 815, 822
    see also Leibniz; Newton
calendar 508, 700, 716, 843, 896, 1146
Cambridge University 173, 215, 257, 258, 738, 823, 915, 1628
    Colleges 106, 194, 197, 240, 244
    education 215, 251, 258, 1537
    libraries 106, 172, 197, 205
    mathematics 738, 791, 823
    medicine 251, 269, 1534, 1537
    science 268, 269, 1210
Campanella, Tommaso 400, 1395
Cape of Good Hope, see Africa, Southern
Cardano, Girolamo 400b, 483, 601B, 662, 665, 934, 1554, 1559
    algebra 730, 803, 819
cartography 198, 906, 1090, 1097, 1116, 1128
    bibliographies 5, 1055, 1061, 1067b, 1115; of history 1060, 1062, 1064, 1065A, 1067
    catalogues 1045, 1052, 1054, 1058, 1059, 1065, 1099, 1119

history 340, 1037, 1046, 1067*b*, 1070, 1082, 1088, 1101, 1104, 1109, 1123, 1127, 1146

map making 1102, 1109, 1115, 1116, 1118–1120

Scientific Revolution and 1113, 1639

see also geography; Ptolemy

Catholic Church 53, 312, 314, 349, 667, 1100, 1407

Council of Trent 343, 1624

see also Church (The); Counter-Reformation; Galileo affair; heliocentrism; Index of Prohibited Books; Inquisition; Jesuits; papacy; religion

celestial physics 873, 1586

censorship of books 101, 138, 139, 603, 893

Champier, Symphorien 652

Charleton, Walter 669, 1395

charms 613, 1400

chemistry 49, 69, 575, 647, 690, 709, 976–1006, 1205, 1474

bibliographies 5, 709, 976, 977, 982, 984, 997

catalogues 842, 978–981

education 298, 989, 1001

history 21*b*, 977*b*, 982*b*, 985, 987, 992, 995, 996, 998, 1004; sixteenth century 1004, 1577; seventeenth century 506, 986, 993, 997, 1003, 1005

see also alchemy; atomism; France; iatrochemistry; Scientific Revolution; Boyle

China 64, 406, 482A*b*, 508, 510, 517, 719, 1101

European botanists 1275, 1318, 1492

"Scientific Revolution" problem 1630, 1643, 1644

see also Asia; Jesuits

Christianity *see* Church, The

Christina, Queen of Sweden 191, 321

chronologies 57, 71–74, 76, 78, 81, 82, 84

by country: Africa 73; America 58, 73; Asia 73; Belgium 59; China 64; Czechoslovakia 68; Denmark 62; England 75, 77, 83; Europe 69, 73; France 66; Italy 65; Japan 64; Netherlands 63; Portugal 1114; Rumania 60; Spain 67

by subject: academies 69; anatomy 1434; architecture 69; astronomy 69; biology 69; Cardano 483; chemistry 69, 1434; Church 74–77, 79, 81, 83; Copernicus 871; discoveries 69, 72, 1061; education 75, 77, 82; electricity 918; Enlightenment 79; exploration 76; geography 69; geology 1042; Huygens 549; inventions 1151; magnetism 918; mathematics 69, 80; medicine 69, 1386, 1390, 1393; music 74; Newton 84; Paracelsus 1521; philosophy 71, 74; physics 1434; printing 129, 133; public hygiene 1393; religion 74–77, 79, 81, 83; Renaissance 84; science 84; technology 57, 74, 76, 1145, 1152; zoology 69

Chuquet, Nicolas 818

Church (The) 74–77, 79, 81, 83, 222, 237, 517, 599

Dissenting Academies 250, 252, 274

16th century England 310, 628, 640

17th century England 307, 309, 313, 315, 322, 330, 334–337, 339, 350, 354, 640

see also Catholic Church; Protestantism; Puritanism; Reformation; social relations of science

Clavius, Christoph 817, 891

Clusius, Carolus *see* L'Escluse, Charles de

College of Physicians, London 149, 168, 194, 264, 362, 363, 1536, 1541

Collegio Romano 302A, 500, 508, 567, 578, 580

colonies 10*b*, 1065A*b*, 1067A*b*, 1294, 1320, 1324, 1326, 1330, 1346, 1348

see also Africa; America; Asia

Columbus, Christopher 1037, 1089, 1321
Comenius, John Amos 222, 226, 230*b*,
        375, 400*b*
comet(s) 534, 704, 853, 862, 867, 895, 911,
        917
    bibliographies 8, 603, 687, 842, 852
Commandino, Federico 155, 156, 817,
        821, 934, 946
Copenhagen 1218, 1240, 1273
Copernicanism 453, 704, 850, 862, 875,
        887
    history 462, 848, 851, 869, 963
    reactions 349, 849, 883, 897, 1580,
        1611
    see also cosmology
Copernicus, Nicolas 400*b*, 521, 580, 594,
        837*b*, 891, 908, 1030
    biographies 837, 860, 872, 898
    library 193, 851, 871
    quincentenary 871, 872, 879, 884
    Rheticus and 837, 863
    Scientific Revolution 596, 1580,
        1586, 1591, 1610
    works 837, 875, 893, 898, 900, 1641
    see also heliocentrism
corpuscular theory 415, 449, 459, 931
    see also optics
cosmography 147, 171, 178, 211, 886,
        1037
cosmology 406, 412, 432, 509, 584, 917A,
        1038, 1579
    Milton 547, 850
    16th century 874, 899, 910, 1580
    17th century 892, 899, 1558, 1572,
        1621
    space 426, 438, 1620, 1623
    see also world picture; Kepler
Counter-Reformation 138, 343, 681, 1624
Culpeper, Nicholas 1495

De Caus, Salomon 1159, 1160   Suster et.
Dee, John 634, 637, 641*b*, 655, 662, 671,
        674*b*, 682*c*, 890 J. D. on Astronomy···
    mathematics 723, 774, 821
    natural philosophy 649, 665, 674b
dentistry 1459, 1460, 1466
Descartes, René 291, 349, 489, 680, 929,
        1583, 1621

bibliographies 98, 396, 398, 400,
        1557
    mathematics 753, 757, 769
    medicine and 436, 444, 1557
    music 964, 965
    philosophy 422, 436, 439, 444, 769,
        1395
    Scientific Revolution and 1597,
        1621, 1639
design 709, 1183
dictionaries 95, 406, 407, 470, 554, 977,
        1369
    bibliographies 19, 47, 49, 397, 833,
        1137
    see also encyclopedias; philosophy;
        Renaissance
Diophantus 154, 779, 819
    see also Greek texts
Dioscorides 154, 1327, 1342, 1487
    see also Greek texts
discoveries 10, 69, 72, 327, 1075, 1145,
        1174
    see also voyages of discovery
Donne, John 491, 503, 1025
Dürer, Albrecht 655, 775, 795, 799, 1343

early printed books 92*b*, 93, 106*c*, 108*b*,
        110*c*, 114*b*, 117*b*, 145
    astronomy 108*b*, 827*b*, 828*b*, 834*b*
    mathematics 108*b*, 113*b*, 711*c*, 717*b*,
        722*b*, 726*b*
    medicine 92*b*, 108*b*, 121*b*, 125*b*, 145,
        984*c*, 1373*c*, 1374*c*, 1377*c*,
        1382*c*, 1385*c*
    reference 91*b*, 94*b*, 95*b*, 121*b*, 125*b*,
        126, 504
    see also Euclid; herbals
earth 534, 1016, 1029, 1030, 1032, 1039,
        1041, 1622
    see also geodesy; geology; geo-
        physics; magnetism
earth sciences 1007–1042
Edict of Nantes 248, 345, 1165
Edinburgh 1273, 1291, 1413
education 75, 77, 82, 211–303, 855, 966,
        1527
    bibliographies 5, 31, 98, 211–213,
        273, 294

geography 211, 249, 287, 1118
science and technology 229, 249,
253, 256, 288, 290, 922
see also botanical gardens; botany;
chemistry; history of education; history of universities;
Jesuits; mathematical education; medical education; Reformation; Scientific Revolution; Scotland; teaching; universities
education in England 222, 237, 259–261,
275, 280
history of 31*b*, 226, 248–250, 259,
276, 279, 280, 413
mathematics 247, 253, 274, 723, 738,
750, 751, 791, 1093
Puritans 250, 252, 263, 274
schools 245, 247, 249, 262, 270, 271,
273
universities 214*b*, 215b, 217*b*, 220b,
221*b*, 262, 265, 266, 271, 273
see also Cambridge University;
medical education; Oxford
University
education in France 98, 119, 222, 237,
289, 292, 293
Collège de France 283, 284
science and technology 288, 291,
922
universities 219*b*, 240, 286, 291, 292,
294, 462; botanical gardens
285, 989, 1001, 1302, 1303,
1331, 1337
see also history of universities;
Jardin des Plantes; medical
education; Montpellier University
education in Germany 233, 237, 295, 297,
368
universities 212*b*, 216*b*, 240, 296–
298, 1302
education in Italy 4*b*, 229, 237, 299, 302,
368, 820, 951
universities 240, 299, 300, 368, 433,
589, 1302, 1340; mathematics
239, 299, 755, 787, 820;

medicine 239, 301, 1409, 1528,
1539; philosophy 239, 240, 300
see also Padua
educational books 211*b*, 213*b*, 294, 302,
1203, 1233
arithmetic 211*b*, 722*b*, 735, 737,
742*b*, 744*b*, 745, 779
in England 127, 132, 134, 245, 252,
273
sciences 211*b*, 714, 835, 922
educationists 222, 226, 227, 235, 237, 246,
254, 272, 1394
by name: Ascham 227; Aubrey 272;
Castiglione 227, 246; Dury
226, 267, 316; Elyot 227, 1394;
Franckc 226; Hartlib 226, 267,
316, 375; Hoole 226; Ignatius
Loyola 235; La Salle 226, 231;
Luther 237; Melanchthon 227;
Pepys 274, 522; Rabelais 289,
544, 1556; Ramus 228, 462,
507, 891; Ratke 222, 226, 233;
Recorde 253, 274, 723, 750;
Vives 227, 246; Ward 265;
Webster 204, 265; Wilkins 265,
373
see also humanists; Jesuits; Comenius; Erasmus; Locke; Milton
electricity 918, 925, 950
embryology 1476, 1534
encyclopedias 48, 222–224, 393, 406, 504,
607, 1153
bibliographies 16, 19, 47, 49, 833,
1135, 1137
see also dictionaries
engineering 340, 505, 1132*c*, 1142*b*, 1144,
1161, 1172, 1178
see also technology
England 31, 1025, 1169, 1391, 1511, 1628
bibliographies 10–13, 87, 109–111,
114, 117, 474, 984, 1081, 1364
chronologies 75, 77, 83
English Revolution 335, 341, 1615
geography 1078, 1081, 1116, 1119,
1120
medicine 1364, 1396, 1408, 1428,
1511; fevers 1412–1414; folk
1400, 1421; obstetrics 1469,

1480; physicians 168, 264, 548, 1419, 1530, 1543, 1547; physiology 1450, 1451
military science 1130, 1138, 1143
occult science 613b, 614b, 621, 632, 640, 641, 655, 804, 1000
Royal Greenwich Observatory 1225, 1240
science 136, 309, 310, 342b, 421, 422, 469b, 548, 926, 1364, 1567b
scientific movement 272, 322, 1405, 1585
societies and guilds 361, 1234
see also almanacs; astrology; astronomy; book trade; Cambridge university; College of Physicians; education in England; educational books; history of medicine; Huguenots; industries; libraries (pre-1700); library history; literature; London; mathematicians; medical education; natural history; naval and maritime arts; Oxford University; physicians; physics; Puritanism; Restoration; Royal Society; schools; science and religion; social relations of science
Enlightenment 79, 1113
epidemics and epidemiology 1321, 1506–1513, 1540
Erasmus, Desiderius 227, 237, 246, 289, 354, 407, 791
L'Escluse, Charles de (Clusius) 1276, 1285, 1323
Euclid 154, 710b, 713b, 720b, 723b, 726b, 774, 821
see also Greek texts
experimental science 317, 325, 365, 484, 498, 582, 924, 959, 970
Scientific Revolution and 1562, 1578, 1585, 1612
exploration 1091, 1079, 1097, 1111, 1296, 1312, 1328
bibliographies 10, 1045c, 1054c, 1057

history 76, 1075, 1083, 1105, 1107, 1110
see also botanical expeditions; geography; plant collecting

Fabricius of Aquapendente, Girolamo 1380, 1476
fauna 517, 1270, 1312, 1343
Fermat, Pierre de 753, 780, 793, 800
Flamsteed, John 186, 704, 882, 1039
flora 1270, 1281, 1282, 1298, 1300, 1334, 1343, 1347
Asia 517, 1320, 1346, 1353, 1482, 1492
see also Africa, Southern; herbals; medicinal plants; plant collecting
Florence 182, 185, 365, 1302
Accademia del Cimento 355, 365, 369, 370, 484
Fludd, Robert 521, 638, 654, 665, 675, 1211
folklore 1282, 1400, 1421
fortifications 8b, 1133b, 1175, 1181, 1193, 1195
Fracastoro, Girolamo 400, 1071, 1074
France 66, 538, 544, 652, 680, 1001, 1105, 1305, 1467, 1523
book trade 141, 144
science 119, 705, 922, 1009, 1049; botany 1273, 1276, 1278, 1295, 1303, 1333; chemistry 986, 995, 1005, 1006
Strasbourg 141, 1273, 1487
see also academies; education in France; history of medicine; history of universities; Huguenots; libraries (pre-1700); literature; mathematicians; medical education; Montpellier University; Paris; physics
Francesca, Piero della 812, 946
freemasonry 606, 611, 645, 679
French pox see syphilis

Galen 146b, 154, 1442, 1445b, 1520, 1534, 1536

Renaissance editions 130, 1327, 1402, 1425b, 1452
Galilei, Galileo 153c, 340, 343, 400b, 433, 526b, 859, 860, 973
    astronomy 889, 903; heliocentrism 343, 347, 351, 863, 963
    biographies 418, 947
    influences 443, 508, 567, 589
    instruments 484, 1204, 1211, 1632
    mechanics 484, 932, 934, 943, 949, 951
    methodology 437, 484, 534, 1597
    philosophy of science 418, 419, 437, 442, 1597
    Scientific Revolution 1572, 1597, 1600, 1607, 1612, 1624, 1632
    symposia 443, 521, 526, 527, 936
    works 534, 539, 932, 1593
    see also Aristotelianism; Galileo affair; physics
Galileo *see* Galilei, Galileo; *and* Galileo affair
Galileo affair 349, 502, 566, 863, 963
    reappraisals 343, 347, 351, 451, 568, 580, 963
    see also Collegio Romano
Gans, David 1633
gardens 1265b, 1292, 1300A, 1318, 1347; England 134, 1262b, 1290, 1315, 1350 *see* also natural history
Gassendi, Pierre 400, 454, 457, 1600
Gasser, Achille Firmin 1555
Gastaldi, Giacomo 1074
Gaymans, Antoni 1351
geodesy 1016, 1026, 1029, 1036, 1088
    see also earth; geology
geography 127, 132, 134, 332, 487, 1048, 1069
    bibliographies 5, 8, 1044, 1045c, 1049, 1054c, 1056c, 1067A, 1078, 1081
    history 69, 1044b, 1068, 1075, 1078, 1081, 1129
    Renaissance 340, 1037, 1091, 1118
    see also cartography; education; exploration; libraries (pre-1700);
Henry the Navigator; Ptolemy; Ramusio
geology 480, 1008c, 1009b, 1010b, 1012c, 1028, 1035, 1038
    history 1010b, 1015b, 1019, 1022, 1040, 1042
    see also earth; geodesy
geomagnetism 1018, 1020, 1024
geometry 211, 448, 716, 753, 759, 763, 772, 778, 799
    17th century 714, 749, 761
    see also Euclid
geophysics 1013 *see also* earth
Germany 101b, 203, 453, 461, 851, 997, 1000, 1062, 1069
    astronomy 831, 834, 851
    book fairs 101, 123, 128
    botanical gardens 298, 1273
    materia medica 298, 1491, 1493
    observatories 1218, 1235, 1240
    Thirty Years' War 222, 240, 367
    see also astrology; education in Germany; libraries (pre-1700); mathematics; Reformation
Gesner, Conrad 124, 125, 130b, 311, 462, 473b, 524b, 552, 597b
Gilbert, William 798, 911, 924, 1020, 1585, 1597
Glanvill, Joseph 623, 1395
glass-making 1146, 1156, 1195
gravitation theory 559, 865, 927, 931, 952
Greek texts 70, 97, 792, 945, 1589
    Bessarion and 131, 196
    manuscripts 154c, 155c, 162c, 175c, 188, 196c, 198, 394, 786
    medicine 1360, 1423, 1456
    see also Archimedes; Aristotle; Diophantus; Dioscorides; Euclid; Galen; Hero; Hippocrates; Ptolemy
gunpowder 1180, 1188
gynaecology 1471c, 1474, 1475, 1477, 1478, 1480
    see also obstetrics

Halley, Edmund 189, 876, 1034
Harriot, Thomas 537, 543, 557, 563, 705A, 798, 802, 803

Hartlib, Samuel 226, 267, 316, 375
Harvey, William 1395, 1402, 1423*b*,
　　1429*b*, 1431, 1437, 1443, 1621
　　anatomy 1449, 1455
　　circulation of blood 1430, 1449; *De
　　　motu cordis* 1429, 1444, 1448
　　physiology 1438, 1449, 1450, 1597
heliocentrism 347, 351, 521, 863, 872,
　　1586
　　Catholic Church and 343, 349, 893,
　　　963, 1580
　　see also Copernicanism
Helmont, Johannes Baptista van 400*b*,
　　990, 1360*b*, 1395, 1402, 1430,
　　1553, 1558, 1560, 1597
Henry the Navigator 1072, 1077, 1099,
　　1114, 1122
herbals 1252*b*, 1288*b*, 1298, 1495
　　bibliographies 8, 1253, 1268*c*
　　history 1255*b*, 1258*b*, 1262*b*, 1300*b*,
　　　1344*b*
herbaria 1259*b*, 1272, 1288*b*, 1307, 1320,
　　1351
　　see also plant collecting
Hermann, Paul 1280, 1310, 1320, 1330
hermeticism 154, 380, 606*b*, 617*b*, 618*c*,
　　631, 654, 667, 984*b*
　　Newton and 651, 659, 665
　　Renaissance and 667, 677
　　Scientific Revolution and 651, 1616,
　　　1618, 1641
　　see also occult science
Hernández, Francisco 1318, 1328, 1490,
　　1494
Hero, of Alexandria 154, 792, 1211
　　see also Greek texts
Heurnius, Justus 1280, 1310
Hippocrates 146, 154, 1367*b*, 1368*b*
　　see also Greek texts
historiography of science 21, 999, 1354,
　　1357, 1603, 1635
　　see also Merton thesis; Scientific
　　　Revolution
history of astronomy 69, 833*b*, 836*b*–
　　839*b*, 841*b*, 843, 844
　　Renaissance 340, 484, 505, 509, 521,
　　　525, 832, 834, 1583
　　16th century 340, 771, 891, 910, 911

17th century 892, 897, 906, 911, 915
history of education 222*b*, 225, 227, 237,
　　238
　　Renaissance 227, 232, 279, 302, 391,
　　　462
　　Reformation 232, 237–239, 243, 262
　　16th century 227, 228, 239, 273
　　17th century 245, 264, 265, 294, 319
　　see also history of universities;
　　　teaching
history of mathematics 69, 80, 505, 731,
　　734, 740, 743, 758, 761, 776,
　　778, 789
　　bibliographies 715, 719, 724, 728,
　　　729
　　Cambridge 738, 791
　　China 510, 719
　　notations 746, 748, 760, 802, 803
　　perspective 775, 795
　　probability 447, 732, 762, 784, 785,
　　　793, 806, 818, 824
　　see also *mathematical disciplines*
history of medicine 551, 1138, 1387, 1396,
　　1397, 1399, 1401, 1406, 1454,
　　1504, 1506, 1507
　　bibliographies 21, 1354–1357, 1362,
　　　1363, 1365, 1366, 1369, 1371*c*,
　　　1379*c*, 1381
　　chronologies 69, 1386, 1390, 1393
　　social 1399, 1405, 1529
　　by period: Renaissance 462, 525, 643,
　　　1402, 1417, 1525, 1540, 1617;
　　　16th century 119, 340, 1407–
　　　1410, 1413, 1414, 1418; 17th
　　　century 1395, 1405, 1408, 1411,
　　　1413, 1419
　　by place: England 1364, 1396, 1421;
　　　France 119, 1525; Geneva
　　　1388; Italy 340, 445, 572, 1362,
　　　1528; Netherlands 1366, 1468,
　　　1557; Spain 1407, 1410, 1411,
　　　1486, 1490, 1494
　　see also anatomy; materia medica;
　　　medical education; physi-
　　　cians, surgeons, syphilis
history of science 21, 484, 495, 554, 596,
　　1573, 1576, 1581

bibliographies 17, 19, 22, 23, 25, 44c,
478, 1371c, 1381c; current 26,
27, 27A; guides 16, 18, 24
from a Marxist view 533, 556, 1596
journals 16, 21, 56
surveys 485, 515, 551, 600, 1566,
1589
see also chronologies; technology
history of technology 1146–1149, 1152,
1153, 1156, 1171
bibliographies 1132c, 1134–1136,
1141, 1142, 1144
chronologies 57, 74, 76, 1145, 1152
essays 1141, 1150, 1173
see also engineering; glassmaking,
military science; shipbuilding
and shipping
history of universities 14b, 214b–221b,
273b, 286, 298
Aberdeen 100, 282, 1538
Paris 219, 283, 284, 291, 1531
Pisa 239, 1302
by period: Renaissance 217, 240, 241;
16th century 239, 240, 243, 266,
273, 300; 17th century 240,
255–257, 266–269, 291, 294, 355
see also history of education; Jardin
des Plantes; Padua; and Cambridge, Leiden, Oxford,
Montpellier universities
Hobbes, Thomas 959, 1583, 1621
Hooke, Robert 189c, 210c, 430, 588, 669,
881, 935, 990, 1041, 1208
Huguenots 248, 304, 316, 344, 345, 1165,
1483, 1549
see also Hartlib; immigrants in England; Papin
humanism 136, 237, 289, 457, 592, 786,
975
bibliographies 99, 103, 105, 112,
120, 122
education and 232, 237, 238, 289
medicine and 239, 1416, 1418, 1425,
1537
Renaissance science and 239, 338,
340, 354, 525, 545, 552, 808,
1327

see also botany; manuscripts;
Netherlands; Renaissance
humanists 99b, 103b, 112b, 122b, 188, 338,
487, 786
by name: Benivieni 160; Caius 194,
1537; Gasser 1555; Holste 165;
Linacre 1536; Münzer 163;
Nostradamus 573; Pico della
Mirandola 162; Pinelli 198;
Ramus 228, 462, 507, 891;
Rheticus 835, 837, 863; Valla
799, 946
see also educationists; Bessarion;
Cardano; Erasmus
Huygens, Christian 199c, 479b, 489, 499,
549, 569, 793, 1213, 1638
hydrology 1027

iatrochemistry 632, 986, 1517, 1574
ideas, history of 406, 630
see also philosophy
illustrations 506, 550, 633, 654, 1104
alchemy 657, 668, 676
instruments 1207, 1223, 1226, 1238,
1242
medicine 1392, 1397, 1398, 1403,
1505, 1531; anatomy 1270,
1424, 1440, 1474, 1475; gynaecology 1474, 1475, 1477;
pharmacy 1484, 1486; surgery
1461, 1463, 1464
natural history 1254, 1266, 1270,
1298, 1343, 1347;
immigrants in England 1169, 1234
see also Huguenots
Index of Prohibited Books 101, 138, 603
India 61, 482Ab, 1182, 1346, 1353
see also Asia; materia medica
Industrial Revolution 1142, 1148
industries 1146, 1156, 1169, 1170, 1173,
1195
Inquisition 138, 584, 625, 681, 893
see also Galileo affair
instrument makers 817, 1120, 1215, 1228,
1231, 1234, 1245
instruments 340, 484, 521, 549, 599, 1146,
1196–1247

bibliographies 16, 21, 166, 1196–
    1200
collections 1066, 1205, 1207, 1210,
    1233
drawing 1233, 1245
gnomonic 833, 1246
mathematical 198, 817, 1120, 1197,
    1203, 1205, 1233, 1234, 1237
meteorological 1206, 1212, 1219,
    1221, 1231
microscopes 423, 503, 1198, 1208,
    1222, 1335, 1612
optical 1198, 1244
surveying 1203, 1205, 1207, 1216,
    1226, 1233, 1243
thermometers 1201, 1202, 1204,
    1211, 1221
see also astronomical instruments;
    illustrations; navigational in-
    struments; surgery; time mea-
    surement
intellectual communities 357, 366–368,
    502
intellectual history 11*b*, 13*b*, 15*b*, 307,
    313, 997, 1592, 1594
inventions 1145, 1147, 1151, 1168, 1585
    patents 1162, 1163, 1166
    see also technology
Invisible College 366
Ireland 215, 266, 364
Italy 4*b*, 53, 65, 203, 302A, 340, 589, 951,
    1273
    Council of Trent 343, 1634
    fortifications 1133*b*, 1175
    historiography of science 27A, 484,
        525, 729
    philosophy 400, 408, 709, 941
    Pisa 239, 1302
    science 302A, 484, 485, 525, 578,
        773, 934, 1133
    Venetia 138, 445, 572, 681
    witchcraft 625, 681
    see also academies; algebra; Aris-
        totelianism; arithmetic, *libri
        d'abaco*; Bologna; book trade;
        botanical gardens; education
        in Italy; Florence; Galileo af-
        fair; history of medicine; hu-

    manism; Inquisition; libraries
    (pre-1700); library history;
    mathematicians; mathematics;
    mechanics; medical education;
    Padua; physics; Renaissance;
    Rome; Urbino; Venice

Japan 64, 482A*b*, 508, 517, 1240, 1324
    see also Asia
Jardin (Royal) des Plantes, Paris 285,
    989, 1001, 1273, 1303, 1331,
    1333, 1337
Jesuits 324, 715, 719, 851, 861, 1275, 1289,
    1624
    in China 472, 508, 854, 1275, 1492
    in education 222, 229, 235, 289, 290,
        299, 287
    see also Catholic Church; Collegio
        Romano
John Paul II, Pope 580, 963
journals 16, 56, 355, 595A
    see also science
Jungius, Joachim 461, 997, 1629

kabbala *see* cabbala
Kepler, Johannes 349, 448, 594, 789, 873,
    879, 885, 1641A
    astrology 665, 698, 701
    astronomy 502, 503, 865, 870, 907,
        911, 1586
    bibliographies 400, 836, 838, 839,
        879
    biographies 858–60, 902
    cosmology 448, 704, 909, 972
    musical theory 968, 972, 973
    philosophy 461, 873, 962
    physics 873, 929, 945, 954, 958, 962,
        1597
Komensky, Jan Amos *see* Comenius

La Brosse, Guy de 1001, 1331, 1337
Le Boë (Dubois), François de, *called*
    Sylvius 1360, 1430
Leeuwenhoek, Antoni van 589, 1335
Leibniz, Gottfried Wilhelm 199, 400*b*,
    402*b*, 416, 783, 790, 800, 811,
    815, 1213
    calculus 583, 781, 797, 801, 815, 822

Newton and 583, 601A, 783, 801
Renaissance mathematics and 765,
    777
Leiden University 147c, 192c, 303, 941,
    1218, 1240, 1320, 1446
    botany 1273, 1276, 1285, 1293, 1314,
        1323
Leonardo da Vinci 495, 521, 541, 575,
    807, 1035, 1187, 1600
    architecture 530, 1192, 1193
    bibliographies 400, 477, 482c, 1192
    biographies 505, 529, 1453
    library 184, 579
    manuscripts 201, 541, 1453
    physical sciences 946, 1453, 1577
    technology 1168, 1172, 1176, 1193
    see also natural history; physiology
Leyden *see* Leiden
libraries (modern) 36, 37, 200
    catalogues 42–45, 90b, 100, 102,
        104b, 106, 109, 110, 114, 116b,
        1254
    see also under special subjects
libraries (pre-1700) 38b, 96b, 115b, 118b,
    135, 145, 189
    by country: England 106, 115, 146,
        148–151, 164, 167, 168, 171–
        173, 176, 178–181, 186, 189,
        194, 195, 197, 200, 202, 204,
        205, 208–210, 682, 823, 1281;
        France 45, 169, 207, 1337, 1531;
        Germany 101, 163, 165, 183,
        190, 199, 1370, 1379; Italy 153–
        156, 160, 162, 175, 182, 184,
        185, 187, 196, 198, 201, 206,
        210A, 579, 1378, 1380;
        Netherlands 147, 152, 159, 192,
        199; others 100, 166, 191, 193,
        203, 851, 854, 871, 1085
    by owner—institutional: Aberdeen
        100; Cambridge 106, 197; Flo-
        rence 185; Leiden 147; Milan
        201; Oxford 146, 164, 200, 202;
        Paris 45, 169, 207; Pei Thang
        854; Rome 155, 156, 175;
        Venice 154, 206; Wolfenbüttel
        1379

by owner—private: Ashmole 189;
    Bachmann 1370; Barrow 823;
    Benivieni 160; Brown (T. and
    E.) 150; Burton 208; Caius 194;
    Canevari 1378; Christina of
    Sweden 191; Colbert 207;
    Copernicus 193, 851, 871;
    Cranmer 171; Dee 682;
    Deighton 167; Fabricius 1380;
    Federigo da Montefeltro 155,
    156; Flamstseed 186; Galileo
    153; Gaurico 187; Goodyear
    1281; Gostling 173; Halley 189;
    Hartgill 179; Herrera 166;
    Holste 165; Hooke 189, 210;
    Huygens 199; La Brosse 1337;
    Leibniz 199; Leonardo da
    Vinci 184, 579; Leoniceno
    210A; Locke 176, 181; Lorkyn
    158, 1547; Lumley 171;
    Mazarin 207; Moore 706, 826;
    Münzer 163; Newton 195, 823;
    Peucer 190; Pickering 1547;
    Pico 162; Pinelli 198; Plempius
    159; Ralegh 180; Ray 189; Read
    1448; Richelieu 207; Scaliger
    192; Scarburgh 148; Scheubel
    183; Séguier 207; Sloane 209;
    Stafford 178; Sturmy 1085;
    Thomas 1547; de Thou 207;
    Webster 204; Winthrop 174
by subject: astronomy 96, 154–156,
    179, 185–190, 193, 195, 205,
    854; chemistry 150, 174, 984;
    geography 96, 147, 150, 151,
    163, 165, 166, 171, 180, 205;
    humanist 146, 149–151, 154,
    160, 162–165, 171, 175, 178,
    180, 185, 188, 189, 191, 192,
    198, 208, 210A; mathematics
    96, 148–152, 154–156, 165, 173,
    178, 179, 183, 185, 188, 190,
    192, 195, 198, 204, 205, 706,
    823; medicine 45, 96, 146–152,
    154–160, 163–168, 171, 174,
    176, 178, 179, 185, 190, 192,
    194, 204, 205, 210A, 1379, 1448,
    1547; natural history 189,

1281, 1337; occult science 162,
166, 174, 205, 682, 984; philosophy 96, 146, 147, 150, 154,
165, 166, 176, 180, 185, 190,
192; science 152–154, 162, 164,
166, 175, 176, 179, 182, 184,
186, 189, 191, 195, 197, 199,
201, 205, 210, 1085
library history 38*b*, 104*b*, 113*b*, 161, 169,
170, 203, 303
scientists 118*b*, 135, 145, 189*c*
England 96, 172; Cambridge 106,
173, 197, 205, 1547; Oxford
164, 202, 208, 254
France 169, 207, 1531
Italy 131, 162, 185, 188, 196, 201, 206
Lilly, William 697
Linacre, Thomas 1536
literature 463
English 87, 109–111, 695, 696, 700
French 98, 119, 538
science and 481, 491, 503, 504, 1025,
1588, 1598
see also astronomy; Milton; Shakespeare
Locke, John 176*c*, 181*c*, 222, 225, 235,
400*b*, 449, 1551
logic 395*b*, 729*b*
see also methodology
London 177, 361, 1225, 1240, 1511
see also College of Physicians; Royal
Society
longitude 584, 588, 589, 1095

magic 166*c*, 621, 627, 652, 659, 673*b*, 678,
1184
bibliographies 602, 603, 606, 611,
613, 615*c*, 617
history 596, 622, 626, 633, 640, 656,
663, 666, 673
in the Renaissance 340, 627, 644,
653, 665, 667, 677
mathematics and 804, 818
religion and 640, 1129
science and 663, 665, 666, 1606,
1641
see also occult science; Bruno; Dee

magnetism 462, 798, 918, 1018, 1020,
1024, 1224
see also earth
Malpighi, Marcello 1430
manuscripts 90*b*, 147, 394
collectors 154, 191, 192, 196, 198,
209, 321
humanistic 97*b*, 102*c*, 104*b*, 154*c*–
156*c*, 175*c*, 188, 191, 196*c*, 198,
786
navigation 1050, 1058, 1099
scientific 116*b*, 201, 377, 443, 525,
1078, 1172; astronomy 686,
830, 831; occult 610*c*, 668, 686
see also Greek texts; humanism;
mathematics; medicine;
Leonardo da Vinci; Newton
maps *see* cartography
Marcgrave, Georg 1326
Marciana Library, Venice 131, 196, 206
marine science 1043–1129
see also naval and maritime arts
materia medica 8*b*, 302A, 977*b*, 978*c*,
981*c*, 1342, 1481–1495
abortifacients 1493
America 1311, 1486, 1490, 1494
apothecaries 361, 1489
history 502, 1481, 1487, 1488, 1493,
1556
India 1311, 1313, 1482
pharmacopoeias 8*b*, 991, 1067A*b*,
1483, 1485
teaching 298, 1302, 1489
see also botanical gardens; Germany; Jardin des Plantes;
medicinal plants
mathematical education 211, 249, 281,
283, 295, 502, 559, 714
Jesuits and 229, 290, 299
teachers 820, 1093
see also arithmetic, teaching; education in England; education in
Italy; educational books; history of education; practical
mathematics; universities
mathematicians 743, 753, 754, 795, 804,
808, 819, 1131
arithmetic 722*b*, 735, 737, 742, 746

bibliographies 715, 719, 721, 724, 725, 729
England 723, 751, 791, 804–826, 1093, 1234
France 283, 284, 720
Italy 725*b*, 755, 786, 805, 812, 817, 820
Netherlands 722*b*, 733, 734, 1203, 1234
Spain 467, 747
see also Barozzi; Barrow; Bombelli; Bovelles; Cardano; Dee; Descartes; Dürer; Fermat; Harriot; Leibniz; Leonardo da Vinci; Moore; Napier; Newton; Pacioli; Ramus; Recorde; Regiomontanus; Stevin; Stiefel; Tartaglia; Viète; Wallis; Wren
mathematics 49, 425, 505, 706–826, 873, 1095, 1233
analysis 734, 743, 770, 789
bibliographies 5, 108, 708, 709, 712, 714, 716, 725, 745
catalogues 751, 830, 842, 1131, 1379
Germany 721, 735, 736 742, 746, 765
Greek 154, 188, 198, 786, 819
Italy 445, 525, 572, 708, 729, 730, 772, 787, 792
logarithms 712*b*, 731, 733, 745, 764, 789, 797
manuscripts 710, 725, 726, 766, 779, 782, 786, 830, 1233
Netherlands 709, 771
see also algebra; arithmetic; Bologna; calculus; Cambridge University; early printed books; education in England; education in Italy; geometry; history of mathematics; libraries (pre-1700); magic; naval and maritime arts; practical mathematics; Scientific Revolution; trigonometry; Newton
mathematization of science 425, 778, 898
matter 406, 409, 415, 424, 439

Maurolico, Francesco 818, 946, 958
mechanical philosophy 349, 423, 454, 669, 928, 1451
chemical theory and 596, 986
genesis 380, 411, 415, 436, 452, 1578, 1581
Scientific Revolution and 1562, 1606, 1608, 1627
mechanics 121*b*, 709*b*, 771, 922, 939, 951, 959, 1612, 1641
catalogues 43, 44, 55, 100, 1131, 1132
history 506, 556, 921, 928, 930, 943, 949, 960; Renaissance 589, 946, 961
in Italy 484, 578, 589, 773, 932, 934, 963
see also gravitation theory; physics
mechanization of design 1183
mechanization of the world picture 1571, 1572, 1584, 1627
medical education 236, 282, 298, 1394, 1418, 1432, 1436, 1524–1540
England 1530, 1534, 1536, 1537; Cambridge and Oxford 251, 255, 269, 276, 1283; London 264, 1455, 1536
France 285, 294, 1489, 1525, 1531, 1532, 1535
Italy 239, 301, 302A, 445, 1302, 1462, 1476, 1479, 1528, 1539
Netherlands: Leiden 303, 1446, 1552
see also anatomy; materia medica; physiology
medicinal plants 584, 1482, 1490, 1492–1494
see also herbals; materia medica
medicine 49, 119, 174, 298, 444, 462, 702, 1004–1138, 1354–1422, 1492, 1518
bibliographies 5*b*, 108, 121*b*, 1355*b*, 1358–1360, 1364, 1367–1369, 1496
biographies 1441, 1541, 1543, 1545; bibliographies 1359, 1362, 1363, 1365, 1366, 1369, 1527; catalogues 1371, 1381, 1542, 1544

book trade 127, 132, 134, 145b, 572
catalogues 43, 981, 1371–1379,
    1381–1385, 1472, 1542, 1544
manuscripts 154, 1372, 1384, 1398,
    1429, 1453, 1551
ophthalmology 1389, 1415
paediatrics 1361, 1394, 1534
plague 8b, 1392, 1511
popular 599, 643, 1400, 1409, 1421
practice 361, 1399, 1418, 1525, 1529,
    1530, 1543, 1547
scientific 1401, 1406
see also [1] Cambridge University;
    early printed books; England;
    humanism; illustrations; li-
    braries (pre-1700); literature;
    Oxford University; religion;
    Scientific Revolution;
    witchcraft; [2] anatomy; den-
    tistry; epidemics; gynaecol-
    ogy; history of medicine; ma-
    teria medica; medical educa-
    tion; midwifery; obstetrics;
    physiology; psychiatry; psy-
    chology; surgery; syphilis; [3]
    Boerhaave; Cardano;
    Descartes; Fabricius; Galen;
    Gasser; Harvey; Belmont;
    Hippocrates; Locke;
    Paracelsus; Paré; Rabelais;
    Servetus; Shakespeare; Taglia-
    cozzi; Vesalius
medieval science 14b, 478b, 501, 653
Mercator, Gerhard 723, 1102, 1118
Mersenne, Marin 411, 455, 479, 489, 665,
    968, 973
Merton thesis 308, 318, 350, 352
metals 994, 1002, 1008, 1014, 1023, 1146,
    1177, 1191, 1195
meteorology 1007b, 1013b, 1017, 1025,
    1212
    see also instruments
methodology 422, 1452, 1570
    of science 417, 430, 437, 484, 590
    see also logic; Bacon; Galilei
Mexico 1328, 1341, 1490
    see also America
microscopes see instruments

Milan: Biblioteca Ambrosiana 201
military science 166, 723, 771, 1117,
    1146, 1233
    bibliographies 8, 10–13, 709, 1130,
        1131, 1137, 1138, 1143
    see also ballistics; England; history
        of technology
Milton, John 226, 503, 504, 847, 850
mineralogy 480, 1028, 1033, 1038
mining 1014, 1146, 1188
"modern" subjects, teaching 249, 250
Monardes, Nicolas Bautista 1311, 1341,
    1486, 1494
Montpellier University 977, 1276, 1288,
    1305, 1316, 1352, 1464, 1489,
    1532, 1535
    botanical garden 1273, 1278
Moore, Sir Jonas 706, 826
More, Henry 593
Müller, Johannes see Regiomontanus
Münster, Sebastian 1103
Münzer, Hieronymus 163
museums 570, 1312, 1315, 1432, 1526
music 8b, 53, 74, 154, 171, 211, 709b, 964–
    975, 992
    Scientific Revolution and 974, 1588
    theory 340, 549, 578, 967, 968, 973,
        975; Descartes 964, 965; Kepler
        968, 972, 973; science and 967,
        970, 971

Napier, John 745, 797
natural history 38, 108, 211, 302A, 311,
    340, 467, 842, 1089, 1248–1353
    bibliographies 1067A, 1256, 1264,
        1267, 1270A, 1379c
    in England 486, 548, 1251b, 1264b,
        1281, 1282, 1296, 1307, 1338
    history 1261, 1270A, 1271, 1338
    idea of nature 406, 412, 440, 1338
    Leonardo da Vinci 575, 1266, 1347
    naturalists 1279, 1289, 1296, 1338
    ornithology 1251, 1325
    Renaissance 484, 505, 525, 1025,
        1251
    see also botany; gardens; illustra-
        tions; Netherlands; Scientific
        Revolution; zoology

natural philosophy 49*b*, 121*b*, 385, 575, 599, 908, 917A, 960, 1583, 1645
   see also nature; physics; Aristotle; Dee; Descartes; Newton
nature 406, 412, 440, 486, 1025, 1338
   see also sky
naval and maritime arts 340, 521, 527, 584, 1089, 1090, 1123–1126, 1162
   bibliographies 8, 709, 1054*c*, 1056*c*, 1063, 1066
   in England 1085, 1093, 1095, 1098, 1106, 1151, 1155
   history 1057*b*, 1086, 1091, 1095, 1106, 1108, 1112, 1137*b*
   mathematics 723, 729, 1093, 1095
   in Netherlands 709, 771, 904, 1055*b*, 1066, 1067A
   sailing 1046, 1065A, 1076, 1080, 1110, 1158, 1180
   see also manuscripts; navigational instruments; Portugal; ship-building and shipping
naval history 1054*c*, 1056*c*, 1057*b*
   architecture 578, 1154, 1155
   see also military science; shipbuilding and shipping
navigational instruments 588, 1065A, 1085, 1090, 1125
   catalogues 1066, 1099, 1217, 1225, 1227
   history 1093, 1095, 1127, 1220, 1229, 1238, 1243
   see also instruments
Neoplatonism 651, 675, 908, 962, 975, 1033, 1417, 1570
Netherlands 59, 63, 219, 240, 344, 589, 709*b*, 771*b*, 1055*b*, 1070, 1234, 1326
   Amsterdam 1066, 1190, 1273
   cartography 1055*b*, 1070, 1102
   Dutch East India Co. 1067A, 1300A, 1320, 1324, 1330, 1346
   Dutch West India Co. 1326
   humanism 103*b*, 112*b*, 122*b*, 289, 407
   natural history 1275, 1300A, 1350, 1351

science 558, 564, 709*b*, 771*b*, 904, 1162
   see also Belgium; history of medicine; Leiden University; libraries (pre-1700); mathematicians; mathematics; naval and maritime arts; physicians
New World 1046, 1269, 1289, 1297, 1329, 1494
   see also America
Newton, Sir Isaac 84, 400*b*, 468*b*, 475*b*, 576, 587, 593, 598, 704, 956, 957
   biographies 513, 528, 553, 562, 565, 601
   essays on 346, 520, 532, 583, 596
   library 195, 823
   manuscripts 330, 519, 553, 555, 559, 1641A
   mathematics 559, 753, 767, 789, 797, 800
   natural philosophy 596, 929, 939, 940, 955–957, 1637
   non-scientific thought of 330, 555
   Principia 475, 533, 555, 576, 601A, 1593, 1619, 1641A; background 519, 916, 957; editions 856, 857, 866, 876
   religion 311, 330, 334, 335, 349, 559
   scientific method 406, 957, 1583
   Scientific Revolution 1609, 1612, 1614, 1619, 1620, 1621, 1637
   see also alchemy; calculus; Leibniz; optics
Newtonian science 349, 561, 929, 1609, 1614
nutrition 575, 1277, 1300A, 1408, 1420

observatories 842, 1218, 1235, 1240
   see also astronomical instruments
obstetrics 1392, 1469*b*, 1470*b*, 1472*c*, 1473, 1475, 1480, 1526
   see also gynaecology
occult science 162*c*, 340, 420, 584, 602–682
   bibliographies 38*b*, 602–606, 611, 617
   numerology 665, 992

see also alchemy; astrology; cabbala; England; freemasonry; hermeticism; magic; Scientific Revolution; witchcraft
oceanography 1034, 1073
optics 406, 588, 649, 729*b*, 823, 940, 942, 955, 1562
   history 521, 929, 933, 937, 944, 945, 953, 1198*b*, 1577
   Newton 475, 576, 957
   see also corpuscular theory; instruments; perspective; vision; Kepler
Orta, Garcia d' 1311, 1313, 1482
Oxford 542, 570, 1209
Oxford University 172, 239, 258, 570, 951, 1273, 1283, 1628
   colleges 164, 200, 240, 254, 277, 951
   history 215*b*, 251, 254, 276
   libraries 172, 200, 254, 276; Bodleian 146*c*, 202, 208*c*
   medicine 251, 269, 276, 1283, 1450, 1534
   science 239, 254, 268, 269, 339, 1283; astronomy 915, 1205, 1240
   see also universities

Pacioli, Luca 735, 812, 819, 820A, 946
Padua 198, 301, 433, 1273, 1302, 1340, 1409, 1587
papacy 53, 305, 580, 963
   see also Catholic Church; Vatican Library
Papin, Denis 316, 1160, 1194
Paracelsianism 574, 632, 1000, 1001, 1519, 1523
Paracelsus 462, 990, 995, 1402, 1519–1522, 1527
   bibliographies 400, 1514–1516, 1522
   medicine 1409, 1502, 1517, 1518, 1534
Pardies, Ignace Gaston 479, 940
Paré, Ambroise 1467
Paris 240, 489, 783, 977, 1240, 1305
   Académie des Sciences 355, 371, 1319
   Collège de France 283, 284

see also history of universities; Jardin des Plantes
Pascal, Blaise 98, 544, 753, 793, 816, 1600
Pepys, Samuel 274, 522
perspective 578, 795, 818, 1117
   see also optics
Petty, Sir William 226, 492, 669, 1394
Peuerbach, Georg 131, 552
pharmacists 361, 1489
pharmacology *see* materia medica
philosophy 389–462, 1580, 1585, 1599, 1608
   bibliographies 98, 390, 393, 400, 403, 404, 709, 1139, 1140
   dictionaries 49, 393*b*, 405, 406
   history 71, 74, 406, 408, 456, 458, 461*b*, 462
   Renaissance 390*b*, 391*b*, 404*b*, 408, 432, 434, 442, 462
   of science 410, 461, 551, 873, 1570; Bacon 413, 414, 450; medicine 551, 1539; occultism 655, 674
   in Western Europe 53, 300, 461, 709*b*, 941
   see also Aristotelianism; libraries (pre-1700); scepticism; science; Scientific Revolution; technology; universities; Aristotle; Galilei; Kepler
physic gardens 1292, 1295
   see also botanical gardens
physical sciences 21*b*, 512, 863, 918*b*–920*b*, 946, 1453
physicians 1359*b*, 1362*b*, 1388, 1527, 1542*b*, 1548, 1554–1556
   England 1541, 1543, 1547, 1549, 1551
   Netherlands 1404, 1545, 1552
physics 918–963, 971, 1434, 1577
   bibliographies 108, 709, 842*c*, 918–920, 1379*c*
   England 798, 890, 924, 956, 1205; Newton 939, 952, 957
   France 283, 291, 294
   history 298, 556, 925, 933, 941, 948, 950, 956
   Italy 484, 525, 556, 575; Galilei 936, 943, 947, 963, 1597

see also mechanics; natural philosophy; optics; Scientific Revolution; Kepler; Newton
physiology 436, 596, 624, 1423*b*, 1424–1455, 1505, 1539, 1553
education 1283, 1450, 1526, 1534
history 1261, 1387, 1430, 1434, 1438*b*, 1441*b*, 1451, 1454
Leonardo da Vinci 530, 575, 1453
see also Harvey
Pico della Mirandola, Giovanni 162, 655
Piero della Francesca 812, 820A, 946
pilots' books and tools 1055, 1058, 1238
planetary theories 868, 877, 888, 892, 894
bibliographies 827, 833, 841
of individuals 506, 881, 883, 894, 907, 1580, 1586
plant collecting 584, 1273, 1276, 1296, 1307, 1308, 1318, 1335, 1348, 1350
America 1294, 1311, 1341; Mexico 1328, 1490
Asia 1275, 1311, 1317, 1324
Cape of Good Hope 1275, 1280, 1300A, 1310, 1320, 1330
by collector: Anguillara 1276; Barlow 1318; Bartholinus 1280; Bauhin 1276; Belon 1276, 1284; Boel 1318; Browne 1318; Cavelier de la Salle 1333; Cheler 1276; Cleyer 1324; Constantin 1276; Cunninghame 1318; Dourez 1276; Flacourt 1333; Foxe 1280; Hermann 1280, 1310, 1320, 1330; Hernandez 1318, 1328, 1490; Heurnius 1280, 1310; Kaempfer 1318; Kamel 1318; Kiggelaer 1280; Lobel 1276; Meister 1324; Monardes 1311, 1341, 1486; Oldenlandus 1280; Orta 1311; Pena 1276; Platter 1276; Plumier 1318; Rauwolf 1272, 1276, 1317; Raynaudet 1276; Reed 1318; Robin 1274; Rumpf 1318, 1653; Sloane 1318; Solier 1276; Starrenburgh 1280;

Tournefort 1333; Tradescant 1281, 1315, 1318, 1339
see also botanical expeditions; exploration; flora; herbaria
Platonism 299, 406, 434, 448, 527, 787
Platter, Felix 1276, 1288, 1527
poetry and poetics 307, 432, 474*b*, 509, 655, 873, 1025
scientific 119, 538, 594, 613*b*
polyhedra 448, 810
popes *see* papacy
Portugal 1082, 1096, 1100, 1114, 1275
naval and maritime arts 1050*b*, 1065A*b*, 1090, 1124; history 1077*b*, 1080, 1084, 1087, 1092, 1095
see also Henry the Navigator
practical mathematics 599, 750, 782, 1063, 1093, 1233
see also arithmetic
printing 86*b*, 107, 113*b*, 129, 133, 139*b*, 487, 1146, 1162, 1168, 1531
Copernicus 893, 900
impact of 136, 137, 139, 142, 143, 517
see also book trade; early printed books; Index of Prohibited Books; technology; Venice
prognostication(s) 132, 603, 684, 685, 693, 695, 703
see also almanacs; astrology
Protestantism 171, 207, 219, 237, 625
Calvinism 354A, 628, 1574
modern science and 312, 314, 341, 1574
see also Erasmus; Reformation
psychiatry 624, 1496, 1502, 1503
psychology 1496*b*–1500*b*, 1501, 1502, 1505, 1592
Ptolemy 154, 850, 883
Geography 1043*b*, 1047*b*, 1051*b*, 1053*b*, 1070, 1113, 1118, 1121
see also astronomy; Greek texts
Puritanism 250, 263, 274, 333, 655
science and 317, 320, 322, 349, 352, 1585, 1594
see also Protestantism; religion; science and religion

Quakers 309

Rabelais, François 289, 544, 1556
Ralegh, Sir Walter 180
Ramus, Peter 228, 462, 507, 891
Ramusio, Giovanni Battista 1071, 1074, 1094, 1121
Ratichius *see* Ratke
Ratke, Wolfgang 222, 226, 233
Rauwolf, Leonhard 1272, 1277, 1317
Ray, John 189, 1262, 1299, 1345
Recorde, Robert 253, 274, 723, 750
Reformation 203, 207, 383, 407, 625, 1173
    education and 219, 222, 237–239, 243, 262
    science and 312, 314, 325, 349, 507, 1574, 1611
    Thirty Years War 222, 240, 367
    see also Protestantism; religion; science and religion
Regiomontanus, Johannes 131, 552, 788, 808, 914, 946
religion 53, 307, 319, 399, 420, 794, 992
    history 5*b*, 11*b*, 15*b*, 31*b*, 75–77, 81, 83
    Judaism 1620, 1633
    medicine and 1405, 1407, 1417, 1518, 1529
    refugees 248, 345, 1165, 1234
    tolerance 315, 322, 1314
    see also Catholic Church; Church (The); magic; Puritanism; Reformation; science and religion; science and society; theology; witchcraft; Newton
Renaissance 84, 203, 302A, 487, 511, 521, 525, 599
    art and architecture 53, 506, 825, 945
    bibliographies 100*c*, 103, 105, 122, 390
    biographies 52, 53, 390, 407, 420, 796, 1161, 1172
    dictionaries 52, 53, 95, 397, 406, 504
    methodology 417, 484
    music 966, 969
    see also Aristotelianism; early printed books; Greek texts;

hermeticism; history of education; history of universities; humanism; humanists; libraries (pre-1700); magic; occult science; philosophy; Durer; Leonardo da Vinci
Renaissance science 328, 466, 476, 485, 515, 1564, 1566
    essays and symposia 511, 521, 525, 535, 540, 599
Restoration 13, 210, 383, 704, 823, 1585
    science 342*b*, 826, 1567, 1582
revolution(s) 260, 331, 416, 1619, 1631
    see also Scientific Revolution
Rheticus, Georg Joachim 835, 837, 863
Ries, Adam 721, 746
Rome 138, 191, 321, 603, 625, 667
    Accademia dei Lincei 355, 369, 484, 618
    libraries 165, 615, 1378
    see also Catholic Church; Collegio Romano; Vatican Library
Rondelet, Guillaume 1285, 1316
Rosicrucians 603, 604, 611, 616, 638, 645, 655
Royal Botanic Garden, Edinburgh 1291
Royal College of Physicians, London *see* College of Physicians
Royal Greenwich Observatory 1225, 1240
Royal Society of London 46, 177, 372–388, 956, 1534
    fellows 174, 209, 376, 384, 386, 522, 766, 1106
    history 355, 372*b*, 377–387, 388*b*; index (Birch) 381
    "History of trades" program 492, 571
    Oldenburg 581, 585

Sagres, School of 1072, 1077
sale catalogues 148–152, 189, 192, 210, 706
"Salomon's House" 365, 387
San Marco library, Venice 131, 154, 196, 206
Santorio, Santorio (Sanctorius) 1204, 1211, 1452

Savonarola, Giovanni Michele 1479
Saxton, Christopher 1119
scepticism 420, 436, 441, 617, 625, 646,
    1003, 1601
schools 213, 243, 245, 257, 273, 297, 302
    England 247, 249, 270, 271, 273
    see also education; teaching
science 463–601
    bibliographies 5b, 8, 39, 43–46, 98,
        342, 463–482A
    chronologies 57, 69, 72, 74, 76, 84,
        586, 1061, 1145, 1152
science and religion 312, 314, 319, 325,
        334, 346, 349, 352, 353, 354A
    England 309, 310, 313, 317, 320, 322,
        333, 335, 337, 339, 341, 354
    Scientific Revolution 1595, 1601,
        1627; Judaism 1620, 1633; Pu-
        ritanism 320, 1585, 1594;
        Reformation 1574, 1611
    see also Puritanism; Reformation;
        religion; theology
science and society 319, 326–328, 338,
        340
    England 315, 320, 323, 329, 333, 341,
        342, 643
    technology 308, 318, 350, 352
    see also religion; social relations of
        science
scientific instruments *see* instruments
scientific method 514, 924, 1583, 1590,
        1597
    see also methodology; Newton
Scientific Revolution 84, 596, 1562–1646
    arts and 506, 974, 1588, 1598
    bibliographies 346, 1564, 1567,
        1568, 1606
    concept of 1613, 1625, 1626, 1630,
        1631, 1634–1636, 1639, 1641
    education and 237, 278, 1418, 1628,
        1629, 1641
    essays and surveys 346, 1562, 1564–
        1568, 1581, 1582, 1595, 1597,
        1605, 1606, 1612, 1626A, 1641,
        1642
    historiography 1565, 1599, 1603,
        1613, 1625, 1631, 1634, 1635,
        1644

history of 1573, 1575, 1576, 1589,
        1593, 1596, 1602, 1604, 1606,
        1626
philosophy and 506, 1570, 1571,
        1578, 1580, 1599, 1601, 1608,
        1619, 1620, 1633
scientific method 1583, 1590, 1597
social and cultural aspects 820,
        1569, 1585, 1592, 1632, 1636
source books 1562, 1563, 1565, 1602
interactions
    mathematics 1570, 1584, 1595, 1612,
        1628, 1638, 1641
    medical sciences 1419, 1422, 1574,
        1612, 1617, 1639, 1640
    natural sciences 1580, 1586, 1620,
        1623, 1640; astronomy 699,
        890, 1230, 1612, 1632, 1639;
        chemistry 996, 1001, 1562,
        1612, 1641; natural history
        1001, 1578, 1641
    occult science 641, 656, 661, 665,
        1606, 1607, 1612, 1616, 1618,
        1641
    technology 143, 506, 536, 1599, 1605
    others 974, 1113, 1183, 1578, 1612,
        1639
other aspects: China and 1630, 1644;
        ecology and 1621; women and
        591, 1621
    see also astrology; cartography; ex-
        perimental science; hermeti-
        cism; mechanical philosophy;
        music; science and religion;
        universities; Bacon; Coperni-
        cus; Descartes; Galilei; New-
        ton
scientific revolutionaries 1631
scientific societies 355–388, 1562
    see also academies; College of
        Physicians, London; Royal So-
        ciety of London
Scotland 100, 679, 1291, 1400, 1413
    education 215, 266, 281, 282, 1538
Servetus, Michael 400, 1433, 1435, 1574
Shakespeare, William 486, 655, 657, 694,
        1025, 1282

medical knowledge 1501, 1505, 1546, 1550, 1561
shipbuilding and shipping 12*b*, 534, 578, 1054*b*, 1065A*b*, 1067A*b*, 1096, 1123, 1146, 1176, 1176A, 1639
sailing ships 1110, 1158, 1180
ibn Sina (Avicenna) 1539
sky 904, 905, 912, 1035
Sloane, Hans 209, 1305, 1307, 1318, 1320
smallpox 1321, 1511
Snell (Snel, Snellius), Willebrord 818, 1016
social relations of science 304–354, 366, 452, 551, 820, 1592, 1595
   see also religion; science and religion; science and society; Scientific Revolution; technology
Spain 17*b*, 67, 203, 240, 464, 467, 924, 1328
astrology 1090, 1407
mathematics 467, 747
naval and maritime arts 584, 1126
Philip II 166, 584, 1490
technology 166, 547, 584
   see also history of medicine
Stahl, George Ernst 990, 1360, 1502
statistics 794, 1534
Steno *see* Stensen
Stensen, Niels 480, 577, 1021, 1028
Stevin, Simon 735, 769, 771, 795
Stiefel, Michael 746, 768
students 266, 277, 294, 301, 542, 1532
surgeons 1359, 1379, 1458, 1463, 1464, 1545
surgery 1392, 1456–1468, 1526
bibliographies 8, 1359, 1456–1458, 1465
history 1387, 1458, 1461, 1463, 1465, 1468
instruments 167, 1461, 1477
surveying 906, 1011, 1117, 1146, 1203
   see also cartography; instruments
Sylvius (Le Boë, François de) 1360, 1430
syphilis 8*b*, 660, 1321, 1506*b*, 1508, 1512, 1540
history 1507, 1509–1511, 1513
   see also Fracastoro

Tagliacozzi, Gaspare 1462
Tartaglia, Niccolò 556, 578, 934, 946, 1176, 1600
teaching 82, 231, 249, 250, 302
   see also anatomy; arithmetic; botanical gardens; education; materia medica; mathematical education; medical education; physiology; practical mathematics; schools; students
technology 288, 309, 406, 429, 558, 578, 771, 922, 1130–1195, 1639
bibliographies 108, 121, 983, 1140, 1142, 1167
chronologies 57, 74, 76, 586, 1145, 1152
European expansion and 517, 1174, 1182
Huguenots 304, 344, 345, 1165
philosophy of 551, 1139, 1140
Renaissance 505, 530, 1138, 1143, 1176, 1176A, 1184, 1186, 1233
   see also education; engineering; history of technology; inventions; manuscripts; printing; science; Scientific Revolution; Leonardo da Vinci
telescopes *see* astronomical instruments
textile industry 1168, 1169
theatre 637, 1157, 1192
theology 49, 332, 349, 383, 390*b*, 796
science and 306, 311, 331, 336, 348, 353, 1129, 1601
   see also Galileo affair; religion; science and religion
theses 29, 101
bibliographies 5, 28, 31, 977; history 30, 32, 33; history of science 34, 35; medicine 1360, 1383, 1385, 1394, 1423
time measurement 406, 431, 549, 1213, 1239
instruments 1162, 1207, 1209, 1220, 1224, 1226, 1234, 1237, 1246; mechanical 1125, 1146, 1196, 1236
   see also instruments
tobacco 1254

Tournefort, Joseph Pitton de 1305, 1306, 1333
trades, history of 492, 571
Tradescants 1281, 1315, 1318, 1339
translations: from Arabic 89, 686; from Greek 97, 720, 723, 1425
travel 53, 134, 1045c, 1091, 1135b
Trent, Council of 343, 1624
trigonometry 718b, 731, 741, 754

universities 161, 212b, 220, 241, 256, 257, 266, 273, 294b
  Scientific Revolution and 278, 1628, 1641
  by country: Ireland 215, 266, 364; Netherlands 240, 589 *see also* Leiden; Scotland 100, 215, 266, 281, 282, 1538; Spain 240; Switzerland 212, 1352, 1527
  by subject: astronomy 462, 891, 892, 915; botany 298, 1303, 1314, 1323, 1352; mathematics 281, 299, 787, 820; music 966; philosophy 239, 240, 294, 300, 941, 1539; science 239, 298, 355, 941
  see also botanical gardens; education; education in England, in France, in Germany, in Italy; history of universities; medical education
Urbino 155, 156, 817
Ursus (Bär, Nicolai Reymers) 902

Valla, Giorgio 799, 948
Van Helmont *see* Helmont
Vatican Library 155, 156, 175, 191, 203
Venice 301, 445, 578, 681, 1176A
  Biblioteca Marciana 131, 154, 196, 206
  printing 138, 140, 572
Vesalius, Andreas 521, 1426b, 1427b, 1430, 1436, 1442, 1527
Viète, François 739, 769, 780, 789, 819, 883

Vinci, Leonardo da *see* Leonardo da Vinci
vision 406, 423, 945
Vives, Juan Luis 227, 246, 1502
voyages of discovery 1045c, 1054c, 1056c, 1065Ab, 1067Ab, 1118, 1289, 1312, 1328, 1589
  see also discoveries

Wales 215, 1119
Wallis, John 383, 753, 766, 769
war and science 926, 1138b, 1143b, 1180
Werner, Johann 799, 837, 1068
Weyer, Johann 624, 1502
Willis, Thomas 1412, 1413
windmills 1146, 1162
witch-craze 607, 608b, 628, 656, 660, 664, 672
witch hunting *see* witch-craze
witchcraft 127, 406, 607, 623, 639, 642, 664, 673, 681
  bibliographies 8, 603c, 606–608, 611c, 612c, 617, 673, 979c
  Europe 599, 636, 644, 646, 648, 665, 681; England 620, 623, 628, 639, 640
  medicine and 624, 660, 1502
  religion and 628, 640, 681
  trials 620, 625, 636
  see also occult science
Wittich, Paul 910
women in science 591, 1621
world picture 494, 1578, 1579, 1606, 1621
  mechanization of 1571, 1572, 1554, 1627
  see also cosmology
Wren, Christopher 471, 531, 880, 881, 1553

Yagel, Abraham ben Hananiah 678

zoology 69, 121b, 1256b, 1270, 1300, 1309, 1312, 1343, 1375c, 1439

nb Gouk   no entries

nb Gouk   no entries

nb Gouk   no entries